Springer Proceedings in Physics 67

Springer Proceedings in Physics

Managing Editor: H. K. V. Lotsch

44 *Optical Fiber Sensors*
Editors: H. J. Arditty, J. P. Dakin, and R. Th. Kersten

45 *Computer Simulation Studies in Condensed Matter Physics II: New Directions*
Editors: D. P. Landau, K. K. Mon, and H.-B. Schüttler

46 *Cellular Automata and Modeling of Complex Physical Systems*
Editors: P. Manneville, N. Boccara, G. Y. Vichniac, and R. Bidaux

47 *Number Theory and Physics*
Editors: J.-M. Luck, P. Moussa, and M. Waldschmidt

48 *Many-Atom Interactions in Solids*
Editors: R .M. Nieminen, M. J. Puska, and M. J. Manninen

49 *Ultrafast Phenomena in Spectroscopy*
Editors: E. Klose and B. Wilhelmi

50 *Magnetic Properties of Low-Dimensional Systems II: New Developments*
Editors: L. M. Falicov, F. Mejía-Lira, and J. L. Morán-López

51 *The Physics and Chemistry of Organic Superconductors*
Editors: G. Saito and S. Kagoshima

52 *Dynamics and Patterns in Complex Fluids: New Aspects of the Physics–Chemistry Interface*
Editors: A. Onuki and K. Kawasaki

53 *Computer Simulation Studies in Condensed Matter Physics III*
Editors: D. P. Landau, K. K. Mon, and H.-B. Schüttler

54 *Polycrystalline Semiconductors II*
Editors: J. H. Werner and H. P. Strunk

55 *Nonlinear Dynamics and Quantum Phenomena in Optical Systems*
Editors: R. Vilaseca and R. Corbalán

56 *Amorphous and Crystalline Silicon Carbide III, and Other Group IV-IV Materials*
Editors: G. L. Harris, M. G. Spencer, and C. Y.-W. Yang

57 *Evolutionary Trends in the Physical Sciences*
Editors: M. Suzuki and R. Kubo

58 *New Trends in Nuclear Collective Dynamics*
Editors: Y. Abe, H. Horiuchi, and K. Matsuyanagi

59 *Exotic Atoms in Condensed Matter*
Editors: G. Benedek and H. Schneuwly

60 *The Physics and Chemistry of Oxide Superconductors*
Editors: Y. Iye and H. Yasuoka

61 *Surface X-Ray and Neutron Scattering*
Editors: H. Zabel and I. K. Robinson

62 *Surface Science: Lectures on Basic Concepts and Applications*
Editors: F. A. Ponce and M. Cardona

63 *Coherent Raman Spectroscopy: Recent Advances*
Editors: G. Marowsky and V. V. Smirnov

64 *Superconducting Devices and Their Applications*
Editors: H. Koch and H. Lübbig

65 *Present and Future of High-Energy Physics*
Editors: K.-I. Aoki and M. Kobayashi

66 *The Structure and Conformation of Amphiphilic Membranes*
Editors: R. Lipowsky, D. Richter, and K. Kremer

67 *Nonlinearity with Disorder*
Editors: F. Abdullaev, A. R. Bishop, and S. Pnevmatikos

68 *Time-Resolved Vibrational Spectroscopy V*
Editor: H. Takahashi

69 *Evolution of Dynamical Structures in Complex Systems*
Editors: R. Friedrich and A. Wunderlin

Volumes 1–43 are listed on the back inside cover

F. Abdullaev A. R. Bishop
S. Pnevmatikos (Eds.)

Nonlinearity with Disorder

Proceedings
of the Tashkent Conference,
Tashkent, Uzbekistan, October 1-7, 1990

With 88 Figures

Springer-Verlag
Berlin Heidelberg New York
London Paris Tokyo
Hong Kong Barcelona
Budapest

Professor Fatkulla Abdullaev

Thermal Physics Department, Uzbek Academy of Sciences
Tashkent-135, Chilanzar, Katartal Str. 28, Uzbekistan

Dr. Alan R. Bishop

Los Alamos National Laboratories, Los Alamos, NM 87545, USA

Dr. Stephanos Pnevmatikos †

ISBN-13:978-3-642-84776-9 e-ISBN-13:978-3-642-84774-5
DOI: 10.1007/978-3-642-84774-5

Softcover reprint of the hardcover 1st edition 1992

Typesetting: Camera ready by authors/editors
54/3140 - 5 4 3 2 1 0 - Printed on acid-free paper

Preface

The Tashkent Conference was the second in a series begun in Los Alamos (May 1988) devoted to the emerging field of *combining* effects of disorder and nonlinearity. Since the Los Alamos meeting, appreciation of the field has grown considerably and attracted focused attention from both mathematicians and physicists. The Tashkent meeting was therefore lively and interesting: many new results were presented and new projects proposed. Yet it is clear that we are still largely at a stage of collecting phenomena and special analytical cases – the comprehensive picture will take some time to emerge, and along the way we can expect many surprises and paradigms to emerge. Certainly the importance of the field to real materials and devices is beyond doubt in an age when science is facing the challenge of "novel" materials for unusual and special-purpose applications. Such materials intrinsically involve noise, disorder, nonlinearity, competing interactions, reduced dimensionality and geometries, nonequilibrium processes, and so on. In some cases we need to use these features, in others to avoid them, but in all cases we must learn to understand and then control them.

The central issues discussed at Tashkent involved the same basic components as at Los Alamos (Springer Proceedings in Physics, Vol. 39). On the one hand, *nonlinearity* on its own can lead to spatially and temporally coherent structures (such as "solitons" or "polarons" and their cousins) *or* to intrinsic spatio-temporal chaos. On the other hand, noise or disorder alone can also promote stochasticity *or* localization (e.g., Anderson localization), with "disorder" arising from many sources, including impurities, random spatial structures, thermal and quantum fluctuations, and stochastic applied fields. After several decades of intensive study, for each of the above phenomena we have achieved a fair (although far from complete) understanding and we are able to begin putting them together – as nature so often does. Classifying the regimes where these influences combine constructively, destructively or largely independently, and developing appropriate analytical and numerical approaches accordingly, is the challenge before us.

The articles included in this volume are organized into (I) Mathematical and Statistical Aspects, and (II) Physical Applications. Much of (I) addresses effects of disórder, noise, and external forces on soliton or near-soliton systems, since these are very well controlled starting points. Examples include continuum and discrete sine-Gordon, nonlinear Schrödinger, Korteweg-de Vries, and ϕ-four equations, which capture many of the basic types of nonlinearity found in nature. The physical applications found in (II) also reflect the diversity of this field. Systems range from Indo-European languages to DNA to Josephson junctions to liquid crystals to optical fibers and magnets. The propagation of energy, mass, magnetic flux, etc., is a dominant theme where the interplay of coherence and chaos is clearly a paramount concern.

We benefited from much local assistance in Tashkent, particularly from the staff of the Thermal Physics Department of the Uzbek Academy of Sciences. Organizing a meeting in this exotic but remote location was a complex logistical exercise. However, those who were able to make the trip will always treasure the experience and remember the generous hospitality.

There remains one very sad comment left to us as members of the Organizing Committee. Shortly after the Tashkent Conference our friend and colleague Stephanos Pnevmatikos was senselessly and tragically murdered in Crete. His loss is greatly felt by family and friends, and his enthusiasm and creativity will be a loss to nonlinear science in many ways. He was not able to complete the editing of this volume but we humbly dedicate these proceedings to his memory. We hope that a little of his passion for "nonlinearity *with* disorder" will emerge and infect future generations of young researchers.

Los Alamos
February 1992

F. Abdullaev
A.R. Bishop

Якка тўлкин эрур солитон,
ўзлигини саклар мукобил...
Хохи ишон, хохи майли тон,
Фалак тилсимлари тохануз ботил.

Дунё нима? – Борлик мавжудот.
Харакат не? – Олам икболи.
Дунёни ўрганмок яшашдан мурод,
Хар кимнинг кўлида ўз истикболи.

Манбаи не жахон равнакин,
Не уни бошкарар келажак томон?
Абадий кураш бу, йўк ўзга кохин,
Инсоний килмокчун келгуси замон!

Якка тўлкин эрур солитон,
ўзлигини саклар мукобил...
Поймол этса хак ишни ёлгон,
Керак эмас хеч ўзга котил...

* * *

Soliton is a solitary wave,
What does preserve itself immutable...
One can believe, one can reject,
That many mysteries have not been yet investigated.

What is the world? – Everything that exists.
And motion is its happiness.
The sense of this life consists in world study,
But everyone has got his own fate.

What things cause the worlds?
What things define their future?
Continuous struggle, but not the will of the Providence,
The expected century of Mankind will come.

Soliton is a solitary wave,
What does preserve itself immutable...
If justice is suppressed by evil,
The world will not need another executioner...

M. R. Djumaev

Contents

Part I Mathematical and Statistical Aspects

Part II **Physical Applications**

Part I

Mathematical and Statistical Aspects

Linear Chaos

B.V. Chirikov

Institute of Nuclear Physics, 630 090 Novosibirsk, Russia

Abstract. The controversial phenomenon of quantum (wave) chaos is reviewed using a simple analogy with classical linear waves in cavities. Estimates for the main statistical properties of wave dynamics are evaluated and discussed. The transient nature of wave chaos is emphasized and explained in detail.

The main purpose of my talk is to present a fairly new conception of the so-called quantum chaos. Formally, it is a part of quantum dynamics, rather surprising but not at all exotic. On the contrary, the quantum chaos turns out to be a typical phenomenon which was overlooked until recently. The understanding is now gradually coming from the classical mechanics where the phenomenon of dynamical chaos was well ascertained and studied in great detail (see, e. g., Ref. [1]). For this reason I am going first to remind you of the classical dynamical chaos (Section 2).

In what follows I restrict myself to the conservative (energy-preserving) Hamiltonian (nondissipative) dynamics as a more fundamental one. To illustrate the general theory I will use simple models of classical and quantum billiards as well as waveguides and cavities.

1. Simple Models

Consider free motion of a particle in the domain of d dimensions surrounded by a $(d-1)$-dimensional perfectly reflecting wall. For $d = 2$ the model is called billiard, and it was extensively studied by many authors (see, e. g., Refs [2, 3]). For special shapes of the wall the motion is completely integrable, or regular, which means that there is the full set of d motion integrals in involution (see, e. g., Ref. [4]). The simplest example is a rectangular parallelepiped with sides a_i. Then motion integrals are momenta $|p_i|$ or the actions

$$n_i = \frac{a_i\,|p_i|}{\pi}\,. \tag{1}$$

The corresponding Hamiltonian $H(\theta_i, n_i)$ in action-angle variables is given by

$$H^2 = \pi^2 \sum_i \frac{n_i^2}{a_i^2} + m^2, \tag{2}$$

where m is particle's mass, and we put the speed of light $c = 1$.

Springer Proceedings in Physics, Vol. 67 — Nonlinearity with Disorder
Editors: F. Abdullaev · A.R. Bishop · S. Pnevmatikos

The completely integrable motion is quasi-periodic, i. e. it has discrete Fourier spectrum with d basic frequences $\omega_i = \pi^2 n_i / Ha_i^2$ (for model (2)). This is a nonlinear oscillation since $d\omega_i / dn_i = \pi^2 / Ha_i^2 \neq 0$. Being regular (by definition!) the motion is unstable in phases θ_i just because of nonlinearity: $\delta\theta_i \sim t\delta n_i$. Yet, this unavoidable instability is weak, only linear in time.

Now, introduce a small perturbation, that is deform the wall in such a way to completely destroy integrability and to convert the motion into chaotic one. This always can be done [2, 3], and we shall characterize the perturbation strength by a small angle $\varepsilon \ll 1$ between two surfaces of the wall, unperturbed and corrugated.

In quantum mechanics the same model is described by the linear Schrödinger equation. For boundary condition $\Psi = 0$ (infinitely high potential wall) the actions n_i are now integers.

Instead of quantum Ψ waves we may consider any classical linear waves (sound, electromagnetic, etc.) inside a cavity with appropriate boundary conditions. Then the classical problem (e. g., Eq. (2)) corresponds, as is well known, to the limit of geometrical optics with the rays as dynamical trajectories. The Hamiltonian is now the local dispersion relation (see, e. g., Ref. [5]):

$$H \equiv \omega\,(k_l, x_l) = \omega\,(n_i, \theta_i) \tag{3}$$

where x_l are Cartesian coordinates, wave numbers $k_l = p_l$, and we put $\hbar = 1$. Particularly, for electromagnetic waves in a medium with refraction index $n(x_l)$ the Hamiltonian

$$\omega(k_l, x_l) = \frac{k}{n(x_l)} \,. \tag{4}$$

Finally, we consider a straight waveguide in vacuum with the perturbation independent of longitudinal coordinate x_3. Then $k_3 = \text{const}$, and the dispersion relation is the same as Hamiltonian (2)

$$\omega^2 = k_1^2 + k_2^2 + m^2 \,. \tag{5}$$

It describes "heavy" photons with "mass" $m = |k_3|$.

2. Nonlinear Ray Chaos

As is well known the classical dynamical chaos is only possible in nonlinear systems, that is those whose equations of motion are nonlinear [1]. Such are, for example, the simple models described above, but not a linear (harmonic) oscillator. On the other hand, the main condition for chaos is linear (local) instability of motion discribed by the linearized equations of motion [1]. Moreover, the instability must be strong (exponential) and not a linear one characteristic for a regular motion (Section 1).

The rate of local instability is given by the Lyapunov exponent Λ of the linearized equations of motion. In a billiard or cavity

$$\Lambda \sim \ln \frac{a}{R}, \tag{6}$$

where $R \sim \alpha / \varepsilon$ is the local curvature radius of the wall corrugation with characteristic linear size α. The quantity

$$h \sim \Lambda d \sim d \cdot \ln \frac{a}{R} \tag{7}$$

is called metric entropy, or KS-entropy (after Kolmogorov and Sinai). It is the most important characteristic of chaotic motion in the modern ergodic theory (see, e. g., Ref. [6]). Roughly, the condition

$$h \geq \Lambda > 0\,; \qquad \frac{a}{R} \sim \frac{\varepsilon\, a}{\alpha} \geq 1\,. \tag{8}$$

is necessary for chaos to occur.

I will not go into details of this condition. Rigorous results can be found in Refs [2, 3]. Extensive studies of ray chaos in cavities were performed by Abdullaev and Zaslavsky [7] (see also Refs [5, 8]). For my purposes the rough but transparent estimates are sufficient to expose the nature of dynamical chaos.

Let me just mention that for a big perturbation parameter $\varepsilon \gtrsim 1$ the chaos occurs even for $\alpha \sim a$. Such is a very popular billiard model, the "stadium": two semicircles connected by two straight lines of arbitrary finite length [3]. Again, to simplify the presentation I consider small perturbation $\varepsilon \to 0$, hence $\alpha \to 0$ (8). The perturbation in this case is like tiny ripples.

Now, what is the role of nonlinearity in chaos? The point is that instability (8) is necessary but not sufficient condition for chaos. The other condition is boundedness of the motion, at least in some unstable variables. For example, unstable motion of a linear system (the so-called hyperbolic motion) is unbounded and perfectly regular. Nonlinearity makes the energy surfaces closed and compact even for an unstable motion. In a billiard, for example, the motion is trivially restricted by the wall.

So, what is really required for chaos is the boundedness of the motion whatever the cause could be. As to the nonlinearity it depends on the description chosen. As is well known there is an equivalent description in terms of the phase space density, or distribution function, $f(x_1, p_1, t)$ which obeys linear Liouville equation

$$\frac{df}{dt} = \frac{\partial f}{\partial t} + [H, f] = 0\,, \tag{9}$$

where [,] are Poisson brackets. This form of dynamical equation is especially convenient for comparison mith quantum mechanics. Of course, Eq. (9) does not change the boundedness of the motion. The density f itself is even the motion integral (9).

How the local instability is described in this presentation? Consider the characteristic function of a small domain that is the initial density concentrated there. In case of the complete chaos the local motion instability results in indefinite expansion of the domain over some $(d-1)$-dimensional manifold and, what is more important, in also indefinite contraction on the complementary manifold, also of $d-1$ dimensions. Notice the energy conservation and zero Lyapunov exponent along a trajectory, which reduces the expansion/contraction dimensions by 2.

Now, if the motion is bounded (in any representation) the expanding manifold has eventually to mix up and to form the so-called foliation with ever increasing number of sheets whose width is exponentially decreasing. This highly intricate structure of the fine-grained (exact) phase density is necessary to provide the time reversibility of density evolution in agreement with trajectory reversibility. However, any coarse-graining (averaging) of the density results in the loss of this reversibility in apparent contradiction with the reversible dynamics.

Note, that the evolution of the coarse-grained density is the same in both directions of time because the only dynamical difference is interchange of the two manifolds, expanding and contracting. Hence, there is no need for the conception of "time arrow".

The time evolution of some coarse-grained density, say, $\bar{f}(n_i, t)$ (usually averaged over phases θ_i) is described by the kinetic equation which, in principle, can be completely derived from the equations of motion or from Liouville equation (9) without any statistical hypothesis. Note, that coarse-graining as well as the use of the phase density are not hypotheses, but a particular method of adequate description to reveal the essential features of the motion.

In case of small perturbation the kinetic equation takes the form of the Focker–Planck–Kolmogorov (FPK) equation that is a diffusion equation. In our examples of ray dynamics the diffusion rate in spatial angles φ_i and in actions n_i is

$$D_\varphi \sim \varepsilon^2 \frac{\mathrm{v}}{a}; \qquad D_n \sim (\varepsilon n)^2 \frac{\mathrm{v}}{a}, \tag{10}$$

where we dropped sup i ($a_i \sim a$, $\varphi_i \sim \varphi$), and where

$$\mathrm{v} \sim a\dot{\theta} \sim a\frac{\partial \omega}{\partial n}; \qquad \frac{\mathrm{v}}{a} \sim \Omega \tag{11}$$

are the velocity of a "particle" and its unperturbed oscillation frequency, respectively.

The diffusion leads to the statistical relaxation

$$\bar{f}\,(n_i, 0) \to \bar{f}_s\,(n_i) = \delta\,(H_0\,(n_i) - E) \equiv \bar{f}_c\,. \tag{12}$$

Here $\bar{f}_s$ is the steady-state, or statistical equilibrium density, E is the energy, and H_0 is unperturbed Hamiltonian of the completely integrable system. The latter relation (12) means that $\bar{f}_s$ is microcanonical distribution fe on an energy surface. This is called ergodicity, which is the weakest statistical property of motion.

In simple dynamical systems like our models density $\bar{f}_s \neq f_c$ is typically different from microcanonical one. The important statistical property is relaxation itself, no matter what would be the final state $\bar{f}_s$. Loosely speaking, the relaxation is equivalent to a continuous motion spectrum. The exponential instability is a sufficient condition, because the discrete spectrum can provide the linear instability at most (see above). Yet, the exponential instability is much stronger a property as compared with the mixing and relaxation. On the other hand, the latter seems to be quite enough to develop a relevant statistical theory. The point is that relaxation, or correlation decay, provides the property of statistical independence, which is the ultimate basis of the probability theory. These considerations, not yet completely understood, are most important in analyzing the conception of wave chaos below (Section 3).

Coming back to our simple models let me mention an obvious estimate for the relaxation time t_c in case of motion ergodicity. From Eqs (10, 11) we have

$$t_c \sim \frac{1}{D\varphi} \sim \frac{n^2}{D_n} \sim \varepsilon^{-2} W^{-1}. \tag{13}$$

This is the principal result in classical, or ray, dynamics to compare below with quantum, or wave, mechanics.

3. Linear Wave Chaos: Statistical Relaxation in Discrete Spectrum

The motion of a quantum particle in billiard is described by Schrödinger equation. This and other quantum equations are linear with respect to quantum dynamical variable, the Ψ-function, which represents the whole closed quantum system. Moreover, all those equations have purely discrete spectrum for any bounded motion. Hence, a common belief that dynamical chaos is impossible in quantum mechanics. But this conclusion is in apparent contradiction with the fundamental correspondence principle, which requires the complete transition to the classical limit, chaos including.

The quantum evolution equations are wave equations, and they are formally equivalent to any other (classical) wave equations like sound, elastic, electromagnetic etc. ones. Of course, the latter must be linear. For nonlinear wave equations there is no problem with dynamical chaos (see, e. g., Ref. [1]). This is why the quantum chaos is also called the wave chaos [9]. For classical waves it is especially clear that the ray dynamics contains nothing beyond the wave equation, and the nonlinear ray chaos must be present somehow in linear wave equations.

To the best of my knowledge, nobody considered so far how the ray chaos is represented and/or modified in the classical wave equations, for example, in cavities or waveguides.

Before discussing this central topic of my talk let us consider first the following question: why the wave (particularly, quantum) spectrum is discrete? The reference to linearity of the equations would be superficial. Indeed, the linear Liouville equation can have a continuous spectrum as well, which means that there are no nonsingular eigenfunctions. The latter is explained just by the arising of indefinitely dense foliations described above, which is obviously an aperiodic process.

It would be more correct to say that linear equations may be qualitatively different. But what is the nature of the difference, and how to recognize it in a given equation seems still to remain an open question.

A constructive solution of this problem could be the attempt either to determine the type of spectrum (discrete or continuous) or to find the ray approximation described by some ordinary differential equation.

The latter is apparently impossible for the Liouville equation. In case of a wave equation the rays, or characteristics, are embedded in the "phase space" of double dimensions as compared to the wave space. In the limit of geometrical optics or of classical mechanics all variables of the former are independent whereas for the wave equation each pair of variables is interconnected by a Fourier transform, and, hence, is restricted by the uncertainty relation.

To describe this situation in a different way one can say that the ray phase space (k_i, x_i) is discrete, the size of a cell being of the order of unity $(\Delta k_i \cdot \Delta x_i \sim 1)$. This is true for any finite $k \to \infty$ and qualitatively different from the very ray limit with its continuous phase space. Hence, the transition to the limit is singular and the implementation of the correspondence principle is far from trivial.

Our solution of this problem [10] (see also review [11]) is based upon introducing characteristic time scales of the wave dynamics on which the latter is close to the ray dynamics, the scales increasing indefinitely towards the ray limit $k \to \infty$. Let me show how this approach works in our simple models.

The most important relaxation time scale t_R is determined by the mean density ρ of frequency (energy) eigenvalues

$$t_R \sim \rho = \Delta^{-1}, \tag{14}$$

where Δ is the average eigenvalue spacing. This is a direct implication of the uncertainty relation $\Delta t \cdot \Delta\omega \sim 1$. Indeed, while $t \sim \Delta t \lesssim \rho$ the discreteness of spectrum is not resolved as $\Delta\omega \sim t^{-1} \gtrsim \Delta$. Hence, the perturbation can and actually does act as one with a continuous spectrum. Particularly, this produces diffusion and relaxation in the discrete spectrum!

Then, an important parameter of wave dynamics is the ratio (see Eq. (13))

$$\lambda = \frac{t_R}{t_e}, \tag{15}$$

which I call the ergodicity parameter. If $\lambda \gtrsim 1$ the ergodic microcanonical steady state $\bar{f}_e$ is reached (12) like in the ray limit. For $\lambda \lesssim 1$ we would expect a different steady state which will be discussed below.

Now, let us estimate parameter λ for our models. To this end, we observe that the total number of eigenvalues up to $n_i \sim n$ is roughly $N(\omega) \sim n^d$. Hence, density ρ and relaxation time scale t_R are

$$t_R \sim \rho = \frac{dN}{d\omega} \sim n^{d-1} \left| \frac{\partial\omega}{\partial n} \right|^{-1}. \tag{16}$$

In combination with Eq. (13), we obtain

$$\lambda \sim \varepsilon^2 n^{d-1} = (\varepsilon n)^2 n^{d-3}. \tag{17}$$

Note, that λ does not depend on dispersion relation $\omega(n_i)$ that is on a particular mechanics of waves.

But the condition $\lambda \geq 1$ is still not sufficient for the wave ergodicity. The point is that the change in action n per collision with the wall is $\Delta n_1 \sim \varepsilon n$. Since in wave mechanics all n_i are integer the transitions between unperturbed states n_i, hence, any diffusion is only possible if

$$\Delta n_1 \sim \varepsilon n \geq 1. \tag{18}$$

This condition allows another simple interpretation if we compare the scattering angle ε of a ray with the minimal diffraction angle $\mu \sim \lambda\!\!\!^{-}/a \sim 1/n$, where $\lambda\!\!\!^{-} \sim a/n$ is the wave length. Then, condition (18) means that diffraction can be neglected: $\mu \lesssim \varepsilon$. This is, of course, the well-known condition for geometrical optics to hold. Yet, this is also not sufficient, generally, because there is another condition (17). The relation between them depends on spatial dimensions. If $d \geq 3$, usual condition (18) is crucial for ergodicity. However, for $d=2$ the ergodicity depends on λ only while the diffraction parameter $\varepsilon n \geq n^{1/2}$ must be very big for $n \gg 1$.

In any event the wave dynamics becomes ergodic for sufficiently large n if ray dynamics is chaotic. This implies that wave eigenfunctions are also ergodic in agreement with Shnirelman theorem announced in 1974 and proved in 1990 [12]. From Eqs (17) and (18) the ergodicity border in n or in ω is

$$n_e(\omega_e) \sim \begin{cases} \varepsilon^{-2}; & d=2\,, \\ \varepsilon^{-1}\,; & d \geq 3\,. \end{cases} \tag{19}$$

The transition to the ray limit depends on two parameters, wave number $n \to \infty$ and time interval $t \to \infty$. The complexity of transition is in that the result depends on the order of two limits. If, first, we take formally the limit $n \to \infty$, then for any finite $t \to \infty$ the wave dynamics becomes the ray dynamics. However, if we fix n, no matter how large, the wave behaviour remains close to the ray diffusion while $t \leq t_R(n)$ only. Hence, the ray limit is singular, indeed.

Above the ergodicity border (19), as $n \to \infty$, almost every eigenfunction, in Wigner's representation, approaches microcanonical distribution $\bar{f}_e$ (12) [12]. Moreover, the fluctuations in chaotic eigenfunctions were shown to be Gaussian [20] (see also Ref. [10]). Yet, they are not completely random. The most important elements of their microstructure are so-called "scars", that is some enhancements above the average microcanonical density along unstable periodic rays. The scars had been discovered in quantum billiard models by Heller [21] and were subsequently confirmed by many others (see, e. g., Ref. [22]). The question how to reconcile scars with Shnirelman's theorem is not completely clear so far. Apparently, all the scars are of the minimal width comparable with an elementary wave cell (Section 3). If so, the scars would not affect any quasi-classical integral relation which is Shnirelman's wave ergodicity.

4. Wave Chaos: Diffusion Localization

A more interesting question is what happens if $n \lesssim n_c$ or $\omega \lesssim \omega_c$? For $d \gtrsim 3$ the answer is very simple: nothing happens as all the transitions are suppressed, so that the initial state of wave field doesn't change at all. It is also called perturbative localization, because the wave perturbation theory is applicable just under condition $\varepsilon n \lesssim 1$. In this case the eigenfunctions remain close to the unperturbed ones.

For $d = 2$ the problem is much more difficult. The first guess could be as follows. The diffusion proceeds during time interval $t_R \sim \rho$, so that the angular size of the wave packet becomes $(\Delta\varphi)^2 \sim D_\varphi t_R \sim \lambda \lesssim 1$.

However, there are at least two questions. First, would the diffusion stop for $t \gtrsim t_R$? This plausible conjecture was well confirmed numerically, indeed [10, 11]. Second, would the density ρ remain unchanged if $\Delta\varphi \lesssim 1$? Apparently, it will not as now only a part of the unperturbed states contributes to ρ. A plausible estimate for new density and relaxation time scale is

$$\tilde{\rho} \sim \rho\, \Delta\varphi \sim \tilde{t}_R\,. \tag{20}$$

Then, from the diffusion law $(\Delta\varphi)^2 \sim D_\varphi \tilde{t}_R$, we obtain

$$(\Delta\varphi)_s \sim \varepsilon^2 n; \quad (\Delta n)_s \sim D_n' \sim l_s\,, \tag{21}$$

which is much smaller than the first guess. In the second estimate (21) $D_n' \sim (\varepsilon n)^2$ is the diffusion rate per collision with wall. This estimate was also well confirmed numerically in somewhat different but related models [10, 11].

The size l_s is called also the localization length as it characterizes suppression, or localization, of the ray diffusion. Note, that condition (18) must be satisfied, otherwise the perturbative localization occurs with minimal

$$(\Delta\varphi)_p \sim \frac{1}{n}; \quad (\Delta n)_p \sim 1. \tag{22}$$

For $\varepsilon n \gtrsim 1$ a very peculiar steady state is formed whose size is given by estimate (21). As in the ray limit this steady state is a result of statistical (diffusion) relaxation $\bar{f}(n, 0) \to \bar{f}_s(n)$ but it differs from the limiting microcanonical distribution $\bar{f}_e(n)$ in two major ways.

First, the wave steady state depends on the initial state as the former is a result of diffusion localization of the latter. Hence, the steady state is always spread around the initial state. Second, the steady state is composed of a finite number M of eigenfunctions owing to the discreteness of the wave phase space. Roughly, $M \sim l_s$ (21), hence, statistical fluctuations are finite, $\sim M^{-1/2} \sim \sim (\varepsilon n)^{-1}$. One can say also that wave phase density $\bar{f}(n, t)$ represents always a finite ensemble of $\sim M$ systems [11]. Fluctuations of the classical electromagnetic field in wave-guides were discussed, e. g., in Ref. [16], without any relation to the wave chaos though.

This remains true for an ergodic steady state as well, for any λ and d, when $M \sim n^{d-1}$. This estimate is inferred from the only restriction $\omega(n_i) \approx$ const for the ergodic steady state and eigenfunctions. In turn, it implies a finite width $\Delta\omega$ of the energy shell occupied by an ergodic state. Using the last estimate and Eq. (16), we obtain

$$\Delta\omega \sim \Omega \,. \tag{23}$$

As $n \to \infty$ the relative width $\Delta\omega/\omega \to 0$ because $\omega(n) \to \infty$ in our units ($\hbar = 1$), and the energy shell becomes an energy surface of the ray dynamics. Estimate (23) is also valid for nonergodic steady states.

Even on the relaxation time scale t_R the wave behaviour is qualitatively different from the limiting ray dynamics in that the former is perfectly stable whereas the latter is strongly unstable. It was proved in numerical (computer) experiments via time reversal in both cases [13]. For the ray dynamics the diffusion is immediately restored due to a fast growth of computation errors while for waves the "antidiffusion" proceeds down to the initial state which is recovered with surprisingly high accuracy.

At a first glance, it seems to contradict to the correspondence principle. The resolution is in that there exists another characteristic time scale t_{BZ} on which the wave dynamics is also unstable, yet $t_{BZ} \ll t_R$. It was discovered and explained by Berman and Zaslavsky [14] (see also Refs [10, 11]). The instability manifests itself in rapid spreading of a narrow wave packet which follows for a while the beam of rays according to the Ehrenfest theorem. In case of exponential instability the corresponding time scale is very short. Roughly (see Eq. (7)),

$$t_{BZ} \approx \frac{\ln n}{\Lambda} \ll t_R \,. \tag{24}$$

Nevertheless, $t_{BZ} \to \infty$ as $n \to \infty$ in accordance with the correspondence principle. Yet, this transition, as well as that for t_R discussed above, is singular, being also a double limit in n and t.

A nice picture of the initially unstable wave evolution can be found in Ref. [15] where the formation of the foliation in phase space, described above in Section 2, is clearly seen.

5. The Nature of Wave and Quantum Chaos

The principal distinction of the wave chaos is its transient character. In other words, the wave dynamics remains close to the ray chaotic motion on a finite time scale only [10]. Moreover, particular statistical properties of the wave evolution correspond to rather different time scales. The strongest ones related to the exponential instability persist on the shortest time scale (24) which is almost independent on integer wave numbers n. Yet, it is important that such significant statistical behaviour as diffusion and relaxation continues much longer, on time scale (16), thus providing a relevant statistical description of wave dynamics. This is very important in many applications as it allows for a fairly simple statistical representation of the essential features of otherwise highly complicated phenomena.

From mathematical point of view the problem of wave chaos requires a fundamental generalization of the contemporary ergodic theory [6] which was developed for the ray chaos, and which is essentially asymptotic in time ($|t| \to \infty$). Note the considerable simplification of the theory due to that asymptotic approach.

To emphasize the finite-time nature of the wave chaos we call it pseudochaos as distinct from the true ray chaos. The ultimate origin of wave chaos limitations is in the discrete phase space as explained above (Section 3).

The same is true, by the way, for the computer simulation of any dynamical systems. Moreover, in the digital computer any quantity is "quantized" that is represented by an integer number, whereas in the wave mechanics only the product of each pair of canonically conjugated variables does so. As a result any dynamical trajectory in computer eventually becomes periodic as compared with almost periodic evolution of waves. Note, that it is the effect of round-off "errors" which are not random at all. This should be taken into account in computer simulations of ray dynamics. The shortest time scale (24) is of the order of length n_m (the number of bits) of computer word (mantissa), and it is negligible in most cases. Fortunately, the relaxation time scale (16) grows like some power of the maximal computer number ($\sim 2^{n_m}$) which is typically not a serious restriction (for details see, e. g., Refs [10, 17]).

Coming back to wave dynamics I would like to stress that even though from formal mathematical point of view all linear wave equations have similar properties the physics of quantum waves is fundamentally different. While most classical linear waves are simply a low-amplitude approximation (an important exception is electromagnetic waves) the linear quantum mechanics is as yet the most fundamental and universal theory. Hence, the "true" classical chaos is but a limiting pattern, very important in the theory but never realized, strictly speaking, in nature.

On the other hand, the evolution of Ψ wave is only a part, and a simpler one, of the quantum dynamics as a physical theory. The other part, much more difficult and vague, is the quantum measurement with its misterious Ψ collapse. It is not excluded that the latter is the most spectacular example of the true quantum chaos (for discussion see, e. g., Refs [11, 18, 19]).

Acknowledgements. It is a pleasure to acknowledge a long-term fruitful collaboration in quantum chaos with my friends Giulio Casati, Joe Ford, Italo Guarneri, Felix Izrailev and Dima Shepelyansky. My thanks are also due to S.S. Abdullaev, G.P. Berman and G.M. Zaslavsky for stimulating discussions.

References

1. A. Lichtenberg and M. Lieberman, Regular and Stochastic Motion, Springer, 1983; G.M. Zaslavsky, Chaos in Dynamic Systems, Harwood, New Jork, 1985.
2. Ya.G. Sinai, Usp. Mat. Nauk, **25**, N 2 (1970) 141.

3. L.A. Bunimovich, Zh. Eksper. Teor. Fiz., **89** (1985) 1452.
4. V.I. Arnold and A. Avez, Ergodic Problems of Classical Mechanics, Benjamin, 1968.
5. J. Wersinger, E. Ott, and J. Finn, Phys. Fluids, **21** (1978) 2263.
6. I. Kornfeld, S. Fomin and Ya. Sinai, Ergodic Theory, Springer, 1982.
7. S.S. Abdullaev, G.M. Zaslavsky, Zh. Exper. Teor. Fiz., **80** (1981) 524; **85** (1983) 1573; **87** (1984) 763.
8. A.V. Chigarev, Yu. V. Chigarev, Akustich. Zh., **24** (1978) 765.
9. P. Šeba. Phys. Rev. Lett., **64** (1990) 1855.
10. B.V. Chirikov, F.M. Izrailev and D.L. Shepelyansky, Sov. Sci. Rev. C **2** (1981) 209; Physica, D **33** (1988) 77.
11. B.V. Chirikov, Time-Dependent Quantum Systems, Proc. Les Houches Summer School on Chaos and Quantum Physics, Elsevier, 1990.
12. A.I. Shnirelman, Usp. Mat. Nauk, **29**, N6 (1974) 181; On the Asymptotic Properties of Eigenfunctions in the Regions of Chaotic Motion, Addendum to the book: V.F. Lasutkin. The KAM Theory and Asymptotics of Spectrum of Elliptic Operators, Springer, 1990.
13. D.L. Shepelyansky, Physica, D **8** (1983) 208; G. Casati, B.V. Chirikov, I. Guarneri, and D.L. Shepelyansky, Phys. Rev. Lett., **56** (1986) 2437.
14. G.P. Berman and G.M. Zaslavsky, Physica, A **91** (1978) 450.
15. K. Nakamura, A. Bishop, and A. Shudo, Phys. Rev., B **39** (1989) 12 422.
16. S.S. Abdullaev, Zh. Tekhn. Fiz., **51** (1981) 697.
17. F. Rannou, Astron. Astroph., 31 (1974) 289; Y. Levy. Phys. Lett., A **88** (1982) 1; J. McCauley, ibid., **115** (1986) 433.
18. B.V. Chirikov, Wiss. Z., **24** (1975) 215.
19. M. Gell-Mann and J. Hartle, Quantum Mechanics in the Light of Quantum Cosmology. Proc. 3rd. Intern. Symp. on The Foundations of Quantum Mechanics in the Light of New Technology, Tokyo. 1989.
20. F.M. Izrailev, Phys. Lett., A **125** (1987) 250.
21. E. Heller, Phys. Rev. Lett., **53** (1984) 1515.
22. R. Jensen et al., Phys. Rev. Lett., **63** (1989) 2771.

Second-Order Perturbations for Solitons

D.J. Kaup

Department of Mathematics and Computer Science, and
Institute for Nonlinear Studies, Clarkson University,
Potsdam, NY 13699-5815, USA

Abstract. A singular perturbation theory for nonlinear Schrödinger solitons is extended into second order. There are certain subtle points which occur in second-order, and these are detailed and described.

Recently, this author has presented a singular perturbation theory for soliton propagation in optical fibers.[1] This work was an extension of previous soliton perturbation theories [2–8], and gave explicit formulas for calculating first-order effects. In fact, the analytic results were shown to very nicely compliment known numerical calculations [1,9].

Also recently Smith and Mollenauer have experimentally observed a long range phase-independent interaction occurring between soliton pairs [10]. One suggestion for a possible source of this interaction was the continuous spectrum (radiation) emitted by the perturbed solitons. This would be a second-order interaction, where the nonlinearity couples the emitted radiation from one soliton to the other soliton. Such an effect has never been calculated before, so no real estimates exists. The purpose of this paper is to provide the basic equations required for such a calculation.

Of course in statistical studies, it is the second-order terms which are the most interesting. They are the terms which in general will not average to zero. When one first looks at the general equations for second order soliton perturbations, it is a bit surprising to see how rapidly the number of possible terms have multiplied. Usually one does not need to evaluate every possible term; but one does need some way to isolate the important terms. And it is not easy to know a prior exactly which second-order terms are the important ones. Consequently it is of considerable value to have simply a complete list such as we give here. By studying this list (after determining what are the important first-order terms), one could isolate the important second-order terms.

In the process of detailing these second-order equations, certain apparent secular terms arose. We have traced the origin of those secularities and found a method for handling them. This is discussed in Appendix B, with the result that the final expansion is free of secular terms.

As shown in Ref 1, one may expand solutions of the perturbed NLS

$$i\partial_t q = -\partial_x^2 q - 2q^* q^2 + \epsilon R[q, q^*] \tag{1}$$

in the singular perturbation expansion about a single soliton as

$$q = q_0 + \epsilon q_1 + \epsilon^2 q_2 + \cdots \tag{2}$$

Springer Proceedings in Physics, Vol. 67 Nonlinearity with Disorder
Editors: F. Abdullaev · A.R. Bishop · S. Pnevmatikos

where

$$q_0 = \frac{Ae^{i\alpha}}{cosh\theta} \tag{3}$$

$$\theta = 2\eta(x - \bar{x}) \tag{4}$$

$$\alpha = -2\xi(x - \bar{x}) + \bar{\alpha} \tag{5}$$

In the above, A is the amplitude of the soliton, η is the inverse of the width, $\bar{x}$ is its position, and $\bar{\alpha}$ is its phase at its center. As in Ref 1, we allow the soliton amplitude and width to be independent variables. However if the perturbations is small, we do expect A to be very close to the value of 2η, which is the pure soliton value.

We also introduce the multiple time scales whence

$$\tau_n = \epsilon^n t \tag{6a}$$

$$\frac{\partial}{\partial t} = \frac{\partial}{\partial \tau_0} + \epsilon\frac{\partial}{\partial \tau_1} + \epsilon^2\frac{\partial}{\partial \tau_2} + \cdots \tag{6b}$$

We take $A = A(\tau_1, \tau_2, \cdots)$ and $\eta = \eta(\tau_1, \tau_2, \cdots)$. For simplicity and to avoid certain secular terms, we also take $\bar{x}$ and $\bar{\alpha}$ to be independent of $\tau_0(= t)$ and to be dependent only on the slow time scales. Consequently, in order to have the proper zeroth-order behavior, we expand $\bar{x}$ as

$$\bar{x} = \bar{x}_{-1}\frac{1}{\epsilon} + \bar{x}_0 + \bar{x}_1\epsilon + \cdots \tag{7}$$

and similarly for $\bar{\alpha}$.

For the higher order parts of the solution, define the column matrix

$$v = \begin{bmatrix} e^{-i\alpha}(q_1 + \epsilon q_2 + \cdots) \\ e^{i\alpha}(q_1^* + \epsilon q_2^* + \cdots) \end{bmatrix} \tag{8}$$

then the evolution of v is given by

$$i\partial_t v + 4\eta^2 Lv = F \tag{9}$$

where the operator L is

$$L = \sigma_3(\partial_\theta^2 - 1) + \frac{2}{cosh^2\theta}(2\sigma_3 + i\sigma_2), \tag{10}$$

with σ_1, σ_2 and σ_3 being the Pauli spin matrices. The eigenstates and closure of L are summarized in appendix A.

The Source F is of the form

$$F = \begin{bmatrix} \mathcal{R} \\ -\mathcal{R}^* \end{bmatrix} \tag{11}$$

where from all of the above, we obtain

$$\epsilon\mathcal{R} = \epsilon e^{-i\alpha}R - \frac{i}{cosh\theta}\partial_t A - \left(\frac{A\theta}{\eta cosh\theta} + \frac{\epsilon\theta}{\eta}\mu\right)\partial_t\xi$$

$$+(4\eta^2 - A^2)\left[\frac{2A}{cosh^3\theta} + \frac{2\epsilon}{cosh^2\theta}(2\mu + \mu^*)\right]$$

$$+i\theta\left[\frac{Asinh\theta}{\eta cosh^2\theta} - \frac{\epsilon}{\eta}\partial_\theta\mu\right]\partial_t\eta$$

$$+\left[\frac{2\xi A}{cosh\theta} - 2iA\eta\frac{sinh\theta}{cosh^2\theta} + 2i\eta\epsilon\partial_\theta\mu + 2\epsilon\xi\mu\right](\partial_t\bar{x} + 4\xi)$$

$$+\left(\frac{A}{cosh\theta} + \epsilon\mu\right)(\partial_t\bar{\alpha} - 4\eta^2 - 4\xi^2) - 2\epsilon^3(\mu^2\mu^*)$$

$$-\frac{2A\epsilon^2}{cosh\theta}(\mu)^2 - \frac{4\epsilon^2 A}{cosh\theta}(\mu^*\mu) \tag{12}$$

where μ is the first component of v, which by (8) is

$$\mu = e^{-i\alpha}(q_1 + q_2\epsilon + q_3\epsilon^2 + \cdots) \tag{13}$$

The terms of order ϵ in (12) were given in Ref 1. Here we have given the general expression which can be expanded to any order desired. One would only have to use (6), (7), and (13) in order to expand (12) to the order desired.

One should note that the coefficients of $\partial_t\xi$ and $\partial_t\eta$ contain terms secular in θ at second-order since, in general, μ will approach a plane wave for large θ. The origin of these terms and their handling is described in Appendix B.

As discussed in Appendix A (and B), the correct expansion for v is in terms of the continuous eigenfunctions of L. Thus we take

$$v = \eta\int_{-\infty}^{\infty} dk\,[\ g(\eta k + \xi, t)|\psi, k> + \bar{g}(\eta k - \xi, t)|\bar{\psi}, k>\ ] \tag{14}$$

Consequently v, to all orders, is orthogonal to the four states $|\phi_e>, |\phi_0>$, $|\chi>$, and $|\theta\phi_e>$. Inner products of (12) with these states determine the evolution of the soliton parameters. In first order [1], these are

$$\partial_t\xi = \frac{\epsilon\eta}{2A} < \phi_0|\sigma_3|F_{ext} > \tag{15}$$

$$4\partial_t A - 2\frac{A}{\eta}\partial_t\eta = -i\epsilon < \phi_e|\sigma_3|F_{ext} > \tag{16}$$

$$\partial_t\bar{x} + 4\xi = -\frac{i\epsilon}{4A\eta} < \theta\phi_e|\sigma_3|F_{ext} > \tag{17}$$

$$\partial_t\bar{\alpha} - 4\eta^2 - 4\xi^2 = \frac{\epsilon}{2A\eta} < \chi|\sigma_3|F_{ext} > \tag{18}$$

where

$$|F_{ext}> = \begin{bmatrix} Re^{-i\alpha} \\ -R^* e^{i\alpha} \end{bmatrix} \tag{19}$$

is the external forcing. The density of the radiation, g and $\bar{g}$, is given in first-order by

$$\partial_t g - 4i\eta^2(k^2+1)g = \frac{1}{2\pi\eta a^2} < \bar{\phi}|\sigma_3|F_{ext} > \tag{20}$$

$$\partial_t \bar{g} + 4i\eta^2(k^2+1)\bar{g} = \frac{-1}{2\pi\eta a^2} < \phi|\sigma_3|F_{ext} > \tag{21}$$

where a is given by (A.19) and $< \phi|$ and $< \bar{\phi}|$ by (A.11) and (A.12).

If we take (9) and expand it to second-order, and then take the inner product of the resultant with the basis vectors $| \psi >, | \bar{\psi} >, | \phi_e >, | \theta\phi_e >, | \phi_o >$, and $| \chi >$, one then will obtain the following six equations

$$2\frac{A}{\eta}\xi_t = \epsilon\langle\phi_o \mid \sigma_3 \mid F_{ext}\rangle - \epsilon^2\langle\phi_o \mid \sigma_3 \mid U\rangle + (\bar{\alpha}_t - 4\eta^2 - 4\xi^2)\,\epsilon\langle\phi_o \mid v\rangle$$

$$+i\epsilon\,(\bar{x}_t + 4\xi)\,\langle\phi_o \mid \sigma_3\,(2\eta\partial_\theta - 2i\xi\sigma_3) \mid v\rangle + \epsilon\eta_t \int_{-\infty}^{\infty} dk\langle\phi_o \mid \sigma_3\,\{g \mid \eta, k\rangle + \bar{g} \mid \bar{\eta}, k\rangle\}$$

$$+\epsilon\xi_t \int_{-\infty}^{\infty} dk\langle\phi_o \mid \sigma_3\,\{g \mid \xi, k\rangle + \bar{g} \mid \bar{\xi}, k\rangle\} \tag{22a}$$

$$i\,(\bar{x}_t + 4\xi)\,4\eta A = \epsilon\langle\theta\phi_e \mid \sigma_3 \mid F_{ext}\rangle - \epsilon^2\langle\theta\phi_e \mid \sigma_3 \mid U\rangle + \epsilon\,(\bar{\alpha}_t - 4\eta^2 - 4\xi^2)\,\langle\theta\phi_e \mid v\rangle$$

$$+i\,(\bar{x}_t + 4\xi)\,\epsilon\langle\theta\phi_e \mid \sigma_3\,(2\eta\partial_\theta - 2i\xi\sigma_3) \mid v\rangle$$

$$+\epsilon\eta_t \int_{-\infty}^{\infty} dk\langle\theta\phi_e \mid \sigma_3\,\{g \mid \eta, k\rangle + \bar{g} \mid \bar{\eta}, k\rangle\}$$

$$+\epsilon\xi_t \int_{-\infty}^{\infty} dk\langle\theta\phi_e \mid \sigma_3\,\{g \mid \xi, k\rangle + g \mid \bar{\xi}, k\rangle\} \tag{22b}$$

$$4iA_t - i2\eta_t\frac{A}{\eta} = \epsilon\langle\phi_e \mid \sigma_3 \mid F_{ext}\rangle - \epsilon^2\langle\phi_e \mid \sigma_3 \mid U\rangle$$

$$+i\,(\bar{x}_t + 4\xi)\,\epsilon\langle\phi_e \mid \sigma_3\,(2\eta\partial_\theta - 2i\xi\sigma_3) \mid v\rangle$$

$$+\,(\bar{\alpha}_t - 4\xi^2 - 4\eta^2)\,\epsilon\langle\phi_e \mid v\rangle + \epsilon\eta_t \int_{-\infty}^{\infty} dk\langle\phi_e \mid \sigma_3\,\{g \mid \eta, k\rangle + \bar{g} \mid \bar{\eta}, k\rangle\}$$

$$+\epsilon\xi_t \int_{-\infty}^{\infty} dk\langle\phi_e \mid \sigma_3\,\{g \mid \xi, k\rangle + \bar{g} \mid \bar{\xi}, k\rangle\} \tag{22c}$$

$$2A\,(\bar{\alpha}_t - 4\eta^2 - 4\xi^2) = \epsilon\langle\chi \mid \sigma_3 \mid F_{ext}\rangle - \epsilon^2\langle\chi \mid \sigma_3 \mid U\rangle$$

$$+4A\,(A^2 - 4\eta^2) - (\bar{x}_t + 4\xi)\,4A\xi + \epsilon\,(\bar{\alpha}_t - 4\eta^2 - 4\xi^2)\,\langle\chi \mid v\rangle$$

$$+i\epsilon\,(\bar{x}_t + 4\xi)\,\langle\chi \mid \sigma_3\,(2\eta\partial_\theta - 2i\xi\sigma_3) \mid v\rangle + \epsilon\eta_t \int_{-\infty}^{\infty} dk\langle\chi \mid \sigma_3\,\{g \mid \eta, k\rangle + \bar{g} \mid \bar{\eta}, k\rangle\}$$

$$+\epsilon\xi_t \int_{-\infty}^{\infty} dk\langle\chi \mid \sigma_3\,\{g \mid \xi, k\rangle + \bar{g} \mid \bar{\xi}, k\rangle\} \tag{22d}$$

$$2\pi i\epsilon\eta a^2\left[\bar{g}_t + 4i\eta^2\left(k^2+1\right)\bar{g}\right] = \epsilon\langle\phi \mid \sigma_3 \mid F_{ext}\rangle$$
$$-\epsilon^2\langle\phi \mid \sigma_3 \mid U\rangle - i\eta\langle\phi \mid \phi_e\rangle\partial_t(A/\eta)$$

$$+2A\left(4\eta^2 - A^2\right)\langle\phi \mid \sigma_3 \mid \nu\rangle + \left(\bar{\alpha}_t - 4\eta^2 - 4\xi^2\right)\epsilon\langle\phi \mid v\rangle$$
$$+i\epsilon\left(\bar{x}_t + 4\xi\right)\langle\phi \mid \sigma_3\left(2\eta\partial_\theta - 2i\xi\sigma_3\right) \mid v\rangle$$

$$+\epsilon\eta_t\int_{-\infty}^{\infty} dk'\langle\phi \mid \sigma_3\left\{g' \mid \eta, k'\rangle + \bar{g}' \mid \bar{\eta}, k'\rangle\right\}$$

$$+\epsilon\xi_t\int_{-\infty}^{\infty} dk'\langle\phi \mid \sigma_3\left\{g' \mid \xi, k'\rangle + \bar{g}' \mid \bar{\xi}, k'\rangle\right\} \tag{22e}$$

$$-2\pi i\epsilon\eta a^2\left[g_t - 4i\eta^2\left(k^2+1\right)g\right]$$

$$= \epsilon\langle\bar{\phi} \mid \sigma_3 \mid F_{ext}\rangle - \epsilon^2\langle\bar{\phi} \mid \sigma_3 \mid U\rangle - i\eta\langle\bar{\phi} \mid \phi_e\rangle\partial_t(A/\eta)$$

$$+2A\left(4\eta^2 - A^2\right)\langle\bar{\phi} \mid \sigma_3 \mid \nu\rangle + \left(\bar{\alpha}_t - 4\eta^2 - 4\xi^2\right)\epsilon\langle\bar{\phi} \mid v\rangle$$
$$+i\epsilon\left(\bar{x}_t + 4\xi\right)\langle\bar{\phi} \mid \sigma_3\left(2\eta\partial_\theta - 2i\xi\sigma_3\right) \mid v\rangle$$

$$+\epsilon\eta_t\int_{-\infty}^{\infty} dk'\langle\bar{\phi} \mid \sigma_3\left\{g' \mid \eta, k'\rangle + \bar{g}' \mid \bar{\eta}, k'\rangle\right\}$$

$$+\epsilon\xi_t\int_{-\infty}^{\infty} dk'\langle\bar{\phi} \mid \sigma_3\left\{g' \mid \xi, k'\rangle + \bar{g}' \mid \bar{\xi}, k'\rangle\right\} \tag{22f}$$

where the states $\mid U >$ and $\mid \nu >$ are given by

$$\mid U >= \frac{2A}{\cosh\theta}\begin{bmatrix} 2\mu^*\mu + \mu^2 \\ -2\mu^*\mu - \mu^{*2} \end{bmatrix} \tag{23a}$$

$$\mid \nu >= \frac{1}{\cosh^2\theta} \mid \phi_e > \tag{23b}$$

The states $\mid \eta >, \mid \bar{\eta} >, \mid \xi >$ and $\mid \bar{\xi} >$ are described in Appendix B.

Of course the first-order terms of these six equations will be the same as in (15)-(21). What is most obvious is the presence of a tremendous amount of cross-coupling between the various modes. For example, deviations of $\bar{x}$ from its unperturbed value ($\bar{x}_t + 4\xi = 0$) coupled into all six equations via the continuous spectra. The general problem is clearly complex and shall have to be treated in pieces. Currently we are investigating the phase-independent part of this interaction and the results will be presented elsewhere.

Appendix A

The operator L was discussed in Ref. 1 and has the following eigenstates

$$L|\psi, k >= (k^2+1)|\psi, k > \tag{A.1}$$

$$L|\bar{\psi}, k> = -(k^2+1)|\bar{\psi}, k> \tag{A.2}$$

$$L|\phi_0> = 0 = L|\phi_e> \tag{A.3}$$

where

$$|\psi, k> = e^{ik\theta}[1 - \frac{2ike^{-\theta}}{(k+i)^2 cosh\theta}] \begin{pmatrix} 0 \\ 1 \end{pmatrix} + \frac{e^{ik\theta}}{(k+i)^2 cosh^2\theta} \begin{pmatrix} 1 \\ 1 \end{pmatrix} \tag{A.4}$$

$$|\bar{\psi}, k> = \sigma_1 |\psi, k> \tag{A.5}$$

$$|\phi_e> = \frac{1}{cosh\theta} \begin{pmatrix} 1 \\ -1 \end{pmatrix}, |\phi_0> = \frac{sinh\theta}{cosh^2\theta} \begin{pmatrix} 1 \\ 1 \end{pmatrix} \tag{A.6}$$

The closure of these states require two more states.[11]

$$L|\chi> = -|\phi_e> \tag{A.7}$$

$$L|\theta\phi_e> = -|\phi_0> \tag{A.8}$$

where

$$|\chi> = \frac{\theta tanh\theta - 1}{cosh\theta} \begin{pmatrix} 1 \\ 1 \end{pmatrix} \tag{A.9}$$

$$|\theta\phi_e> = \theta|\phi_e> \tag{A.10}$$

Although L is self-adjoint, instead of using (A.1) and (A.2) as solutions of the adjoint problem, it is best to use the states

$$< \phi, k| = e^{-ik\theta}[1 - \frac{2ike^{\theta}}{(k+i)^2 cosh\theta]} \widetilde{\begin{pmatrix} 1 \\ 0 \end{pmatrix}} + \frac{e^{-ik\theta}}{(k+i)^2 cosh^2\theta} \widetilde{\begin{pmatrix} 1 \\ 1 \end{pmatrix}} \tag{A.11}$$

$$< \bar{\phi}, k| = < \phi, k|\sigma_1 \tag{A.12}$$

where $\sim$ indicates the matrix transpose. Then

$$< \phi, k|L^A = -(k^2+1) < \phi, k| \tag{A.13}$$

$$< \bar{\phi}, k|L^A = +(k^2+1) < \bar{\phi}, k| \tag{A.14}$$

For the adjoint bound states and closure states, we use the matrix transpose of (A.6), (A.9) and (A.10).

The only nonzero inner products are

$$< \phi, k'|\sigma_3|\bar{\psi}, k> = 2\pi a^2 \delta(k-k') \tag{A.15}$$

$$< \bar{\phi}, k'|\sigma_3|\psi, k> = -2\pi a^2 \delta(k-k') \tag{A.16}$$

$$< \theta\phi_e|\sigma_3|\phi_0 >= 2 =< \phi_0|\sigma_3|\theta\phi^e > \qquad (A.17)$$

$$< \phi_e|\sigma_3|\chi >= -2 =< \chi|\sigma_3|\phi_e > \qquad (A.18)$$

where

$$a = \frac{k-i}{k+i} \qquad (A.19)$$

and the inner product is defined as

$$< u|\sigma_3|v >= \int_{-\infty}^{\infty} d\theta \widetilde{u(\theta)}\sigma_3 v(\theta) \qquad (A.20)$$

with no complex conjugations involved.

Appendix B

Here we will discuss the origin of the secular (in θ) terms in (12), and describe how one can handle them such that the solution for v will have no terms secular in θ.

If one is far away from the soliton (θ large), then (12) reduces to

$$\epsilon\mathcal{R} = \epsilon e^{-i\alpha} R - \frac{\epsilon\theta}{\eta}\mu\partial_t\xi - i\frac{\epsilon\theta}{\eta}(\partial_\theta\mu)\partial_t\eta + \cdots \qquad (B.1)$$

Of course the second and third terms are secular in θ, but these terms are also proportional to either $\partial_t\xi$ or $\partial_t\eta$. Consequently, when η and ξ are stationary, these secularities do not occur. It is not surprising then that these secularities are related to the evolution of η and ξ. In fact, if one traces back to the origin of these terms, he find that the first one originated from the phase factor $e^{-i\alpha}$ in (13) and the second one from taking the independent spatial variable to be θ instead of x. These choices were determined in first order. By introducing the phase factor in (13) and by using θ as the independent spatial variable, then a first-order solution is easily obtained by using separation of variables. But the price of this is seemingly secularity in second-order.

To see how to handle this, first consider only one small-amplitude, simple plane wave emitted by a perturbed soliton. After the emission, it propagates essentially linearly with an unchanging wavelength. However, if we represent our radiation solution as a linear integral over the fundamental solutions (A.4), then since θ is given by (4), if η changes in time, it follows that k must also change in time in exactly the opposite manner, so as to maintain the constancy of the wavelength of the emitted radiation. Once emitted, the radiation is naturally independent of the future history of the soliton. But θ is completely dependent on this history. In other words, by using θ as the independent spatial variable, we are using a variable standard of length which is tied to the evolution of the soliton. To "factor-out" this dependency, we must maintain the product ηk as constant. Since k is really a dummy variable (we shall always integrate over it to obtain physical quantities), it is not very feasible to have k evolve directly. Rather, it is easiest to compensate for this by allowing the density of radiation, g, to evolve appropriately. One can show that these secular terms can be absorbed by taking

$$v = \eta \int_{-\infty}^{\infty} dk \left[g(\eta k + \xi, t)|\psi, k> + \bar{g}(\eta k - \xi, t)|\bar{\psi}, k> \right] \qquad (B.2)$$

where ψ and $\bar{\psi}$ are given by (A.4) and (A.5). By direct calculation, one has

$$-i\partial_t v - \frac{\theta}{\eta}(\partial_t \xi)\sigma_3 v - i\frac{\theta}{\eta}(\partial_t \eta)\partial_t v$$

$$= -i\eta \int_{-\infty}^{\infty} dk[(\partial_t g)|\psi, k> + (\partial_t \bar{g})|\bar{\psi}, k>]$$

$$+(\partial_t \eta) \int_{-\infty}^{\infty} dk[g|\eta, k> + \bar{g}|\bar{\eta}, k>]$$

$$+(\partial_t \xi) \int_{-\infty}^{\infty} dk[g|\xi, k> + \bar{g}|\bar{\xi}, k>] \qquad (B.3)$$

provided $|kg|$ and $|k\bar{g}|$ vanish as $|k| \to \infty$. The states $|\eta, k>, |\bar{\eta}, k>, |\xi, k>$, and $|\bar{\xi}, k>$ are given by

$$|\eta, k> = \frac{2k\theta e^{ik\theta}}{(k+i)^2 cosh^2\theta} \begin{pmatrix} 0 \\ 1 \end{pmatrix} - \frac{2ka}{(k+i)^2}(1 - tanh\theta)e^{ik\theta} \begin{pmatrix} 0 \\ 1 \end{pmatrix}$$

$$+\frac{2i\theta tanh\theta e^{ik\theta}}{(k+i)^2 cosh^2\theta} \begin{pmatrix} 1 \\ 1 \end{pmatrix} - \frac{2ike^{ik\theta}}{(k+i)^3 cosh^2\theta} \begin{pmatrix} 1 \\ 1 \end{pmatrix} \qquad (B.4)$$

$$|\bar{\eta}, k> = \sigma_1 |\eta, k> \qquad (B.5)$$

$$|\xi, k> = -\frac{2a}{(k+i)^2}(1 - tanh\theta)e^{ik\theta} \begin{pmatrix} 0 \\ 1 \end{pmatrix}$$

$$-\frac{2\theta e^{ik\theta}}{(k+i)^2 cosh^2\theta} \begin{pmatrix} 1 \\ 0 \end{pmatrix} - \frac{2ie^{ik\theta}}{(k+i)^3 cosh^2\theta} \begin{pmatrix} 1 \\ 1 \end{pmatrix} \qquad (B.6)$$

$$|\bar{\xi}, k> = -\sigma_1 |\xi, k> \qquad (B.7)$$

Note that these states are nonsecular in θ. Whence the θ - secular terms in (12) can be successfully and effectively summed by taking the k and time dependence of g and $\bar{g}$ to be of the form in (B.2).

References

[1] D.J. Kaup, Institute of Nonlinear Studies preprint #154, (to appear in Phy. Rev A).

[2] D.J. Kaup, SIAM J. Appl. Math. 31, 121-133 (1976).

[3] J.P. Keener and D.W. McLaughlin, Phys. Rev. A 16, 777 (1977).

[4] D.W. McLaughlin and A.C. Scott, Phys. Rev. A 18, 1652 (1978).

[5] V.I. Karpman, Pis'ma Zh. Eksp. Teor. Fiz. 25, 296 (1977) [JETP Lett. 25, 271 (1977)].

[6] V.I. Karpman and E.M. Maslov, Zh. Eksp. Teor. Fiz. 73, 537 (1977) [Sov. Phys. - JETP 46, 281 (1977)].

[7] D.J. Kaup and A.C. Newell, Proc. R. Soc. London Ser. A 361, 413 (1978).

[8] Yuri S. Kivshar and Borsi A, Malomed, Rev. Mod. Phys. 61, 763 (1989).

[9] L.F. Mollenauer, J.P. Gordon and M.N. Islam, IEEE J. Quan. Elec., QE-22, 157-173 (1986).

[10] K. Smith and L.F. Mollenauer, Optics Letters 14, 1284 - 1286 (1989).

[11] D.J. Kaup, J. Math. Anal. Appl. 54, 849 (1976).

Deformation of Solitons in Random Media

M. Wadati

Department of Physics, Faculty of Science, University of Tokyo,
Hongo 7-3-1, Bunkyo-Ku, Tokyo 113, Japan

Abstract. *A partial differential equation which describes nonlinear wave propagations in random media is presented. Based on the equation, behaviors of soliton propagations can be analysed exactly. Under the assumption of Gaussian white randomness, it is shown that the amplitude of a soliton decreases asymptotically as* $x^{-1/2}$*,* x *being the distance of the propagation.*

1. Introduction

A combination of two subjects, nonlinearity and randomness, raises many new problems. For instance, propagation of nonlinear waves in random media or under external noises is a very interesting and important problem. The problem, however, is difficult even in one-dimensional case and seems to require an intensive research on it.

As a model equation for nonlinear waves in random media, the stochastic Korteweg-de Vries(KdV) equation has been proposed [1]. It reads as

$$u_X - 6uu_T + u_{TTT} = \eta(X), \tag{1}$$

where $\eta(X)$ is the random force acting on the system. Here and hereafter, partial differentiations are denoted by subscripts. To be specific, we take $\eta(X)$ to be Gaussian white noise with strength θ,

$$< \eta(X)\eta(X') >= 2\theta\delta(X - X'). \tag{2}$$

It is to be remarked that in order to discuss the spatial evolution the roles of space X and time T are exchanged in eq.(1). An advantage of eq.(1) is that it is analytically tractable. More precisely, the boundary-value problem of eq.(1) is solvable. It can be shown that the average of the soliton solution

$$u(X,T) = -2\kappa^2 \text{sech}^2(\kappa T - 4\kappa^3 X - \kappa T_0), \tag{3}$$

in the limit $X \to \infty$, is expressed as

$$< u(X,T) >= -\frac{\kappa}{\sqrt{3\pi\theta}} X^{-3/2} \exp\left[-\frac{(T - T_0 - 4\kappa^2 X)^2}{48\theta X^3}\right]. \tag{4}$$

This form clearly indicates the deformation of the soliton due to the randomness or the external noise. Similar problems for other soliton systems, such as the nonlinear Schrodinger equation and the Sine-Gordon equation, have been reported (see, for a review, [2,3,4]).

Recently, Ishiwata et al. (IOWT, for short) numerically studied a random Toda lattice [5]. Two kinds of particles with different masses are randomly distributed on a one-dimensional lattice and interact with their nearest neighbors through an identical Toda (exponential) potential. A soliton is sent from one end of the lattice and the deformation of the soliton by the randomness is observed. After many trials , the following "experimental formula" was proposed. Let A_{in} be the amplitude of the incident soliton and A_n the soliton amplitude at the n-th lattice point. During the propagation the amplitude of a soliton decreases as

$$\frac{A_n}{A_{in}} = \frac{1}{(1 + \alpha\sqrt{A_{in}}n)^p}, \quad p = 1.0 \sim 1.2, \tag{5}$$

where α is some constant. As was discussed by IOWT, this result is similar to the prediction (4) by the stochastic KdV equation (1):

$$\frac{A_n}{A_{in}} = \text{const.}(A_{in})^{-1/2} n^{-3/2}, \quad n \to \infty. \tag{6}$$

While the agreement between formulae (5) and (6) is fairly good, some questions remain to be examined. First, the lattice size in the numerical study by IOWT is large (500 lattice sites) but finite. It is difficult to conclude that the formula (5) gives the true asymptotic form. Second, as seen from Fig.3 in [5], the exponent p in eq.(5) looks smaller than $1 \sim 1.2$ for small soliton amplitudes. Third, it is desirable to show that we may treat the randomness of mass as external noise. With these questions in mind, we shall present and analyse a new stochastic equation which describes nonlinear waves in weakly random media.

2. A New Stochastic Equation

We consider a one-dimentional lattice. A force F between two adjacent masses is assumed to be

$$F = -K(\Delta + \alpha\Delta^2), \tag{7}$$

where K is the spring constant, Δ the elongation of the spring and α the strength of nonlinearity. We denote by y_i the displacement of the i-th mass m_i from equilibrium. The equation of motion is

$$\begin{aligned} m_i\ddot{y}_i &= K[y_{i+1} - y_i + \alpha(y_{i+1} - y_i)^2] \\ &- K[y_i - y_{i-1} + \alpha(y_i - y_{i-1})^2]. \end{aligned} \tag{8}$$

We employ the continuum approximation. We write the lattice spacing as h and regard $x = ih$ as a continuous variable. We assume that a distribution of the masses is expressed by

$$m(x) = m_0(1 + \rho(x)), \tag{9}$$

where m_0 denotes a constant part of the distribution. Expanding $y_{i\pm1}$ in Taylor series with $x = ih$ and using them in eq.(8), we have

$$y_{tt} = c^2(x)(y_{xx} + \frac{1}{12}h^2 y_{xxxx} + 2\alpha h y_x y_{xx} + \cdots), \tag{10}$$

$$c(x) = c_0(1 + \rho(x))^{-1/2}, \tag{11}$$

where $c_0 = \sqrt{K/m_0} \cdot h$ is the sound velocity. Within the same approximation, eq.(10) holds for the case where the spring constant K is a function of lattice site i. Note that eq.(10), when terms denoted by dots are discarded, is a generalization of the Boussinesq equation.

We apply the reductive perturbation method [6] to eqs.(10) and (11). We set

$$\epsilon^{1/2} = 2\alpha h, \tag{12}$$

$$\xi = \epsilon^{1/2}(x - c_0 t)/h, \tag{13}$$

$$\tau = \epsilon^{3/2}x/24h, \tag{14}$$

$$\rho(x) = \epsilon\eta(\tau), \tag{15}$$

$$y(x,t) = -h/2 \cdot \phi(\xi, \tau). \tag{16}$$

All new variables are chosen to be dimensionless. We regard ϵ as a smallness parameter. Substituting these changes of variables into eqs.(10) and (11), we get

$$\phi_{\xi\tau} - 12\eta(\tau)\phi_{\xi\xi} + \phi_{\xi\xi\xi\xi} - 6\phi_\xi\phi_{\xi\xi} = 0. \tag{17}$$

Equation (17) with a random function $\eta(\tau)$ is a new stochastic equation. In this derivation, an important condition is eq.(15). That is, randomness of mass is weak. We shall take $\eta(\tau)$ to be a Gaussian white noise,

$$< \eta(\tau)\eta(\tau') >= 2\theta\delta(\tau - \tau'). \tag{18}$$

This assumption is not essential in our theory (other choices of the stochastic function $\eta(\tau)$ could be important), but it makes a further analysis simple.

Formal solution of eq.(17) is obtained as follows. Introducing again changes of variables,

$$X = \xi + 12 \int_0^\tau \eta(\tau) d\tau, \tag{19}$$

$$T = \tau, \tag{20}$$

$$u(X,T) = \phi_\xi(\xi, \tau), \tag{21}$$

we obtain from eq.(17)

$$u_T - 6uu_X + u_{XXX} = 0. \tag{22}$$

This is nothing but the Korteweg-de Vries (KdV) equation and its solvability is known [7,8]. Remark that an independant variable T is the slow space variable as seen from eqs.(14) and (20). It is interesting to notice that a relation (19) implies the fluctuation of the velocity.

3. Deformation of Solitons

We are now in a position to discuss how a soliton changes during its propagation in a weakly random chain. It is known that the Kdv equation (22) has a soliton solution,

$$u(X,T) = -2\kappa^2 \mathrm{sech}^2(\kappa X - 4\kappa^3 T - \kappa X_0). \tag{23}$$

Fourier transform of the soliton solution is given by

$$\begin{aligned}\hat{u}(k,T) &= \int_{-\infty}^{\infty} dX u(X,T) e^{-ikX} \\ &= -4\kappa \frac{\pi k}{2\kappa} \frac{1}{\sinh(\pi k/2\kappa)} \exp[-ik(4\kappa^2 T + X_0)]. \end{aligned} \tag{24}$$

We define

$$W(T) = \int_0^T dT \eta(T), \tag{25}$$

where a Gaussian white noise $\eta(T)$ has been introduced in eq.(18). The following identities hold:

$$< W^2(T) >= 2\theta T, \tag{26.a}$$

$$< e^{i\alpha W(T)} >= \exp(-\frac{1}{2}\alpha^2 < W^2(T) >). \tag{26.b}$$

Then, the average of the soliton solution is expressed as

$$\begin{aligned}< u(X,T) > &=< \int_{-\infty}^{\infty} \frac{dk}{2\pi} \hat{u}(k,T) e^{ikX} > \\ &= \int_{-\infty}^{\infty} \frac{dk}{2\pi} \hat{u}(k,T) e^{i\xi k} < e^{12ikW(T)} > \\ &= \int_{-\infty}^{\infty} \frac{dk}{2\pi} \hat{u}(k,T) \exp(i\xi k - 144k^2 \theta T). \end{aligned} \tag{27}$$

For large T, applying a saddle point method we have

$$\begin{aligned}< u(X,T) > &=< \phi_\xi(\xi,\tau) > \\ &= -\frac{2\kappa}{\pi} \sqrt{\frac{\pi}{a}} \exp(-s^2/4a), \end{aligned} \tag{28}$$

where s and a stand respectively for

$$s = \xi - 4\kappa^2 \tau - \xi_0, \tag{29}$$

$$a = 144\theta\tau + \pi^2/24\kappa^2. \tag{30}$$

This shows that when the initial amplitude $|-2\kappa^2|$ is large it approaches rapidly to the asymptotic form,

$$< u(X,T) > = < \phi_\xi(\xi,\tau) >$$

$$= -\frac{\kappa}{6}(\pi\theta\tau)^{-1/2}\exp[-(\xi - 4\kappa^2\tau - \xi_0)^2/576\theta\tau] \quad (31)$$

We summarize the above result in terms of the original variables x,t and $y(x,t)$. Incident soliton at $x = 0$ has a form

$$y_t(0,t) = -\kappa^{1/2}\epsilon^{1/2}c_0\mathrm{sech}^2(\kappa\epsilon^{1/2}c_0t/h). \quad (32)$$

Due to the random mass distribution, the wave form is deformed during the propagation. Its asymptotic form as $x \to \infty$ is

$$< y_t(0,t) > = A\frac{1}{x^{1/2}}\exp(-B(x-ct)^2/x), \quad (33)$$

where constants A and B are given by

$$A = -\kappa c_0(h/6\epsilon^{1/2}\pi\theta)^{1/2}, \quad (34)$$

$$B = (1 - \frac{1}{6}\kappa^2\epsilon)^2/240\theta\epsilon^{1/2}h, \quad (35)$$

and the velocity c is defined by

$$c = c_0(1 + \kappa^2\epsilon/6). \quad (36)$$

In this expression, ϵ has been assumed to be small.

4. Concluding Remarks

In this paper, starting from a nonlinear lattice with random mass distribution, we have derived a new stochastic equation (17). The equation can be transformed into the KdV equation and the average of wave field can be analytically evaluated. Formula (33) shows that the amplitude of a soliton decreases asymptotically as $x^{-1/2}$ and the width of a soliton increases asymptotically as $x^{1/2}$, x being the distance of the propagation in random media. Extention of this result to the N-soliton case is straightforward since the intervals of solitons increase linearly with time. By using the soliton perturbation theory, effect of dissipation can be included as shown in [9].

In conclusion, a new stochastic equation (17) is a prototype model for nonlinear waves in random media and it should also be useful for the study of the similar problems in hydrodynamics and plasma physics.

Acknowledgements

The author would like to thank Professors F.Kh.Abdullaev, D.J.Kaup, A.V.Mikhailov, Yu.S.Kivshar, St.Pnevmatikos and S.Watanabe for useful discussions and comments. This work is partially supported by Grants-in-aid for Scientific Research Fund from the Ministry of Education, Science and Culture (01540310).

References

[1] M.Wadati: J.Phy.Soc.Jpn. **52** (1983) 2642.
[2] F.G.Bass, Yu.S.Kivshar, V.V.Konotop and Yu.A.Sinitsyen: Phys. Reports **157** (1988) 63.
[3] Yu.S.Kivshar and B.A.Malomed: Rev.Mod.Phys. **61** (1989) 763.
[4] F.Kh.Abdullaev: Phys.Reports **179** (1989) 1.
[5] S.Ishiwata, Y.Okada, S.Watanabe and H.Tanaca: J.Phys.Soc.Jpn. **59** (1990) 3029.
[6] H.Washimi and T.Taniuti: Phys.Rev.Lett. **17** (1966) 996.
T.Taniuti and N.Yajima: J.Math.Phys. **10** (1969) 1369.
[7] C.S.Gardner, J.M.Greene, M.D.Kruskal and R.M.Miura: Phys. Rev.Lett. **19** (1967) 1095.
[8] M.Wadati and M.Toda: J.Phy.Soc.Jpn. **32** (1972) 1403.
[9] M.Wadati and Y.Akutsu: J.Phys.Soc.Jpn. **53** (1983) 3342.

Nonlinear Wave Propagation Through Disordered Media

Y.S. Kivshar

Institute for Low Temperature Physics and Engineering,
Academy of Sciences, 47 Lenin Avenue, 310 164 Kharkov, Ukraine
Present address: Departamento de Física Teórica I,
Facultad de Ciencias Físicas, Universidad Complutense,
E-28040 Madrid, Spain

Abstract. We briefly review recent numerical and analytical results related to the study of the coexistence of nonlinear dynamics and effects of disorder, mostly for soliton bearing systems. In particular, we point out that disorder may give rise to a more intensive development of nonlinear properties via modulational instability, and that localization effects stipulated by disorder in linear systems may vanish in the presence of strong nonlinearity. The examples that we present demonstrate that nonlinearity leads to an actual improvement of the transmission only when it contributes to create soliton pulses.

1 Introduction

Wave propagation in nonlinear disordered media has become an extensive studied subject in recent years [1]. In linear systems, disorder generally originates Anderson localization, a concept that applies to a number of different waves like phonons, acoustic and electromagnetic waves, etc. (see, e.g., Refs.[3],[2] and references therein). Localization means that the transmission coefficient T of a wave decays exponentially with the system length L and, in particular, that a positive finite number exists, the so-called localization length $\lambda(k)$, k being the wave number, so that for $L \gg \lambda(k)$ very little transmission is allowed. Linear wave packets behave in an analogous fashion (see,e.g.,[4]).

The study of the influence of nonlinearity on the transport properties of disorderer systems was started by Payton *et al.* [5], who investigated numerically the role of nonlinearity on the heat flow in a discrete system. They emphasized that the heat flow is enhanced by the introduction of nonlinearity in the system (see also the recent interesting paper by Bourbonnais and Maynard [6] on the same subject). However, at the time they studied the problem, the concept of soliton had not been well established and, as a matter of fact, the relation between disorder and soliton propagation had not been noticed.

In the framework of stationary nonlinear equations, the action of nonlinearity is opposite to that of disorder, and it changes the length dependence of the transmission coefficient, that still tends to zero as the size of the system increases, but following a power law [7]-[9] instead of the exponential one proper to the linear case. Nonlinearity may also produce multistability in the wave transmission through a disordered slab[9]. In nonstationary problems, nonlinearity may lead to modulational instability which is enhanced by disorder[10,11]. Modulational

Springer Proceedings in Physics, Vol. 67 Nonlinearity with Disorder
Editors: F. Abdullaev · A.R. Bishop · S. Pnevmatikos

instability is important because it is a factor that causes appearance of solitons instead of linear waves.

The most remarkable manifestation of influence of nonlinearity on disorder is related to the fact that many nonlinear systems support undistorted propagation of localized waves, solitons. As is well known, there are solitons of three general types: dynamical solitons, envelope solitons, and topological kinks. The amplitude of a *dynamical soliton* (or a dynamical kink) is proportional to its velocity, and the scattering of such an object in a disordered medium is similar to that of a linear wave. This means that the scattering of such solitons may be treated as the scattering of a superposition of linear waves with different wave numbers that usually yields power-like dependence of the transmission coefficient on the system length, and nonlinearity only slightly modifies this dependence (see Refs.[12] and[13]). *Envelope solitons* are two-parametric ones and they demonstrate more complicated behaviour in disordered media. As was demonstrated in Ref.[4], in the case of the envelope soliton localization effects vanish drastically above a certain threshold in a nonlinearity parameter. This means that influence of disorder on propagation of such intensive pulses can become negligible. A similar conclusion follows from the numerical study of the energy transport in anharmonic lattices with disorder: a regime of nonlinear diffusion is characterized by an exponent which depends strongly on anharmonicity but it is insensitive to disorder[6].

Another type of solitary waves is that of *topological kinks*, which may exist only in nonlinear systems with two (or more) equivalent ground states, i.e., when nonlinearity is large. In a number of cases kink dynamics in disordered systems may be described in the framework of a collective-variables approach by an equation for an effective (classical or relativistic) particle, its coordinate being the kink position[14,15]. This situation is similar to that for perturbation-induced dynamics of regular nonlinear systems supporting the kink propagation[16,17].

The present paper aims to demonstrate a number of features in coexistence of nonlinear dynamics and effects of disorder, mostly for soliton bearing systems. Section 2 briefly summarizes scattering of linear waves and linear wavepackets to use these results for comparison with the scattering of nonlinear waves. In Section 3 we present some results obtained for stationary nonlinear models. Section 4 gives examples demonstrating that the typical approach which is usually used in the localization theory of disordered linear systems is not suitable for some nonlinear models due to modulational instability. We point out that modulational instability may be more pronounced in disordered systems, and it gives rise to soliton generation.This is the reason to study soliton propagation. Section 5 reviews the soliton propagation through disordered media for different types of solitons : this propagation has a number of features for each type of solitons. Section 6 concludes the paper and presents some perspectives to study the effects produced by strong nonlinearity in the presence of a disorder. A more detailed and extended version of the present work wil be published in collaboration with Sergei A.Gredeskul [18].

2 Linear systems and disorder

Linear waves in disordered media have been widely studied in recent years, particularly because of the Anderson localization phenomenon . Let us start with a brief summary of the main ideas and results.

Consider, for example, the Schrödinger equation

$$-\frac{d^2u}{dx^2}+\epsilon(x)\,u=k^2u, \tag{1}$$

with a static random potential $\epsilon(x)$ which occurs in a number of physical problems related to wave scattering,k being the wavenumber. We ask the nature of the proper modes of the equation. The most remarkable phenomenon, called Anderson localization (see, e.g., Ref.[2] and references therein) is the following : if disorder is sufficiently strong, i.e. the fluctuations of the potential are sufficiently wild, the proper modes of the equation for a typical realization of the random potential (i.e., in fact, with a probability one) are exponentially localized and the spectrum of the self-adjoint operator is pure point with a dense set of eigenvalues and exponentially decaying wave functions. For one-dimensional systems this occurs for any (arbitrary small) disorder.

Considering now time evolution, Anderson localization implies that for an initial state starting in a finite region of space, the probability to find the particle outside some finite region of space will remain small uniformly in time, for a large enougth region. Such a phenomenon is also naturally believed to imply that an electronic system is an insulator, since electrons are "trapped" by disorder and cannot anymore transport the electric current. Similar phenomena arise for other continuous and discrete linear wave equations when considering propagation in a disordered medium and apply for example to electromagnetic wave propagation in turbulent plasmas, to acoustic waves in ramdom media, etc.[3]. Anderson localization is induced by phase interferences of the waves which are partially reflected by the fluctuations of the medium and interfere together with random phases.

Thus, localization of states implies that the transmission coefficient of a linear wave $T_L(k)$ decays exponentially with the system length L , such that for large L ,

$$L^{-1}\ln T_L(k)\approx-\lambda^{-1}(k), \tag{2}$$

where $\lambda(k)$ is *the localization length*, and hence, if L is sufficiently wider than $\lambda(k)$, it is clear from Eq.(2) that very little transmission will be allowed. If $\epsilon(x)$ describes a set of delta-like point impurities with equal intensities ϵ but random positions in space, which are characterized by the mean distance p^{-1} between separate impurities, the localization length has the form (see,e.g.,Ref.[4])

$$\lambda(k)=\frac{4k^2}{p\epsilon^2}, \tag{3}$$

provided $\epsilon^2\ll k^2$, $p\ll k$,and $k^2\ll\epsilon^2Lp$. As a result, the localization length is a function of the wave number k . This fact means that the transmission coefficient for a linear wave packet which takes into account the spectral density of the packet $P(k)$ may change such a dependence due to *averaging* in all possible wave numbers. However, simple calculations for Eq.(1) indicate that in the case when the spectral function $P(k)$ is a function localized in the vicinity of a certain carrier wave number, say $k = k_0$, the transmission coefficient of the wave packet,

$$\tau = \frac{\int dk\, |P(k)|^2\, T_L(k)}{\int dk\, |P(k)|^2}, \tag{4}$$

defined as an averaged value $< \tau >$ (or for a realization) is qualitatively the same as that of a plane wave, but the exponential dependence for the linear wavepacket is slightly different: the power in the exponent depends on the relation between the parameters (carrier wavenumber k^0 and width) of the wave packet [4,18]. Otherwise, for wide spectral densities the integral in Eq.(4) depends on the real structure of the function $\lambda(k)$.We would like to point out especially the case of the Helmholtz equation, $-u_{xx} + k^2[1+\epsilon(x)]u = 0$, the notations being the same as in Eq.(1), for which $\lambda(k) = 4/p\epsilon^2 k^2$ [cf. Eq.(3)], that yields the power-like dependence of the averaged transmission coefficient[18],

$$< \tau > \sim L^{-1/2}. \tag{5}$$

This power-like dependence is also valid for the transmission coefficient defined for a realization, and it is important that it has arised in result of *averaging* in all wave numbers of the linear wave packet.

3 Stationary nonlinear wave scattering

Let us consider now the influence of nonlinearity on the wave transmission introducing a nonlinear term into Eq.(1), i.e.

$$-\frac{d^2u}{dx^2} + \epsilon(x)u - \alpha|u|^2u = k^2u. \tag{6}$$

Equation (6) may be also obtained as a stationary version of the time-dependent Schrödinger equation (see the corresponding discussion below). The problem of the wave transmission through a nonlinear disordered slab, i.e. when $\epsilon(x) \neq 0$ and $\alpha \neq 0$ only for $0 < x < L$, was investigated in Refs.[7] and [8] (see also Ref.[9]). Summarizing these results, we note that such a nonlinear transmission has some general features in comparison with the linear transmission. First, in the linear case, Anderson localization arises as a result of interference phenomena which are supressed in the nonlinear case due to the absence of the superposition principle. So, it is natural to see [7,8] that the decreasing of the nonlinear transmission coefficient is not exponential but power-like. Another new property of the nonlinear transmission is the existence of *two different problems* that are the same in the linear approach: the fixed output and fixed input problems . The fixed output problem is more simple than the fixed input problem, but, from a physical point of view, the fixed input problem is more important for real situations. For the fixed input problem it was found that for small L, $L \ll L_{cr}$, where $L_{cr} \sim A^{2/3}/\lambda$, A is the wave amplitude and λ is the localization length, the transmission coefficient oscillates between two limit values which decay as $L^{-1/2}$, L being the length of the slab. For large L ($L \gg L_{cr}$), the transmission coefficient recovers the exponential behaviour (2), because for such lengths the influence of nonlinearity is negligible.

The most interesting feature of the fixed input problem is bistability or multistability which occurs when a single input wave with fixed intensity corresponds to a few output solutions. This property of nonlinear transmission is well-known for systems without a disorder (see. e.g., Ref.[19]) and, as was shown in Ref.[8], a weak random potential $\epsilon(x)$ does not change sufficiently degeneracy of nonlinear solutions. The much more important effect induced by disorder is the following. It is well known, bistability arises when nonlinear parameter α exceeds a certain threshold value α_c, where $\alpha_c \sim (kL)^{-1/3}$(see, e.g., Ref.[9]) . When disorder is present, this threshold value is much smaller than in homogeneous case. For example, according to numerical results by Knapp et al.[9], for $max|\epsilon(x)| \sim 0.1$ and $kL = 10^3$, the bistability was observed in 25% of realizations of the random potential when it was taken $\alpha A/k^2 \approx 0.008$. But in the same time bistability does not occur in the similar problem without disorder untill the value $\alpha A/k^2 \simeq 0.3$. So, small disorder may sufficiently change the threshold conditions for the bistability effect, but solutions are changed a little.

4 Modulational instability and wave propagation

Equation (6) was introduced in Refs.[7,8] as a stationary version of the nonlinear Schrödinger equation (NLS)

$$i\phi_t + \phi_{xx} + \alpha|\phi|^2\phi = \epsilon(x)\phi. \tag{7}$$

Indeed, let us look for solutions of Eq.(7) in the form $\phi(x,t) = \exp(-ik^2t)\,u(x)$. Then, substituting of this expression into Eq.(7) yields Eq.(6) for the function $u(x)$. However, unlike the linear case, solutions of nonlinear problems have to be investigated for stability properties. As is well known, the homogeneous NLS equation ($\epsilon = 0$) with $\alpha > 0$ corresponds to modulational instability of a nonlinear plane wave. To demonstrate it, let us consider the exact solution of the equation (7) at $\epsilon(x) = 0$,

$$u(x,t) = A\exp(i\theta),\ \theta = kx - \omega t, \tag{8}$$

for which the following dispersion relation is valid:

$$\omega = k^2 - \alpha|A|^2. \tag{9}$$

To analyze stability of the solution (8), (9), we have to derive equations for small deviations from its exact form. Substituting $\theta = \theta + \pi(x,t)$ and $A = A + a(x,t)$, we may obtain a system of the linear equations for the functions $\pi(x,t)$ and $a(x,t)$, which yields the dispersion relation,

$$(\Omega - kq)(\Omega - 2kq) = q^2(q^2 - 2\alpha|A|^2), \tag{10}$$

Ω and q being the frequency and wave number of the small periodic perturba-

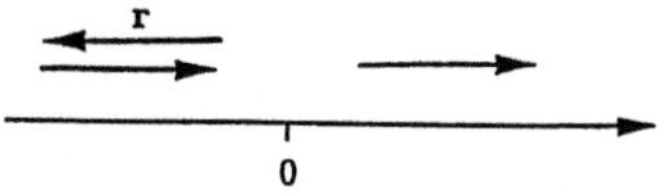

Fig.1. Generation of a counterpropagating wave (r) due to the wave scattering by an impurity ($x = 0$). In the region $x < 0$ interaction between the scattering and reflected waves enhances modulational instability.

tions. For $\alpha > 0$ Eq.(10) has a negative solution for Ω^2 which corresponds to an instability. For example, at $k = 0$ the region of instability, $0 < q^2 < 2\alpha|A|^2$, has the maximum growth rate $\sigma_{cr}^{(0)} = \alpha|A|^2$.

In the case of inhomogeneous systems the problem of modulational instability is more complicated . To understand the main featuares of modulational instability in inhomogeneous media, let us consider the simplest case of a single point impurity (see Fig.1). The impurity generates a reflected wave which interacts with the scattering wave. Therefore, in the region before the impurity position we have *two* coupled nonlinear waves. Modulation instability of two coupled waves has been extensively investigated in nonlinear optics and plasma physics (see, e.g., Ref.[20] for a review and references therein). For example, for two counterpropagating waves with amplitudes A and B of the same order the maximum growth rate is (see, e.g., Ref.[21]),

$$\sigma_{cr} = \frac{1}{2}\alpha\{|A|^2 + |B|^2 + [(|A|^2 + |B|^2)^2 + 6|A|^2|B|^2]^{1/2}\}, \tag{11}$$

i.e. it is *always* larger than the value $\sigma_{cr}^{(0)} = \alpha|A|^2$ for a homogeneous case, and in the case $|A| = |B|$ Eq.(11) yields the result $\sigma_{cr} = 3\sigma_{cr}^{(0)}$ (!). In a general case, the effect of wave scattering includes three simple waves (see Fig.1), but the main effect is stimulated by the interaction betweem the scattering and reflected waves.

So, the arguments presented above demonstrate that modulational instability is always enhanced by inhomogeneities and that effect has a simple physical explanation. To demonstrate an example of this effect, we would like to present some numerical results which were obtained by Peyrard and Bishop [11]. They considered a system of rotators given by the Hamiltonian (in fact, the discrete sine-Gordon model),

$$H = \sum_n [\frac{1}{2} I\dot{\phi}_n^2 + \frac{1}{2}S(n)(\phi_n + \phi_{n-1})^2 + B(1 - \cos\phi_n)], \tag{12}$$

where the coupling constant between rotators $S(n)$ was changed in 2.5 times in 10 central cells (the total chain contained 200 cells). Figure 2 depicts the behaviour of the plane wave in such an inhomogeneous case, the rest of other parameters being the same as in a homogeneous case. It is clear that small disorder generates modulational instability much faster (cf. scales in Fig.2 !).

The effect of modulational instability of nonlinear plane wave suggests that results obtained in Ref.[7] and [8] for Eq.(6) are questionable to be extended for

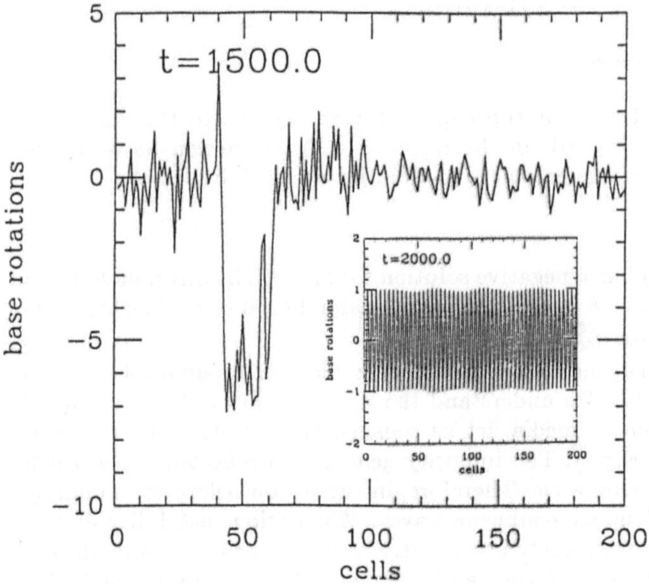

Fig.2 Base rotation ϕ_n along an inhomogeneous chain at $t = 1500$ for an initial wave amplitude of 1.0. For comparison, the homogeneous case at the same parameters but at $t = 2000$ is also presented (Peyrard and Bishop [11]).

the time-dependent case, at least for $\alpha > 0$ in Eq.(7). To analyse the scattering in the framework of the time-dependent NLS equation, Caputo et al.[10] have pe-rformed the scattering experiments in the full time-dependent framework, using the discrete NLS model,

$$i\dot{A}_n + A_{n+1} + A_{n-1} - 2A_n + \alpha|A_n|^2(A_{n+1} + A_{n-1}) = V_n A_n, \qquad (13)$$

the potential V_n is equal to zero for $n < 0$. As is well known, the discrete Ablowitz-Ladik model (13) is exactly integrable at $V_n = 0$ [22]. As follows from the numerical simulations, in the cases $\alpha = 0$ (linear) and $\alpha = -10$ (nonlinear medium without modulational instability) the scatterings of the waves are very similar. On the contrast, for $\alpha = 10$, the field does not decay as fast as for cases mentioned above and, additionally, there are pulses which get created and propagate through the medium, the latter being the solitons of the NLS equation (see Figs.5,6, and 7 in Ref.[10]). Hence, from the comparison between the two cases $\alpha = -10$ and $\alpha = +10$ it is clear that these coherent structures (in fact, solitons) are responsible for the increase in energy transport for the nonlinear case. Therefore, solitons play an important role in nolinear wave scattering and that is why it is necessary to study the soliton scattering in detail.

5 Soliton propagation through disordered media

We will analyse the propagation of solitoins through disordered media for the different types of solitons separately. Our main goal is to point out the qualitative difference between the three types of solitons scattering in a disordered medium.

5.1 Dynamical solitons

One of the systems where dynamical solitons may propagate is a nonlinear lattice of particles with nearest-neighbor interactions between them, which can be described by the Hamiltonian,

$$H = \sum_n [\frac{1}{2} m_n \dot{y}_n^2 + V(y_n - y_{n-1})], \tag{14}$$

where $y_n(t)$ is the displacement of the n-th particle from its equilibrium position, and $\dot{y}_n = dy_n/dt$ and m_n are, respectively, the velocity and the mass of the n-th particle. In the simplest case the interaction potential $V(r)$ is polynomial,

$$V(r) = \frac{Gr^2}{2} + \frac{Ar^3}{3} + \frac{Br^4}{4}. \tag{15}$$

The equation of motion of the n-th particle is

$$m_n \ddot{y}_n = -V'(y_n - y_{n-1}) + V'(y_{n+1} - y_n), \tag{16}$$

and for a homogeneous system it can be transformed into a generalized Boussinesq (GBq) equation (see, e.g., Ref. [23]); from that equation, after some transformations, it is possible to derive the Korteweg-de Vries (KdV) equation (see, e.g., Ref.[24]). For the GBq equation exact kink solutions may be found, however this equation, unlike the KdV equation, is not exactly integrable.

Another example of the system (14) is the Toda lattice, when

$$V(r) = (a/b)[\exp(-br) - 1] + ar, \tag{17}$$

for which the discrete equation (16),(17) becames exactly integrable[25].

Let us assume that a disordered segment embedded between two homogeneous semi-infinite chains has randomly distributed impurity masses $m = \gamma m_0$ with the probability p in the disordered part. A soliton is incident on the disordered segment, and we are interested in the transmission coefficient, which is the transmitted energy E_t over the soliton energy E_{in}. The problem was solved numerically in Ref. [12] for the potential (15). Some results for a quartic nonlinear disordered chain are presented in Fig.3. The behaviour of the soliton may be analytically studied in two limits (see Ref.[12]): the independent scattering approach and linear theory.

In the independent scattering approach we assume that the concentration p of the mass impurities is low, i.e. the mean distance between two nearest-neighbor impurities is larger than the soliton size. In this case we can write

$$\Delta E_{i+1} = E_{i+1} - E_i = -E_i R_i(E_i), \tag{18}$$

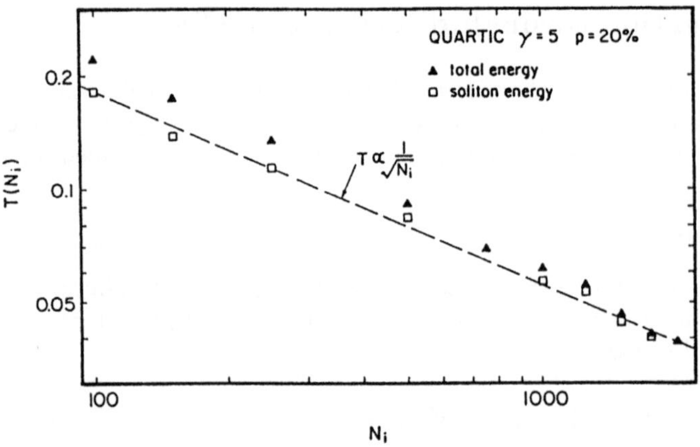

Fig.3 Transmission coefficient of the soliton with the incident energy $E_{in} = 0.66$ as a function of the disordered segment length N_i for the quartic nonlinear chain (Qiming Li et al. [12]).

where $R(E_i)$ is the reflection coefficient for a single impurity, which has been previously calculated, numerically [23] and analytically [24], and it is given by

$$R(E) = a_0 E^{\alpha}(\gamma - 1)^2, \tag{19}$$

where $a_0 \approx 0.66$, $\alpha = 2$ for the quartic potential, and $a_0 = 8.0$, $\alpha = 2/3$ for the cubic potential. Since on the average there are $(\Delta x)p$ impurities in the interval between x and $x + \delta x$, from Eq.(18) a differential equation can be established, taking the continuum limit

$$\frac{dE}{dx} = -pE(x)\, R[E(x)], \tag{20}$$

which, after integrating with the help of Eq.(19), yields

$$T(L) \simeq E(L)/E_{in} = (1 + c_0 L)^{-1/2}, \; c_0 = p\, \alpha\, a_0 E_{in}^{\alpha}(\gamma - 1)^2. \tag{21}$$

Therefore, for length $L \gg 1/c_0$ the independent scattering limit gives for the quartic potential $T(L) \sim L^{-1/2}$, and for the cubic potential $T(L) \sim L^{-3/2}$, i.e. the dependence is implicitely dependent on the type of nonlinearity through the reflection coefficient.

The result $T(L) \sim L^{-1/2}$ is in agreement with numerical simulations for the quartic potential (see Fig.3). As was shown in Ref.[12], the same behaviour follows from the averaged transmission coefficient (4) with the localization length

$$\lambda(k) = \frac{2(1 - p + \gamma p)^2}{(\gamma - 1)^2 p(1 - p)} \cot^2(\frac{k}{2}), \tag{22}$$

i.e. the formula (4) with Eq.(22) yields[12] [cf. Eq.(5)],

$$\tau = \frac{(1-p+\gamma p)}{(\gamma-1)[p(1-p)]^{1/2}} \frac{1}{\sqrt{L}}, \quad L \gg 1. \tag{23}$$

As a result, the independent scattering picture gives $T \sim L^{-1/2}$ for a quartic potential and $T(L) \sim L^{-3/2}$ for a cubic potential. However, for long enough lengths, when nonlinear terms can be neglected, the total transmission must decay as $L^{-1/2}$, i.e. much slower than exponential for T in the case of a plane wave. For the quartic potential these behaviours are the same.

Numerical results for the disordered lattice with the Toda potential (17) heve been obtained by Ishiwata et al.[13] . The authors were interested in the same problem considering mass disorder in a nonlinear chain of particles interacting via the Toda exponential potential. The main result of the paper[13] is the dependence for the soliton amplitude A_n at the n-th particle,

$$\frac{A_n}{A_0} = (1+\alpha\sqrt{A_0 n})^{-p}, \tag{24}$$

where $p = 1 - 1.2$ and $\alpha \simeq 0.009$, which was established numerically. As we can see, the dependence (24) is also power-like.

Therefore, the transmission coefficient of a dynamical soliton decays much slower than for a plane wave, in fact, with a power law. On the other hand, the similar dependences arise for wavepackets in *linear* disordered systems, in particular, for quartic interparticle interaction potential they are the same. So, we may conclude that dynamical solitons in disordered systems propagate as nonlinear wavepackets for which the dependence $T(L)$ is slightly modified due to nonlinearity, but the general dependence is power-like, as for averaged wavepackets in linear theory.

In conclusion of this section, we would like to point out that the similar result, i.e. the power-like decay of the soliton amplitude, follows also from results obtained by Wadati[26] and Wadati and Akutsu[27] (see also Ref.[28]) for the so-called exactly solvable stochastic KdV equation. However, the sense of the averaged soliton calculated in Refs.[26] and [27] is not so clear when results are compared to either the independent scattering approach results or direct numerical simulations, where an unique realisation is actually observed.

5.2 Topological solitons

Another type of solitons is a topological soliton. Such a soliton exists only in a nonlinear system with two (or more) equivalent ground states. The simplest example is the sine-Gordon model,

$$\phi_{tt} - \phi_{xx} + \sin\phi = \epsilon(x)\sin\phi, \tag{25}$$

which in the homogeneous case, i.e. when $\epsilon(x) = 0$, has a kink solution of the form,

$$\phi_k(x,t) = 4\tan^{-1}\exp\left(\frac{x-X}{\sqrt{1-V^2}}\right), \tag{26}$$

X being the kink coordinate.

As is shown in Ref.[17], for a number of cases the kink emission is (exponentially) small, so that kink dynamics may be studied in the so-called collective-coordinate approach. In the framework of such an approach the evolution of the kink's coordinate X is described by the simple equation of classical mechanics ($\dot{X}^2 \ll 1$),

$$\frac{d^2X}{dt^2} = -\frac{dU(X)}{dX}, \tag{27}$$

where the effective potential $U(X)$ is the following,

$$U(X) = \frac{1}{4}\int_{-\infty}^{\infty} dx \, \frac{\epsilon(x)}{\cosh^2(x-X)}. \tag{28}$$

In the paper by Gredeskul et al.[15] Eqs.(27),(28) have been analysed for the case when disorder appears as randomly distributed point impurities with equal intensities, i.e. for $\epsilon(x) = \sum_n \delta(x - x_n)$, x_n being random values. Since the method of the Fokker-Plank equation (which is valid for time-dependent random perturbations studied in the papers [28]-[29], see also [30]) cannot be applied to the problem (notice that the potential (28) is not a Markov one), it was elaborated a statistical procedure to compute the mean characteristics of the kink propagation. As a result, the kink dynamics in disordered systems may be described with the help of an effective random potential, and this approach is in good agreement with direct numerical integration [14].

5.3 Envelope solitons

Let us consider now propagation of envelope solitons in disordered media. We will start, for simplicity, from the NLS equation (7) which in the homogeneous case, i.e. when $\epsilon(x) = 0$, supports the distortionless propagation of localized excitations in the form of envelope solitons,

$$u_s(x,t) = a\frac{\exp\{-(iV/2)x - i[(V^2/4) - a^2]t\}}{\cosh[a(x - Vt)]}, \tag{29}$$

where a is the soliton amplitude and V is its velocity. To study the scattering of the soliton (29) by a random system of point impurities with concentration p and equal intensities ϵ, i.e. $\epsilon(x) = \epsilon \sum_n \delta(x - x_n)$, x_n being random, we will apply the independent scattering approach (see subsection 5.1). However, the soliton (29) is two-parametric pulse unlike a dynamical soliton. Therefore, we have to use two integrals of motion , e.g., the energy

$$E = \int_{-\infty}^{\infty} dx \, [|u_x|^2 + \epsilon(x)|u|^2 - |u|^4], \tag{30}$$

and the "number of quasi-particles",

$$N = \int_{-\infty}^{\infty} dx \, |u|^2. \tag{31}$$

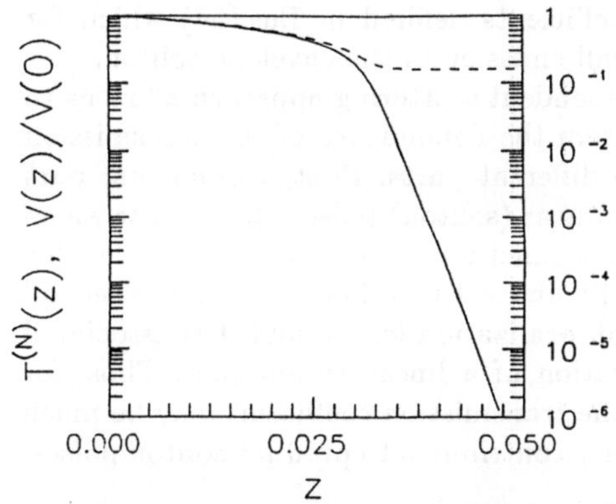

Fig.4 The transmission coefficient $T^{(N)}$ (solid line) and the function $V(z)/V(0)$ (dashed line) when initial conditions are $N(0) = 1.25$, $V(0)01.0$, so that $\alpha(0) = 1.25$ (Kivshar et al. [4]).

Following the line presented in the subsection 5.1, we may obtain the relations [4] ,

$$\Delta E_{j+1} = E_{j+1} - E_j = -E_j R_j^{(E)}(E_j, N_j), \tag{32}$$

$$\Delta N_{j+1} = N_{j+1} - N_j = -N_j R_j^{(N)}(E_j, N_j), \tag{33}$$

where $R^{(E)}$ and $R^{(N)}$ are two reflection coefficients (which are different in nonlinear case!). These reflection coefficients were calculated in Ref. [31] (see also Ref.[24]) and the expressions are

$$R^{(N)} = \frac{\pi\epsilon^2}{64NV}\int_0^\infty dx\, F(x,\alpha), \quad R^{(E)} = \frac{\pi\epsilon^2 V}{256E}\int_0^\infty dx\, x^2 F(x,\alpha), \tag{34}$$

where

$$F(x,\alpha) = \frac{[(x+1)^2+\alpha^2]^2}{\cosh^2[\frac{\pi}{4\alpha}(x^2+\alpha^2-1)]}, \quad \alpha = N/V. \tag{35}$$

The system (32)-(35) was analysed in Ref.[4] in the limit when Eqs.(32),(33) may be reduced to a system of two coupled integro-differential equations. In the limit $a \ll 1$, i.e., small α, the resulting system yields the approximate analytical dependences,

$$T^{(N,E)}(x) = N(x)/N(0) = E(x)/E(0) = \exp(-x/\lambda_0), \tag{36}$$

where $\lambda_0 = V^2(0)/p\epsilon^2$ is the localization length. The dependence (36) was proved by numerical solution of the system [4].On the other hand, in the case $a \sim 1$ the asymptotic change in the transmission coefficients $T^{(N,E)}(x)$ depends crucially on the value of the parameter $\alpha(0)$ (see Fig.4). As was demonstrated in Ref.[4], in this simple model the critical value $\alpha_{cr} \simeq 1.285$ appears, so that above this value the soliton propagation is almost undistorted. This behaviour may be understood analysing the reflection coefficients defined in Eqs.(34) which for large α correspond to exponentially small emission of the envelope soliton.

The results above presented for independent scattering approach allow us to assume that in a general nonlinear system the dependence of the transmission coefficient on the slab length has two different parts. First, a nonlinear part corresponds to a trasformation of a nonlinear (soliton) pulse into a linear wave-packet, and the corresponding length is a function of the pulse amplitude; this length may be extremely large (expecially above a threshold value). The second linear part is characterized by the usual localization length and it is associated with the property of Anderson localization of a linear wavepacket. Thus, for nonlinear systems the decay length of the transmission coefficient may be much longer than for linear ones, if nonlinearity contributes to produce soliton pulses.

6 Conclusions

In conclusion, we have briefly reviewed some problems related to propagation of nonlinear waves through disordered madia. In the case when the nonlinear medium shows modulational instability, the nonlinear constant-amplitude wave decays into a set of solitons. So, nonlinearity leads to an improvement of the transmission only when it contributes to create soliton pulses. Solitons are much robust than linear waves or wave packets and that is why their creation acts against localization effects: solitons have to give non-zero contribution to the transmission coefficient. Moreover, above a certain threshold in amplitude, the envelope soliton described by the Schrödinger equation has very little radiative losses, and it propagates almost undistorted. Using the simplest independent-scattering approach, we have pointed out difference between three types of solitons scattering in disordered media. As a result, topological solitons (kinks) are more robust objects than other solitons, and in typical situations they may be described in the framework of a collective-coordinate approach.

As can be concluded from the paper, one of the main questions in the study of influence of nonlinearity on disorder is modulational instability which may produce solitons to improve transmission. We believe that the study of modulational instability in inhomogeneous media will improve our knowledge about nonlinear effects in disordered systems. As seems to us, another possible line interesting for subsequent studies in this field is direct numerical simulations of *different* soliton bearing systems including disorder and their comparison because the propagation of solitons through a disordered medium essentially depends on type of solitons. In any case, the study of the soliton propagation is very important because solitons give the most relevant contribution to the transmission characteristics of a disordered nonlinear system and, in particular, they may extremely increase the energy transmission in nonlinear systems.

Acknowledgments

I would like to thank Alan Bishop, David Campbell and Sergei Gredeskul for valuable discussions, Michel Peyrard and Stephanos Pnevmatikos (it is very regrettable that he tragically died while I was writing this paper) for a permission to use figures from their papers. I am indebted to James A. Krumhansl who attracted my attention to the paper by Payton et al. [5], probably, the first paper on energy transport in nonlinear disordered systems. I thank Angel Sánchez for critical reading and help in preparation of the manuscript.

References

1. *Nonlinearity and Disorder*, eds. A.R.Bishop, D.K.Campbell and St.Pnevmatikos, Springer Proceedings in Physics no. 39 (Springer, Berlin-Heidelberg, 1989).
2. Lifshitz,I.M., Gredeskul,S.A.,and Pastur,L.A.:*Introduction to the Theory of Disordered Systems.*Wiley, New York (1988).
3. *Scattering and Localization of Classical Waves in Random Media*, ed. Ping Sheng (World Scientific, Sigapore, 1990).
4. Kivshar,Yu.S., Gredeskul,S.A.,Sánchez, A., and Vázquez: Phys.Rev.Lett.**64**, 1693 (1990).
5. Payton,D.N., Rich,H.,and Visscher,W.M.:Phys.Rev. **160**,706 (1967).
6. Bourbonnais,R.,and Maynard,R.: Phys.Rev.Lett. **64**,1397 (1990).
7. Devillard,P., and Souillard,C.M.: J.Stat.Phys. **43**, 423 (1986).
8. Doucot,B., and Rammal, R., Europhys.Lett. **3**, 969 (1987); J.Phys.(Paris) **48**, 509 (1987).
9. Knapp,R., Papanicolaou,G., and White,B., in: Ref.1, p.2.
10. Caputo,J.G., Newell, A.C., and Shelley,M., In:*Integrable Systems and Applications*, eds. M.Balabane, P.Lochak, and C.Sulem, Lecture Notes in Physics, no.342, Springer, Berlin-Heidelberg, 1989, p.49.
11. Peyrard,M., and Bishop,A.R.,In:*Nonlinear Coherent Structures*, eds. M.Barthes and J.Léon, Lecture Notes in Physics No.353 (Springer, Berlin-Heidelberg,1990), p.29.
12. Qiming Li, Soukoulis,C.M., Pnevmatikos,St., and Economou,E.M.: Phys.Rev. B **38**, 11888 (1988).
13. Ishiwata,S., Okada,Y., Watanabe,S., and Tanaca,H.: J.Phys.Soc.Jpn. **59**, 3029 (1990).
14. Pascual,P.J., Vázquez,L., Bishop,A.R., and Pnevmatikos,St., In: Ref.1, p.85.
15. Gredeskul,S.A., Kivshar,Yu.S., Maslov,L.K., Sánchez,A., and Vázquez,L.: Phys. Lett.A (submitted).
16. Fogel,M.B., Trullinger,S.E.,Bishop,A.R., and Krumhansl,J.A.: Phys.Rev.Lett. **36**, 1411 (1976); Phys.Rev.B **15**,1578 (1977).
17. Kivshar, Yu.S., and Malomed,B.A.: Rev.Mod.Phys.**61**, 763 (1989).
18. Gredeskul,S.A., and Kivshar,Yu.,S.:*Propagation and Scattering of Nonlinear Waves in Disordered Systems*, Phys.Rep. (submitted).
19. Delyon,F.,Levy,Y.-E., and Souillard,B.: Phys.Rev.Lett. **57**,2010 (1986).
20. McKinstrie,C.J., and Luther,G.G.: Physica Scripta **30**, 31 (1990).
21. Barday,D., and Remoissenet, M.: Phys.Rev. B **43**, in press (1991).
22. Ablowitz, M.J., and Ladik, J.F.: J.Math.Phys. **17**, 1011 (1976).

23. Qiming Li, Pnevmatikos, St., Economou,E.N., and Soukoulis,C.M.: Phys. Rev. B **37**, 3534 (1988).
24. Kivshar,Yu.S., Kosevich, A.M., and Chubykalo,O.A.: Phys.Rev. B **43**, in press (1991).
25. Toda,M.: J.Phys.Soc.Jpn. **22**, 431 (1967); **23**,501 (1967).
26. Wadati,M.: J.Phys.Soc.Jpn. **52**, 2642 (1983).
27. Wadati,M., and Akutsu,Y.: J.Phys.Soc.Jpn. **53**, 3342 (1986).
28. Bass,F.G., Kivshar,Yu.S., Konotop, V.V., and Sinitsyn, Yu.A.: Phys.Rep. **157**,63 (1988).
29. Pascual,P.J., and Vázquez, L.: Phys.Rev. B **33**, 8305 (1985).
30. Biller,P., and Petruccione,F.: Phys.Rev. B **41**, 2139, 2145 (1990).
31. Kivshar Yu.S., Kosevich, A.M., and Chubykalo,O.A.: Phys.Lett.A **125**, 35 (1987).

Nonlinearity and Disorder in the Statistical Mechanics of Integrable Systems

R.K. Bullough[1,*], *J. Timonen*[2], *and Yu-zhong Chen*[3,**]

[1]MIDIT, The Technical University of Denmark, DK-2800 Lyngby, Denmark
[2]Department of Physics, University of Jyväskylä, SF-40100 Jyväskylä, Finland
[3]UMIST, Manchester M60 1QD, UK
*On leave from UMIST, Manchester M60 1QD, UK
**Currently at Department of Physics, Purdue University, IN, USA

Abstract. Attention is drawn to a theory of the statistical mechanics (SM) of the integrable models in 1+1 dimension — a theory of 'soliton statistical mechanics' classical and quantum [1 − 17]. This SM provides a generic example of *integrable* nonlinearity interacting with disorder. In the generic classical examples, such as the classical SM of the sine–Gordon model, phonons provide disorder in which sit coherent structures — the kink–like solitons. But these solitons are *dressed* by the disorder, in equilibrium, while the breather–like solitons break up to form the disordered structures which are the phonons in thermal equilibrium. On the other hand quantum solitons, dressed by both the vacuum and finite temperature effects, are nevertheless coherent *ordered* structures. In particular, the quantum breathers have a degree of freedom, and a consequent internal structure,which is localised in k–space, and through which the quantum breathers exhibit some features of *quantum chaos*. The disorder of the phonons is eliminated as such in these quantum systems. But the translational motions of these quantum particles continue to be disordered by the Brownian motion thereby contributing to the finite entropies of these particles.

1. Introduction

This short paper is primarily concerned to draw attention to the work on equilibrium statistical mechanics (SM) of the integrable models in one space and one time (1+1) dimensions — a 'soliton statistical mechanics' — which we have reported in a series of papers, e.g. [1 − 16]. The point in the present context is that the work describes a generic situation in which nonlinear coherent structures (solitons) interact with disorder. The purpose of the paper is to try to explain in simple terms how the interplay between nonlinearity, integrability and disorder creates the results it does. These results are typically expressions for the equilibrium free energies of the models.
The *nonlinearity* plays a fundamental role in the analysis and is responsible for a crucial element of it. However, the *integrability* of the models restricts the nonlinear dynamics so that classical motions lie on tori or generalised tori. The existence of the latter, essentially tori with infinite period, ensures that

nonlinearity and integrability together can create the coherent structures which are the solitons.

On the other hand it is the nature of SM that the coherent (ordered) solitons are to move in the disorder of Brownian motion introduced conceptually (or in actuality [7, 12, 16]) by a heat bath at a finite temperature. In the classical examples the action of the Brownian motion is to provide a disordered phonon background which *dresses* the solitons. The phonons derive in the first instance from motions on classical tori described by action–angle variables $P(k)$, $Q(k)$ for each 'mode' k: the phase spaces are cylinders $0 \leq P(k) < \infty$, $0 \leq Q(k) < 2\pi$ and the tori are P=constant, $Q = \omega t$+constant where $\omega = \omega(k)$. But under the action of the heat bath all P's and Q's become accessible and as far as we know the motion is one of infinite dimensional Hamiltonian chaos in which all values of the P's and Q's are *a priori* equally likely (Hamiltonian chaos is ergodic).

In equilibrium these dressed solitons are intrinsically different from free solitons. Indeed the dressed solitons interact through phase shifts between themselves and with the phonons. At first sight this is exactly the nature of solitons in 1+1 dimension. But in the SM these solitons do this even in the action–angle description which the theory adopts. Note that in action–angle description the integrable models form ideal, non–interacting gases first of all.

A generic example is the sine–Gordon (s–G) model with classical equation of motion

$$\phi_{xx} - \phi_{tt} = m^2 \sin \phi \tag{1}$$

where m is a 'mass'. This derives as Hamilton's equations of motion from the Hamiltonian

$$H[\phi] = \gamma_0^{-1} \int_{-\infty}^{\infty} [\frac{1}{2}\Pi^2\gamma_0^2 + \frac{1}{2}\phi_x^2 + m^2(1 - \cos\phi)]dx \tag{2}$$

in which $\gamma_0 > 0$ is a classical coupling constant.

The Π, ϕ are canonical variables satisfying the bracket $\{\Pi, \phi\} = \delta(x - x')$. There is a canonical transformation to action–angle variables [4, 7, 18] in which $H[\phi] \to H[p]$,

$$H[p] = \sum_{i=1}^{N_K + N_{\bar{K}}} (M^2 + p_i^2)^{1/2} + \sum_{j=1}^{N_b} (4M^2 \sin^2\theta_j + \hat{p}_j^2)^{1/2} + \int_{-\infty}^{\infty} \omega(k)P(k)dk, \tag{3}$$

in which $M \equiv 8m\gamma_0^{-1}$ is the 'kink mass', $\omega(k) = (m^2 + k^2)^{1/2}$, and p_i, $\hat{p}_j$ are translational momenta for the kink (antikink) and breather type solutions. It is important to our discussion that the breathers also have an internal action variable $4\gamma_0^{-1}\theta_j$, $0 \leq \theta_j < \frac{\pi}{2}$, which is conjugate to Φ_j, $0 \leq \Phi_j < 8\pi$. The Hamiltonian $H[p]$ describes an ideal, non–interacting gas made up of relativistic massive particales of masses M, 2Msin θ_j: the $\int \omega(k)P(k)dk$ can be identified as a collection of classical *oscillators* of mass m. Under the action of the Brownian motion these oscillators become the 'phonons' which provide the disorder about the coherent solitons. Of course this is an interpretation: tha action of the Brownian motion is to cause all of the 'particles' making up the ideal gas (3)

to interact with each other through phase shifts. Moreover in the classical SM of s–G the breathers are actually *eliminated* as we shall see.
The theory [1 – 16] concernes both classical SM and quantum SM. A point of this paper is to show that the reaction of the coherent structures to disorder is very different in the two cases. Roughly speaking quantum mechanics imposes order on the systems: thus although the classical breathers in the classical SM disappear by becoming chaotic phonons, the quantum breathers in the quantum SM remain, and remain in one of their two degrees of freedom, highly ordered.
To see how all of this may be said to come about we have to sketch first of all the quantum and classical SM of two of the simplest of the integrable models in 1+1. These are the sinh–Gordon (sinh–G) and repulsive NLS models.

2. Interacting phonons: order and disorder in the sinh–G and the repulsive NLS models

The Hamiltonian $H[\phi]$ for the sinh–G model is obtained from (2) for s–G by canonical transformation $\Pi \rightarrow \gamma_0^{-1/2}\Pi$, $\phi \rightarrow \gamma_0^{1/2}\phi$, followed by analytical continuation in γ_0, $\gamma_0 \rightarrow -\gamma_0$. The equation of motion is (1) with $\sin\phi \rightarrow \sinh\phi$. The classical repulsive NLS model has equation of motion

$$-i\phi_t = \phi_{xx} - 2c\phi^*\phi^2, \quad c>0 \tag{4}$$

and ϕ is complex. Both of these models have no coherent structures (solitons). There are phonons only and $H[p] = \int_{-\infty}^{\infty} \omega(k)P(k)dk$ simply with $\omega(k) = (m^2+k^2)^{1/2}$ for sinh–G and $\omega(k) = k^2$ for repulsive NLS. The repulsive NLS is a nonrelativistic limit of the sinh–G model with $m = 1/2$.
Both of these models have been explicitly placed in a heat bath and the quantum master equation solved for the equilibrium density matrix [7, 12, 16]. We would like to follow the dynamics but so far only the equilibrium solution has been calculated. The solution tells us of course to compute the partition function $Z = tr\ exp(-\beta\hat{H})$. It is important to notice that although we are concerned only with oscillators there is a continuous infinity of these. This means that Z is conveniently a functional integral. We note too that $H[p]$ looks like a collection of *linear* oscillators. However, the nonlinearity is still imposed upon the problem.
In [1 – 16] we develop different aspects of two distinct methods for computing both quantum and classical Z. One is indeed a 'method of functional integration' [1, 2, 4, 6]. The other is a variational method which describes the disorder by defining an entropy S and minimising $F = E - \beta^{-1}S$ to determine the actual free energy $F = -\beta^{-1}lnZ$. We call this method the method of 'generalised Bethe ansatz'. We find each method leads to the same expression for F for the chosen model. All of the results [12, 17] take the same general forms as those we give shortly for the sinh–G and repulsive NLS models – namely a series of *linear oscillator* contributions with however excitation energies determined from a set of *nonlinear* coupled integral equations.

Remarkably enough this mathematical structure persists even when there are solitons: the solitons in $H[p]$ eqn.(3) lie on tori of infinite period, e.g. the kinks have p_i = constant and the canonical q_i lies in $-\infty<q_i<\infty$. In effect the classical cylindrical phase space of the oscillator is split open to form infinite sheets coordinatised by p_i, q_i. Motions on these sheets contribute disordered oscillator contributions to F.
For the quantum oscillator contributions to F for the 'pure phonon' systems (sinh–G and repulsive NLS) we find that a typical distinction is between bose oscillators and fermi oscillators. In fact there appear to be anyon [19] oscillator descriptions also [13 – 15]. In the 'pure phonon' systems the bose and fermi descriptions are wholly equivalent. We find for both the quantum sinh–G and quantum repulsive NLS models that, for $L \to \infty$,

$$\begin{aligned} FL^{-1} &= \mu\bar{n} - (2\pi\beta)^{-1}\int_{-\infty}^{\infty} \ell n(1+e^{-\beta\epsilon(k)})dk \\ \epsilon(k) &= \omega(k) - \mu - (2\pi\beta)^{-1}\int_{-\infty}^{\infty} \frac{d}{dk}\Delta_f(k,k')\ell n(1+e^{-\beta\epsilon(k')})dk' \end{aligned} \tag{5a}$$

or *equivalently* (bose–fermi equivalence)

$$\begin{aligned} FL^{-1} &= \mu\bar{n} + (2\pi\beta)^{-1}\int_{-\infty}^{\infty} \ell n(1-e^{-\beta\epsilon(k)})dk \\ \epsilon(k) &= \omega(k) - \mu + (2\pi\beta)^{-1}\int_{-\infty}^{\infty} \frac{d}{dk}\Delta_b(k,k')\ell n(1-e^{-\beta\epsilon(k')})dk' \end{aligned} \tag{5b}$$

in which μ is a chemical potential, Δ_b and Δ_f are boson and fermion phase shifts respectively, and $\Delta_b = \Delta_f + 2\pi\theta(k'-k)$ where $\theta(x)$ is the unit step function. The connection between Δ_b and Δ_f ensures that (5a) (fermion description) and (5b) (boson description) *are* equivalent. Note that the systems conceptually lie in a 'box' of length L and $L \to \infty$ in such a way that $\lim_{L\to\infty} FL^{-1}$ is a well defined free energy density. Thus if there are N particles in the box $\lim_{L\to\infty} NL^{-1} = \bar{n}>0$. This is the *finite density* thermodynamic limit. It is interesting that it is this finite density limit which allows the Brownian motion to induce interactions between the particles of the system.
What happens is this. The thermodynamic limit at finite density is implemented through periodic boundary conditions. For these two simple models a condition on the allowed modes k, which we call $\tilde{k}$, arises. In discretized form this condition is (e.g. [2, 4, 12])

$$\tilde{k}_n = k_n - L^{-1}\sum_{n'}\Delta(\tilde{k}_n,\tilde{k}_{n'})P_{n'} \tag{6}$$

in which $P_n = 0,1$ in fermion description (and $\Delta = \Delta_f$) and $P_n = 0,1,2,...$ in boson description (and $\Delta = \Delta_b$): the Δ are the phonon–phonon S–matrix phase shifts and eqns. (6) are an example of the 'phase shift' equations[1, 2, 17]. It is the conditions (6) which induce the nonlinear integral equations for the excitation energies in eqns. (5). These quantum phonons are disordered in the

usual way. There is an entropy $S = (\partial F/\partial T)_L$ for each fixed L and the limit $L \to \infty$ can be taken.
These quantum phonons are little different from the also disordered classical phonons. The classical SM yields exactly the classical limit of the *bose* expressions eqns. (5b): with $\mu = 0$ and $L \to \infty$ this is

$$FL^{-1} = (2\pi\beta)^{-1} \int_{-\infty}^{\infty} \ell n(\beta\epsilon(k))dk$$
$$\epsilon(k) = \omega(k) + (2\pi\beta)^{-1} \int_{-\infty}^{\infty} \frac{d}{dk}\Delta_c(k,k')\ell n(\beta\epsilon(k'))dk' \tag{7}$$

in which Δ_c is the classical limit of Δ_b. The condition (6) survives but now $\Delta = \Delta_c$ while the P_n prove to be exactly the discretized action variables.
Although the Brownian motion induces the interactions between the phonons in the finite density thermodynamic limit as explained, the phase shifts themselves are a consequence, entirely, of the *nonlinearity*: the phase shifts Δ are essentially linear in the coupling constants. As $\gamma_0 \to 0$ the Δ for sinh–G all vanish and $\tilde{k}_n = k_n$ in (6): $k_n = 2\pi n L^{-1}$ the usual uncoupled k–vector. The situation is directly comparable for s–G. But the coupling constants actually scale the nonlinearity: the Hamiltonian $H[\phi]$ eqn. (2) for classical s–G becomes exactly that for the linear Klein–Gordon model as $\gamma_0 \to 0$. The same is true for sinh–G, and this feature is generic. Thus only the *nonlinear* integrable models have phase shifts and only in these nonlinear models can the Brownian motion induce interactions between the particles. The linear systems remain ideal oscillator gases agitated by the Brownian motion in the usual way while the energies (frequences) conform to the free particle dispersion relations $\epsilon(k) = \omega(k)$.

3. The classical sine–Gordon model and its free energy

The main point of this paper is to contrast the way in which quantum solitons can couple to the disorder of Brownian motion as compared with the way in which the classical solitons can couple to the Brownian motion. The situation is generic for the s–G and attractive NLS models; the Landau–Lifshitz models; the Heisenberg magnet and others[12, 13, 17]. A fine example is the Heisenberg magnet in a longitudinal field [20]. The Toda lattice apparently couples very differently [8]. A physical result of this contrast between the quantum and classical systems is that the Davydov soliton, roughly speaking a classical soliton of the attractive NLS model, will break up into phonons in final equilibrium and will become disordered in that sense — our picture of infinite dimensional Hamiltonian chaos.
We remark that the attractive NLS model has classical equation of motion eqn. (4) with however the coupling constant $c<0$. Unlike the repulsive case this model has soliton solutions of 'breather' type, that is with a translational degree of freedom and an internal oscillatory degree of freedom. The attractive NLS model offers certain difficulties in both the quantum and classical SM: it

is, however, in a well defined sense the non–relativistic form of the s–G model (with mass m=1/2). Therefore we take the s–G model as the generic model. In this section 3 we sketch the classical SM of this model and the way disorder (the phonons) interact with the order (the solitons) breaking up the classical breathers to phonons in equilibrium. Notice that an apposite study for the purposes of this meeting would be to describe the *dynamics* of that break up as the disorder breaks the breather up into its ultimate Hamiltonian chaos. Unfortunately such a calculation has still to be carried out. There is also the interesting question of the 'dynamics' of break up induced in the quantum solitons by turning up the temperature (the disorder). This calculation is not available yet either.
Evidently the sinh–G model is a 'repulsive' form of the s–G model. In the same way analytical continuation in the coupling constant ($c \to -c$) takes the repulsive NLS model to the attractive NLS model and vice versa. As the repulsive models become the attractive models the nonlinearity can admit the solitons.
Our methods for the classical free energies use the action–angle variables as canonical coordinates. Thus for the classical free energy of s-G we appeal to the Hamiltonian $H[p]$ given in eqn. (3). However [1], we use in effect

$$H[p] = \sum_{i=1}^{N_K+N_{\bar{K}}} (M^2 + p_i^2)^{1/2} + \int_{-\infty}^{\infty} \omega(k)P(k)dk \qquad (8)$$

in which (compare (3)) the breather contributions are simply dropped: we have (so far) thus preemptied the argument which shows that the breathers are replaced by phonons in the classical SM.
From (8) our result corresponding to (7) for the sinh–G model is [1] that, for $L \to \infty$,

$$FL^{-1} = (2\pi\beta)^{-1} \int_{-\infty}^{\infty} \ell n(\beta\epsilon(k))dk - 2(2\pi\beta)^{-1} \int_{-\infty}^{\infty} e^{-\beta\tilde{E}(\phi)}dp \qquad (9)$$

while the excitation energies $\epsilon(k)$ and $\tilde{E}(p)$ satisfy the coupled system

$$\begin{aligned} \epsilon(k) &= \omega(k) + (2\pi\beta)^{-1} \int_{-\infty}^{\infty} \frac{d\Delta}{dk}(k,k')\ell n(\beta\epsilon(k'))dk' \\ &\quad + 2(2\pi\beta)^{-1} \int_{-\infty}^{\infty} \frac{d\Delta_s}{dk}(k,p)e^{-\beta\tilde{E}(p)}dp \\ \tilde{E}(p) &= E(p) - (2\pi\beta)^{-1} \int_{-\infty}^{\infty} \frac{d\Delta_s}{dp}(p,k')\ell n(\beta\epsilon(k'))dk' \\ &\quad - 2(2\pi\beta)^{-1} \int \frac{d\Delta_{ss}}{dp}(p,p')e^{-\beta\tilde{E}(p')}dp' \end{aligned} \qquad (10)$$

in which Δ, Δ_s, Δ_{ss} are classical phonon–phonon, phonon–soliton and soliton–soliton phase shifts respectively [1]. (Here the 'soliton' means kink or equivalent antikink solutions of classical s–G).

It is plain that the 'phonons' are the disordered classical phonons of the sinh–G model as described in section 1 (eqn.(7)). Then the solitons appear as independent coherent structures dressed by the coupling to the phonons in a way we make plain shortly.
In this particular description the classical solitons are distinguished from the disordered phonons by densities $\sim e^{-\beta\tilde{E}(p)}$. There is a *semi*classical quantum description [3] in which the soliton densities appear as *fermi* oscillator contributions $ln(1 + e^{\sim\beta\tilde{E}(p)})$. The semiclassical fermi oscillators are disordered in the usual way for fermions: thus the Brownian motion is disordering the classical solitons in the normal way. In this description the 'coherence' of the classical solitons can be identified only by the fermion label which persists in the classical density.
It is possible to *iterate* the integral equations (10) though the expression for FL^{-1} eqn. (9). The result is [1], for $L \to \infty$,

$$FL^{-1} + (F^{(1)} + F^{(2)} + ...) + F_{KG} \tag{11}$$

where

$$\begin{aligned} F^{(1)} &= -\beta^{-1} m(\frac{8}{t\pi})^{1/2} e^{-1/t}[1 - \frac{7}{8}t - \frac{59}{128}t^2 - \frac{897}{1024}t^3 - ...] \\ &- \beta^{-1} m[\frac{1}{4}t + \frac{1}{8}t^2 + \frac{3}{16}t^3 + \frac{53}{128}t^4 + ...]; \end{aligned} \tag{12}$$

$t \equiv (M\beta)^{-1}$, and F_{KG} is *exactly* the free energy density of the linear Klein–Gordon (K–G) equation. There is the problem about the classical (ultraviolet) divergence of the K–G system but this is well understood. The point here is that it is a linear contribution independent of γ_0 (which scales the nonlinearity). The *nonlinear* part is shown explicitly in $F^{(1)}$: it depends only on $t = (8m\gamma_0^{-1}\beta)^{-1}$. When the nonlinearity vanishes as $\gamma_0 \to 0$, $t \to 0$ and $F^{(1)} \to 0$; the contributions $F^{(q)}$ are $0(e^{-q/t})$ and these also vanish when $\gamma_0 \to 0$.
It is a check on our initial choice of Hamiltonian eqn. (8) (no breathers) that we find *exactly* and term by term the results (11), (12) by evaluating the classical partition function Z. This is the functional integral

$$Z = \int \mathcal{D}\Pi\mathcal{D}\phi \; exp(-\beta H[\phi]) \tag{13}$$

in which, for classical s–G, $H[\phi]$ is given by eqn.(2). The functional integral (13) is taken over classical trajectories in the symplectic manifold coordinatised by $\Pi(x,t)$, $\phi(x,t)$. To reach (11) with (12), Z, eqn.(13), is evaluated by the transfer integral method (TIM) [1].
We can now see how the disorder of the phonons dresses the free solitons. Take the free soliton (free kink) from (3) with energy $(M^2 + p^2)^{1/2}$ and form the 1–particle partition function.

$$Z_K = \frac{1}{2\pi}\int_{-\infty}^{\infty} dp \; e^{-\beta(M^2+p^2)^{1/2}} \int_{-1/2L}^{+1/2L} dq. \tag{14}$$

One finds [7, 18], for $L \to \infty$,

$$FL^{-1} = -\beta^{-1}\frac{M}{2\pi}\sqrt{\frac{2\pi}{M\beta}}e^{-\beta M}[1 + \frac{3}{8}(M\beta)^{-1} + ...], \tag{15}$$

while it is easy to see that (12) contains the different expression (to be halved for kinks alone: factor 2 for kinks and antikinks)

$$-2(m\beta)\cdot\beta^{-1}\frac{M}{2\pi}\sqrt{\frac{2\pi}{M\beta}}e^{-\beta M}[1 - \frac{7}{8}(M\beta)^{-1}...]. \tag{16}$$

It is possible to show (e.g. [8]) that the extra factor $2(m\beta)$ in (16) is entirely due to the dressing provided by the disorder from the phonons: the change $3/8 \to -7/8$ in the series expansion depends however also on the disorder provided by the solitons disordered by the phonons disordered by the solitons ... i.e. there is the coupling through the nonlinear integral equations (19) and under the action of the Brownian motion the solitons also phase shift each other.

It will be noticed that $F^{(1)}$ also contains the series $-\beta^{-1}m[\frac{1}{4}t + \frac{1}{8}t^2 + \cdots]$. That this is a dressed i.e. disordered contribution from the classical breathers is intuitive and, indeed, one can compute the free breather contribution as $-\beta^{-1}\frac{M}{2\pi}\cdot 8\pi[(M\beta)^{-2} + \cdots]$ which is apparently dressed in $F^{(1)}$ to $-\beta^{-1}m[\frac{1}{4}(M\beta)^{-1} + \cdots]$. However the dressing is really one of total destruction by the Brownian motion. The breather contributions to $H[p]$, eqn. (3) were dropped in (8) which was then used to compute Z in the form

$$Z = \int \mathcal{D}\mu \exp(-\beta H[p]); \tag{17}$$

in this $\mathcal{D}\mu$ is the proper, and carefully constructed, measure for the new canonical coordinates [1]. These measures are determined by the phase shift equations (for s–G [1]). The simpler phase shift equations for sinh–G were exhibited in (6) earlier. Thus the Brownian motion which induces the phase shift interactions under the finite density condition of the thermodynamic limit determines the measure (the weight) of the trajectories in the classical statistical mechanics.

Now the breathers are deliberately dropped from the calculation because (apparently) they *are* destroyed by the Brownian motion in the approach to equilibrium and become in equilibrium the disordered phonons. In words the mathematics of this is the following [3, 12, 13]:– In the finite density thermodynamic limit the apparent phonon density (called $\rho(k)$) tends to zero as the box size $L \to \infty$. At first sight there are no classical phonons for s–G in thermodynamic limit. But now the classical breathers must make an *extensive* contribution (strictly $\propto L$) if they are to contribute to $\lim FL^{-1}$ as $L \to \infty$. The classical breathers of s–G (and of attractive NLS, of Landau–Lifshitz etc.) have *two* degrees of freedom. Indeed in $H[p]$ eqn.(3) there are the two action coordinates namely $\hat{p}_l$ and Θ_l. Each degree of freedom will provide a contribution $\mathcal{O}(L)$ as $L \to \infty$: the interactions via the phase shift equations (the Brownian

motion) ensure that the breather is *not* like a molecule with an internal vibration agitated by Brownian motion [12]. The two degrees of freedom together provide an $\mathcal{O}(L^2)$ contribution to F and to $L^{-1}F$ this is $\sim L$ as $L \to \infty$. The classical breathers therefore avoid this problem by collapsing in the sense that each angle Θ falls into a narrower and narrower cone of angle $\mathcal{O}(L^{-1})$ as $L \to \infty$. Thus in one sense they become highly ordered: the available phase space is reduced from a plane for the variable Θ, to a line as $L \to \infty$. However in another sense (namely in the other degree of freedom) they are *disordered*: for each label ℓ labelling a breather with translational momentum $\hat{p}_\ell$ the many possible breathers with internal co-ordinates $\Theta_{\ell_1}, \Theta_{\ell_2}, \cdots$ fill this narrowing cone densly in such a way that together they act as a single phonon with the wave number $\tilde{k}_\ell$, and these allowed $\tilde{k}_\ell$ satisfy a phase shift equation like eqn.(6). These 'new' phonons have a finite density $\rho(\tilde{k})$ as $L \to \infty$ in thermodynamic limit [5, 12, 13]. This is how the classical breathers break up to become the disordered classical phonons under the action of the Brownian motion.
We now contrast this break-up of the classical breathers with the behaviour of the quantum breathers faced with the action of the Brownian motion.

4. The quantum sine–Gordon model: the 'quantum chaos' of the quantum solitons

The quantum SM is distinguished [13 – 15] by the fact that the partition function Z eqn. (17) for the classical SM gains a quantum mechanical 'phase factor', the origin of all of the quantum features. Thus the 'action' $-\beta H[\phi]$ for the classical SM is extended in the quantum SM by addition of a phase term $i\Phi[p]$. For the sinh–G and quantum repulsive NLS models $\Phi[p]$ discretizes to

$$\Phi[p_n] = \frac{1}{\hbar}\int_0^{\beta\hbar} dt \sum_n P_n Q_{n,t} \tag{18}$$

in which P_n, Q_n are discretized action–angle variables. Our strategy to evaluate quantum Z is then to replace the quantum phase by *constraints* achieved by setting

$$\frac{1}{\hbar}\int_0^{\beta\hbar} dt P_n Q_{n,t} = \frac{1}{\hbar}\oint P_n dQ_n = 2\pi m_n \tag{19}$$

where m_n is a positive integer: by choosing $m_n = 0, 1$ and $m_n = 0, 1, 2, \ldots$ we find the fermion expression (5a) for $\lim_{L\to\infty} FL^{-1}$ and the boson expression (5b) respectively.
A priori there are many alternative choices for the integers m_n and corresponding different quantum statistics [13–15]: particularly by introducing a winding number $\nu_n \geq 1$ there are apparently *anyon* quantum statistics [13 – 15, 19]. The fact that (19) is Bohr semi-classical quantisation does not trivialise the quantisation of the integrable models: it is simply an exact set of constraints and a deeper quantisation through the quantum groups [11, 12, 15] (which determine

the quantum phase shift equations like (6)) remains imposed on the models. Notice that the only consequence for the pure phonon models like sinh–G is to introduce disorder (entropy) which is the entropy appropriate to the corresponding quantum statistics. Moreover the entropy is unique to the model — despite the alternative quantum descriptions.

In the quantum s–G the order and disorder are very different. The quantum constraints are now constraints imposed on the classical breathers alone: there are no other particles and the kinks and antikinks appear as pairs forming particular quantum breathers. We sketch how this comes about later in this section. These constraints on the classical breathers are that

$$\frac{1}{\hbar}\oint 4(\gamma_0^{''})^{-1}\Theta_\ell d\Phi_\ell = 2\pi m_\ell \tag{20}$$

in which $\gamma_0^{''} = \gamma_0[1-\gamma_0/8\pi]^{-1}$ is the renormalised quantum coupling constant with bare coupling constant $\gamma_0>0$. Since Θ_ℓ = constant the quantum constraints (20) on the classical breathers mean that Θ_ℓ is *discrete*: $\Theta_\ell = \frac{1}{16}m_\ell\gamma_0^{''}\hbar$. If $[8\pi(\gamma_0^{''})^{-1}] = N_b - 1$ where $[x]$ means integral part of x, it proves [12, 17] that there are $N_b - 2$ *distinct* quantum breathers labelled by $1 \le m_\ell \le N_b - 2$ and one kink–antikink pair of corresponding double weight. Thus the free energy is [12, 17], for $L \to \infty$,

$$FL^{-1} = \mu_s\bar{n}_S + \sum_{\ell=1}^{N_b-2}\mu_\ell\bar{n}_\ell - (2\pi\beta)^{-1}\sum_{\ell=1}^{N_b-2}\int d\alpha_\ell M_\ell\cosh(\alpha_\ell)\ell n(1+e^{-\beta\epsilon_\ell})$$
$$- 2(2\pi\beta)^{-1}\int_{-\infty}^{\infty} d\alpha_s M_s\cosh(\alpha_s)\ell n(1+e^{-\beta\epsilon_s}) \tag{21}$$

and this is evidently in fermion description in which rapidities α are used and the μ_ℓ, μ_s are chemical potentials: $M_\ell = 2M\sin(\ell\gamma_0^{''}/16)$, $M_s = M$. The excitation energies ϵ_ℓ, ϵ_S are determined from $N_b - 1$ coupled nonlinear integral equations generalising that in (5a). There are no other contributions to $\lim_{L\to\infty} FL^{-1}$. In particular there are no quantum phonon contributions: for in one sense these are the quantum breathers which have $\ell = 1$. However in another sense we explain shortly the quantum phonons have disappeared.

Thus there are $N_b - 1$ distinct particles stabilised by the action of quantum mechanics on the classical breathers preventing them from breaking up into phonons. The internal actions of the breathers are discretized (localised) by the quantum mechanics and the classical disorder is thereby prevented. This is evidently a form of *quantum chaos.*

The quantum solitons may be said to have a very real internal structure at zero temperature which is not distroyed at finite temperatures. The Bethe ansatz analysis of the bose–fermi equivalent massive Thirring model shows [21 – 23] that the particles of the theory include first of all bare fermions which fill the Dirac sea of negative energy states in the complex rapidity plane. On this is imposed n–particle systems, the n–strings. To conserve the degrees of freedom, introduction of an n–string induces n holes in the Dirac sea and some number of these ($<n$) join the n–string with the same real part of the complex rapidity

to form a complex structure. Simultaneously the remaining Dirac sea particles are rearranged (the back flow of the Dirac sea). There is also a new ordering as the particles in the complex structure adjust to each other, and there is a further adjustment between the different n–particle states which fall into groups defined by the different n. Ultimately one finds the bare particles have indeed disappeared: they renormalise $\gamma_0 \rightarrow \gamma_0''$, m, and the rapidities. What is left for a complete description of the system, is now the n–particle states, dressed with their collection of $<n$ bound holes, and free holes. The free holes finally become the quantum kinks and antikinks. The dressed n–strings become the quantum breathers with the constrained internal action. The Brownian motion induces disorder in one degree of freedom — the translational momentum. But the ordered structure in the other degree of freedom — the internal action Θ_ℓ is not disordered by the Brownian motion and persists in the form of $N_b - 2$ distinct particles. This is the *quantum* chaos the quantum breathers of s–G describe.

The quantum attractive NLS model, the quantum Landau–Lifshitz model, and the quantum ferromagnet apparently show similar features. Quantum mechanics stabilises the chaos of classical phonons as distinct quantum breathers. It is therefore of interest to describe how these internally ordered quantum breathers can become the disordered phonons as the temperature rises to implement the classical SM. That is it is of interest to understand how the quantum SM has the classical SM as the classical limit. We have carried out this calculation for quantum s–G [3] (see also [24]).

The key result is that the $N_b - 2$ quantum breathers become one collective boson [3, 24] in a semi–classical limit $\gamma_0 \rightarrow 0$: this limit also means $N_b \rightarrow \infty$. The rest energies of the bosons ($\propto m$) $\rightarrow 0$ in this limit at fixed $M = 8m/\gamma_0''$. These bosons may lower their energies by falling into their ground states. This 'explains' the 'collapse' of the classical breathers into the classical phonons in the futher classical limit, i.e. why $\Theta = O(L^{-1})$ as L$\rightarrow \infty$ for the classical breathers as explained in the previous section. The free energy density in this classical limit is given exactly by the eqn. (10) together with eqn. (11).

Evidently there is a fascinating interplay of nonlinearity and disorder in the rich structure described by the quantum and classical sine–Gordon model at finite temperatures. It is this we have tried to describe, with a minimum of mathematical detail, in this paper.

References

1. J. Timonen, M. Stirland, D. J. Pilling, Yi Cheng and R. K. Bullough, Phys. Rev. Lett. **56**, 2233 (1986).
2. R. K. Bullough, D. J. Pilling and J. Timonen, J. Phys. A: Math. Gen. **19**, L955 (1986).
3. J. Timonen, R. K. Bullough and D. J. Pilling, Phys. Rev. **B34**, 6525 (1986).

4. R. K. Bullough, D. J. Pilling and J. Timonen, 'Soliton statistical mechanics' in: "Solitons", M. Lakshmanan ed., Springer-Verlag, Heidelberg (1988), pp. 250–281, and refecences.
5. J. Timonen, Yu-zhong Chen and R. K. Bullough, Nucl. Phys. **B** (Proc. Suppl.) **5A**, 58 (1988).
6. R. K. Bullough, Yu-zhong Chen, S. Olafsson and J. Timonen, 'Statistical Mechanics of the NLS Models and their Avatars' in: "Integrable Systems and Applications", Springer Lecture Notes in Physics 342, M. Balabane, P. Lochak and C. Sulem eds., Springer-Verlag, Heidelberg (1989), pp. 12–26.
7. R. K. Bullough, D. J. Pilling and J. Timonen, 'Soliton statistical mechanics and the thermalisation of biological solitons', in: "Nonlinear Coherent Structures in Physics", J. Pouget, M. Remoissenet and R. Ribotta, eds., Journal de Physique C3, **50**, 41 (1989).
8. R. K. Bullough, J. Timonen, Yu-zhong Chen, Yi Cheng and M. Stirland, 'Quantum and Classical statistical mechanics of the integrable models', in: "Nonlinear evolution equations: integrability and spectral methods", A. Degasperis, A. P. Fordy and M. Lakshmanan eds., Manchester University Press, Manchester (1990), pp. 605–617.
9. R. K. Bullough and S. Olafsson, 'Complete integrability of the integrable models: quick review', in: IX Intl. Congress on Math. Phys. 17–27 July, 1988, Swansea, Wales, B. Simon, A. Truman and I. M. Davies eds., Adam Hilger, Bristol (1989), pp. 329–334.
10. R. K. Bullough and S. Olafsson, 'Algebra of Riemann–Hilbert Problems and the Integrable Models — A Sketch', in: "Proc. 17th Intl. Conf. on Diff. Geom. Methods in Theor. Phys.", A. I. Solomon ed., World Scientific, Singapore (1989), pp. 295–309.
11. R. K. Bullough, S. Olafsson, Yu-zhong Chen and J. Timonen, 'Integrability Conditions', in: "Proc. XVIII Intl. Conf. on Diff. Geom. Methods in Theor. Phys." (Lake Tahoe, July 2–9, 1989: NATO ARW Physics and Geometry) Ling-Lie Chau and Werner Nahm eds., Plenum, New York (1990) and references. To appear 1990.
12. R. K. Bullough, Yu-zhong Chen and J. Timonen, 'Soliton statistical mechanics: thermodynamic limits for the integrable models', in: "Proc. IV Intl. Workshop on Nonlinear and Turbulent Processes in Physics", V. E. Zakharov, A. G. Sitenko, N. S. Erokhin and V. M. Chernousenko eds., World Scientific, Singapore (1990) and references. To appear 1991.
13. R. K. Bullough and J. Timonen, 'Quantum solitons on quantum chaos: coherent structures, anyons and statistical mechanics', in: "Microscopic Aspects of Nonlinearity in Condensed Matter", A. R. Bishop and V. Tognetti eds., Plenum, New York. To appear 1991.
14. R. K. Bullough and J. Timonen, 'Quantum and classical integrability: new approaches in Statistical Mechanics', in: "Nonlinear Science: the Next Decade", A. R. Bishop and D. K. Campbell eds., Physica D: Nonlinear Phenomena. To appear 1991.
15. R. K. Bullough and J. Timonen, 'Quantum groups and quantum complete integrability: theory and experiment', in: "Proc. XIX Intl. Conf. on Diff. Geom. Methods in Theor. Phys. (Rapallo, Italy, June 19–24, 1990), U.

Bruzzo and C. Bartocci eds., Lecture Notes in Physics, Springer-Verlag, Heidelberg. To appear 1991.
16. J. Timonen, D. J. Pilling and R. K. Bullough, 'Quantum kinetic Heisenberg models', in: "Coherence, Cooperation and Fluctuations", F. Haake, L. M. Narducci and D. Walls eds., CUP, Cambridge (1986), pp. 18–34.
17. Yu-zhong Chen, 'Classical and quantum statistical mechanics of the 1+1 dimensional integrable models', Ph.D. Thesis, University of Manchester (1989).
18. R. K. Bullough, 'Statistical mechanics of the sine–Gordon field: Part I', in: "Nonlinear Phenomena in Physics", F. Claro ed., Springer-Verlag, Heidelberg (1985), pp. 70–103 and references.
19. A. Zee, 'Semionics, a theory of high temperature superconductivity', in: "High Temperature Superconductivity; Proc. of the Los Alamos Symp. 1989", K. S. Bedell, D. Coffey, D. E. Keltzer, D. Pines and J. R. Schrieffer eds., Addison Wesley, Redwood City, Calif. (1990), pp. 248–298.
20. R. K. Bullough, Yu-zhong Chen, J. Timonen, V. Tognetti and R. Vaia, Phys. Lett. **A145**, 54 (1990).
21. H. B. Thacker, Rev. Mod. Phys. **53**, 253 (1981).
22. S. G. Chung and Yia-Chung Chang, J. Phys. A: Math. Gen. **20**, 2875 (1987).
23. J. Timonen and R. K. Bullough, to be published.
24. Nu Nu Chen, M. D. Johnson and M. Fowler, Phys. Rev. Lett. **56**, 507 (1986); 1427 (E) (1986).

Classical and Quantum Mechanical Analysis of Order and Chaos in the Discrete Self-Trapping Equation

P.L. Christiansen[1], *L. Cruzeiro-Hansson*[2], *H. Feddersen*[3], *R. Flesch*[4], *M. Salerno*[5], *and A.C. Scott*[1]

[1]Laboratory of Applied Mathematical Physics, The Technical University of Denmark, DK-2800 Lyngby, Denmark
[2]Present address: Department of Crystallography, Birkbeck College, University of London, Malet Street, London WC1E 7HX, UK
[3]Present address: Department of Mathematics, Heriot-Watt University, Riccarton, Edinburgh EH14 4AS, Scotland
[4]Present address: Research Center of Crete, P.O. Box 1527, GR-711.10 Heraklion, Crete, Greece
[5]Permanent address: Department of Theoretical Physics, University of Salerno, I-84100 Salerno, Italy

Abstract. Results for the classical discrete self-trapping (DST) equation with few degrees of freedom and varying degrees of nonlinearity are reported. The effect of order and chaos on the corresponding quantum DST equation is demonstrated. Linear and nonlinear localization is found in a many degrees of freedom DST model of a globular protein molecule.

Introduction

Several years ago the discrete self-trapping (DST) equation was introduced as a useful model for several interesting physical phenomena including: dynamics of small molecules, molecular crystals, and biomolecules [1-3]. Subsequently, a quantum theory for the DST equation was developed [4-6]. The present contribution summarizes results obtained in Refs. [7-11] on the relationship between (nonlinear) classical dynamics and (linear) quantum dynamics in the DST system. Also, a recent application [12] of the classical DST model to globular protein is briefly discussed. Here the relationship between (linear) Anderson localization and localization due to anharmonicity is illustrated.

In simplest terms, the classical DST system comprises f anharmonic oscillators (freedoms), which are coupled through linear dispersive interactions:

$$\left[i \frac{d}{dt} - \omega_o\right]\bar{A} + \gamma \,\mathrm{diag}(|A_1|^2, \cdots, |A_f|^2)\bar{A} + \varepsilon M \bar{A} = 0 \,, \tag{1}$$

where $\bar{A}$ is the column vector

$$\bar{A} = \mathrm{col}(A_1, A_2, \cdots, A_f) \tag{2}$$

of complex components. With the anharmonic parameter γ and the dispersive parameter ε both equal to zero, Eq. (1) represents a system of f noninteracting, harmonic oscillators each of frequency ω_o. In the general case ($\varepsilon \neq 0$ and $\gamma \neq 0$) the dynamics of Eq. (1) exhibits the effects of both anharmonicity and dispersion. This interaction can lead to either self-trapping (i.e. local modes or solitons) or chaos or a mixture of self-trapping and chaos depending upon the parameters and initial conditions for $\bar{A} = \bar{A}(t)$ that are chosen.

Springer Proceedings in Physics, Vol. 67 Nonlinearity with Disorder
Editors: F. Abdullaev · A.R. Bishop · S. Pnevmatikos

We note that the DST system is Hamiltonian and has two conserved quantities, the <u>number</u>

$$N = \sum_{i=1}^{f} |A_i|^2 \tag{3}$$

and the <u>energy</u>

$$H = \omega_o N - \frac{\gamma}{2} \sum_{i=1}^{f} |A_i|^4 - \varepsilon \sum_{i \neq j}^{f} m_{ij}\, A_i^* A_j \ , \tag{4}$$

m_{ij} being the elements of the coupling matrix M.

Therefore the quantization of the DST system becomes an easy task: the complex mode amplitudes (A_i^* and A_i) become boson creation and annihilation operators ($\hat{B}_i^\dagger$ and $\hat{B}_i$ satisfying

$$[\hat{B}_i, \hat{B}_j^\dagger] = \delta_{ij} \ .) \tag{5}$$

With full symmetrization of the operator products Eq. (3) becomes the <u>number operator</u>

$$\hat{N} = \sum_{i=1}^{f} \left(\hat{B}_i^\dagger \hat{B}_i + \frac{1}{2}\right) \tag{6}$$

and Eq. (4) becomes the <u>energy operator</u>

$$\hat{H} = \left(\omega_o - \frac{\gamma}{2}\right)\hat{N} - \frac{\gamma}{2} \sum_{i=1}^{f} \hat{B}_i^\dagger \hat{B}_i \hat{B}_i^\dagger \hat{B}_i \ . \tag{7}$$

The $\hat{H}$ and $\hat{N}$ operate upon wave functions of the form $|n_1\rangle|n_2\rangle \cdots |n_f\rangle$. Thus the quantized discrete self-trapping (QDST) equation describes an assembly of bosons.

Two degrees of freedom and higher degrees of nonlinearity

In the continuum limit, $f \to \infty$, the classical DST equation reduces to the nonlinear Schrödinger equation, which is integrable. Also, for $f < 3$ the DST system, having the two conserved quantities, (3) and (4), is integrable. In Refs. [7-8] we have investigated the integrable case $f = 2$ for higher degrees of nonlinearity, σ, the quantum case arising in the boson representation of spin systems. Thus we have, after a gauge transformation, removing the oscillator frequency ω_o in Eq. (1),

$$\left.\begin{aligned} i\dot{A}_1 + \gamma|A_1|^{2\sigma} A_1 + \varepsilon A_2 &= 0 \\ i\dot{A}_2 + \gamma|A_2|^{2\sigma} A_2 + \varepsilon A_1 &= 0 \end{aligned}\right\} \ . \tag{8}$$

The elements of the coupling matrix are $m_{12} = m_{21} = 1$ and $m_{11} = m_{22} = 0$. Note that the degree of nonlinearity in Eq. (1) corresponds to $\sigma = 1$. The conserved quantities become

$$N = |A_1|^2 + |A_2|^2 \tag{9}$$

and

$$H_\sigma = -\frac{\gamma}{\sigma+1}\left[|A_1|^{2\sigma+2} + |A_2|^{2\sigma+2}\right] - \varepsilon\left[A_1^* A_2 + A_1 A_2^*\right] \ . \tag{10}$$

The dynamics of this system can be expressed in terms of various canonical variables. Thus we may introduce polar coordinates by $A_1 \equiv a_1 \exp(i\theta_1)$ and $A_2 \equiv a_2 \exp(i\theta_2)$ and use the angle difference $\theta = \theta_1 - \theta_2$ and the square module $N_1 = a_1^2$ as canonical variables. In terms of the variables introduced by Feynman et al. [13]

$$\left.\begin{aligned} r_1 &= A_1A_2^* + A_2A_1^* \\ r_2 &= i(A_1A_2^* - A_2A_1^*) \\ r_3 &= |A_1|^2 - |A_2|^2 \end{aligned}\right\} \tag{11}$$

the DST equation (8) can be rewritten as "the Feynman top"

$$\dot{\overline{r}} = \overline{\Omega}_\gamma \times \overline{r} \tag{12a}$$

with

$$\underline{r} = (r_1, r_2, r_3) \tag{12b}$$

and

$$\overline{\Omega}_\gamma = \left(-2\varepsilon, 0, \gamma\left(|A_2|^{2\sigma} - |A_1|^{2\sigma}\right)\right) . \tag{12c}$$

The conserved quantities become

$$N = \left(r_1^2 + r_2^2 + r_3^2\right)^{\frac{1}{2}} \tag{13}$$

and

$$H = -\frac{\sigma}{\sigma+1}\left[\left(\frac{N+r_3}{2}\right)^{\sigma+1} + \left(\frac{N-r_3}{2}\right)^{\sigma+1}\right] - \varepsilon r_1 . \tag{14}$$

Using r_3 and its canonically conjugate momentum, p_3, the resulting Hamiltonian equations can readily be integrated. Table 1 indicates the increasing degree of complexity of $r_3(t)$ as σ increases. (Terminology as in Ref. [14]). The conjugate momentum is seen to satisfy a generalized pendulum equation.

In the quantum case the symmetrized number operator and energy operators become

$$\hat{N} = \hat{B}_1^\dagger\hat{B}_1 + \hat{B}_2^\dagger\hat{B}_2 , \tag{15}$$

where 1 has been subtracted for convenience, and

Table 1. Feynman variable and its conjugate momentum versus degree of nonlinearity

Degree of nonlinearity	Feynman variable	Equation for conjugate momentum
σ	$r_3(t)$	$p_3(t)$
0	sinusoid	p_3 = const
1,2	Jacobi elliptic function	Pendulum equation with sinusoid
3,4	Inverse of hyperelliptic integral of 1st kind, 3rd class	Generalized pendulum equation with inverse of elliptic integrals of 3rd kind
5,6	Inverse of hyperelliptic integral of 1st kind 5th class	Generalized pendulum equation with inverse of hyperelliptic integrals of 3rd kind, 2nd class

Table 2. Quantum operators corresponding to $|A|^{2\sigma+2}$ for nonnegative integer values of σ

σ	$\lvert A\rvert^{2\sigma+2}$	$\frac{(2\sigma+2)!}{[(\sigma+1)!]^2}$	$P_\sigma(\hat{B}^\dagger\hat{B})$
0	$\lvert A\rvert^2$	2	$\hat{B}^\dagger\hat{B} + \frac{1}{2}$
1	$\lvert A\rvert^4$	6	$(\hat{B}^\dagger\hat{B})^2 + (\hat{B}^\dagger\hat{B}) + \frac{1}{2}$
2	$\lvert A\rvert^6$	20	$(\hat{B}^\dagger\hat{B})^3 + \frac{3}{2}(\hat{B}^\dagger\hat{B})^2 + 2(\hat{B}^\dagger\hat{B}) + \frac{3}{4}$
3	$\lvert A\rvert^8$	70	$(\hat{B}^\dagger\hat{B})^4 + 2(\hat{B}^\dagger\hat{B})^3 + 5(\hat{B}^\dagger\hat{B})^2 + 4(\hat{B}^\dagger\hat{B}) + \frac{3}{2}$
4	$\lvert A\rvert^{10}$	252	$(\hat{B}^\dagger\hat{B})^5 + \frac{5}{2}(\hat{B}^\dagger\hat{B})^4 + 10(\hat{B}^\dagger\hat{B})^3 + \frac{25}{2}(\hat{B}^\dagger\hat{B})^2 + \frac{23}{2}(\hat{B}^\dagger\hat{B}) + \frac{15}{4}$

$$\hat{H}_\sigma = -\frac{\sigma}{\sigma+1}\; P_\sigma(\hat{B}_1^\dagger\hat{B}_1) + P_\sigma(\hat{B}_2^\dagger\hat{B}_2) \;-\; \varepsilon(\hat{B}_1^\dagger\hat{B}_2 + \hat{B}_2\hat{B}_1^\dagger) \;. \tag{16}$$

In the polynomials, P_σ, given in Table 2, the average of all orderings of $\hat{B}^\dagger$ and $\hat{B}$ requires consideration of $(2\sigma+2)!/[(\sigma+1)!]^2$ terms.

Since the operators $\hat{N}$ and $\hat{H}_\sigma$ commute the eigenvalue problem for $\hat{H}_\sigma$

$$\hat{H}|\psi\rangle = E|\psi\rangle \;, \tag{17}$$

E being the energy, can be solved in the finite dimensional spaces of $\hat{N}$

$$\hat{N}|\psi\rangle = n|\psi\rangle \;, \tag{18}$$

n being the number of quanta in the system. The most general wave function is

$$|\psi\rangle = \sum_{j=0}^{n} c(n,j)\,(n-j,j) \tag{19}$$

where c(n,j) are possibly complex coefficients and (n-j,j) denotes the number state $|(n-j)\rangle|j\rangle$. The c(n,j) are determined by Eq. (17) leading to the following matrix equation for c(n,j)

$$H_n(\sigma)\bar{c} = E\bar{c} \tag{20a}$$

with

$$\bar{c} = \mathrm{col}(c(n,0),\, c(n,1),\, \cdots,\, c(n,n)) \;. \tag{20b}$$

Following [4] the diagonal elements of $H_n(\sigma)$ are readily constructed by means of the first term of Eq. (16) and Table 2. Detailed calculations show that the coefficients of number states (n,0) and (0,n) <u>increase</u> with σ. In this way, increasing anharmonicity leads to increasing localization. In [6] the quantum structure of the unsymmetric classical stationary solution is discussed for $\sigma = 1$. For $\sigma \geq 1$ the two levels of lowest energy are separated by a small energy difference given by

$$\frac{\Delta E}{\varepsilon} \sim \frac{2n}{\Pi_\sigma(n)(n-1)!}\left(\frac{\varepsilon}{\gamma}\right)^{n-1} \tag{21}$$

for small ε/γ and large n. From Table 3 follows that a conservative estimate of the denominator polynomial is

$$\Pi_\sigma(n) \sim n^{(\sigma-1)(n-1)} \;. \tag{22}$$

Table 3. Denominator polynomial appearing in Eq. (20)

σ	$\Pi_\sigma(n)$
1	1
2	$(n+1)^{(n-1)}$
3	$\Pi_{k=1}^{n-1}\left[n^2+\left(\frac{3}{2}-\frac{1}{2}k\right)n+\left(\frac{5}{2}+\frac{1}{2}k^2\right)\right]$
4	$\Pi_{k=1}^{n-1}\left[n^3+(2-k)n^2+(6-k+k^2)n+(5+k^2)\right]$

Thus we see that ΔE becomes very small with increasing σ. This leads to a dramatic increase in the time that the quantum wave packet of an unsymmetric solution remains localized.

Three degrees of freedom

For $f \geq 3$ the classical solution trajectories of Eq. (1) can be either chaotic or quasi-periodic. In Refs. [9-10] the system with $f = 3$ and the dispersion matrix

$$\varepsilon M = \begin{Bmatrix} 0 & 1 & 1 \\ 1 & 0 & 1 \\ 1 & 1 & 0 \end{Bmatrix} \tag{23}$$

is studied in the classical case and the quantum case.

In the classical case the fraction of phase space, ρ_c, occupied by chaotic trajectories is determined. The dimension of the phase space can be reduced by introducing the angle, ϕ_i, and the momentum conjugate, P_i, $i = 1,2,3$. Thus

$$A_1 = \sqrt{P_1}\, e^{-i(\phi_1+\phi_3)}$$

$$A_2 = \sqrt{P_2}\, e^{-i(\phi_2+\phi_3)} \qquad (24a,b,c)$$

$$A_3 = \sqrt{P_3-P_1-P_2}\, e^{-\phi_3} .$$

Here P_3 is in fact the conserved quantity N given by Eq. (3). Thus ϕ_3 becomes an ignorable coordinate. The initial conditions are randomly distributed over the energy surface $H(P_1,P_2,\phi_1,\phi_2) = E$, H being the Hamiltonian for the system. Thus, ϕ_1 becomes a function of P_1,P_2,ϕ_2. Figure 1 shows the Poincaré section of regular - and chaotic - trajectories for particular choices of P_1, N and E. In Figure 2 results for ρ_c, based on 50-100 trajectories (vertical error bars determined as the 95% confidence interval of our estimate), are shown.

In the quantum case studies of energy spectra of quantum systems have shown that the density of spacings, S, between uncorrelated, adjacent energy levels, p(S), can be used as an indicator of chaos or regularity in the corresponding classical systems (see numerous references in [9]). When the classical phase space is partitioned into regular regions and only one chaotic region the following interpolation formula has been proposed [15]

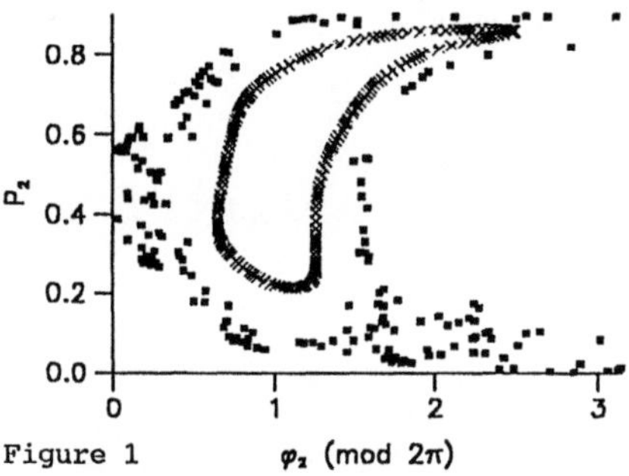

Figure 1

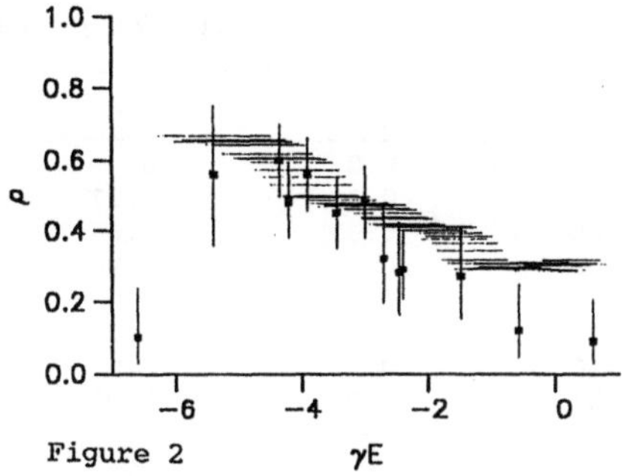

Figure 2

Figure 1. Poincaré section of regular (×) and chaotic (▪) trajectories for $P_1 = 0.1$, $\gamma N = 3$, $\gamma E = -2.4$. (From Ref. [9].)

Figure 2. ρ_c (▪) and ρ_q (−) as functions of energy, γE, for fixed $\gamma N = 3$. Error bounds indicated by vertical and horizontal bars. n = 310 in the quantum DST giving a total of 8164 energy levels. (From Ref. [9].)

$$p_I(S) = \exp\left[-(1-\rho_q)S - \frac{\pi}{4}\rho_q^2 S^2\right] \times \left[(1-\rho_q)^2 \exp\left[\frac{\pi}{4}\rho_q^2 S^2\right] \operatorname{erfc}\left[\frac{\sqrt{\pi}}{2}\rho_q S\right] + 2(1-\rho_q)\rho_q + \frac{\pi}{2}\rho_q^3 S\right], \tag{25}$$

ρ_q being the fraction of the energy shell that is chaotic. Eq. (25) interpolates between the Poisson distribution (for the quantum counterpart of the regular classical system [16]) and the Wigner distribution (corresponding to the chaotic classical system [17]). If it is assumed in the case of DST with f = 3 that all chaotic regions are connected we may determine ρ_q by fitting $p_I(S)$ given by Eq. (25) to the available data. The results are indicated in Figure 2. Each ρ_q value is calculated from a sequence with 2000 levels, the energy E being the average energy and the error being the corresponding standard deviation.

The quantum results confirm the trend of the classical ones. However, the correspondence is not as good as that found in other systems (referenced in [9]). One reason may be that with the value of n used, the density of levels is not large enough (a fact that is expressed by the horizontal error bars). A clear divergence from the classical result is observed in the low γE limit. This is due to the particularly low density of quantum levels in this region. For high values of γE the agreement is not good either. This may be because the system is not generic at extreme values. In the intermediate range of energy, where $\rho_c \simeq \rho_q \simeq 1/2$, the agreement seems to be within the error bars.

Four degrees of freedom

For f = 3 the conservation of number (3) implies that the invariant manifold, on which the dynamics takes place, has dimension four so that the (three-dimensional) KAM tori can disconnect it into more than one regular and one chaotic region. On the other hand, when f > 3 the appearance of Arnold diffusion makes the assumption that the phase space consists of a regular region and a chaotic region exactly valid.

In Ref. [11], we use, for f = 4, a Monte Carlo approach to compute ρ_c in the classical phase space and compare it with the corresponding ρ_q obtained from the interpolation formula Eq. (25) for the nearest neighbour level spacing distribution. In order to distinguish between regular and chaotic trajectories we compute the maximum Liapunov exponent. Table 4 summarizes our results

Table 4. Numerical results for ρ_c and ρ_q. $f = 4$, $\gamma H = -6$

ε	ρ_q	ρ_c
0	0.0	0
0.03	0.25	0.33
0.05	0.33	0.67
0.08	0.41	0.98
0.1	0.56	1.0
0.15	0.85	1.0
0.20	0.95	1.0
0.5	0.97	1.0

calculated in the energy surface $\gamma H = -6$ where

$$H = -\frac{\gamma}{2}\sum_{i=1}^{4} \alpha_i |A_i|^4 - \varepsilon \sum_{i\neq j}^{4} m_{i,j}\, A_i^* A_j \tag{26}$$

with $m_{i,j} = 1 - \delta_{ij}$. Note that we choose

$$\alpha_1 = 1, \quad \alpha_2 = (\sqrt{5}-1)/2, \; \alpha_3 = \sqrt{\alpha_2}, \quad \alpha_4 = 1/\alpha_2 \tag{27}$$

in order to ensure that the level spacings distribution is generic in the $\varepsilon = 0$ limit. The quantum mechanical parameter, ρ_q, shows good agreement with ρ_c for ε close to zero where the nearest neighbour level spacings distribution is Poisson-like ($\rho_q \simeq 0$), and for bigger values of ε, where the distribution is Wigner-like ($\rho_q \simeq 1$). For the intermediate values of ε, i.e. $0.05 \leq \varepsilon \leq 0.15$, the agreement between ρ_q and ρ_c is not so good. For these values of ε the evaluation of the maximal Liapunov exponents, λ, becomes critical because they are very small (i.e. $\lambda \simeq 0.1$ for $\varepsilon = 0.08$) and the time to estimate their convergence is very long (i.e. more than 1500 time units for $\varepsilon = 0.08$). Thus it becomes an enormous computational task to decide the character, chaotic/regular, for the necessary number of trajectories in phase space. This introduces an unavoidable uncertainty in the ρ_c determination. On the other hand, the possibility that the conjectured distribution, Eq. (25), is not correct should not be overlooked [18].

A DST model of globular protein

In Ref. [12] the DST system given by the Hamiltonian

$$H = \sum_{i=1}^{f} \omega_i |A_i|^2 - \frac{\gamma}{2}\sum_{i=1}^{f} |A_i|^4 - \sum_{i\neq j}^{f} m_{ij}\, A_i^* A_j \tag{28}$$

is used (with $f = 194$) to model the dynamics of the enzyme adenylate kinase (ADK). The atomic structure of ADK is fairly well known and can be obtained from the Protein Data Bank, Department of Chemistry, Brookhaven National Laboratory. Thus the interaction matrix with elements m_{ij} has been obtained by means of these data. The ω_i's, $i = 1,2,\cdots,194$, are the site frequencies. These are shown at random from an interval of width 35 cm^{-1} since reliable estimates have not been obtained from the avilable data.

For the so-called stationary solutions of the DST equation given by

$$A_i(t) = \Phi_i\, e^{-i\omega t} \qquad i = 1,\cdots,f \tag{29}$$

a measure of the localization is

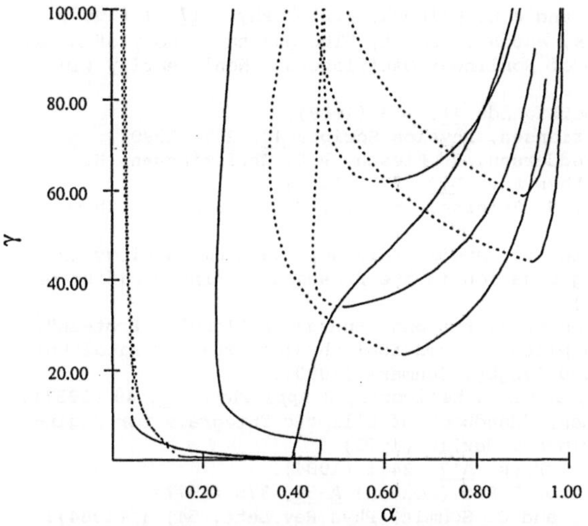

Figure 3. DST-model of ADK. Relationship between γ and α parameters for selected stationary solutions. Full (dotted) curve represents stable (unstable) solution. (From Ref. [12].)

$$\alpha = \sum_{i=1}^{f} \Phi_i^4 \Big/ \sum_{i=1}^{f} \Phi_i^2 \, . \tag{30}$$

Thus α assumes its maximum value, α = 1, for a solution which is completely localized on a single site, and its minimum value, α = 1/f, for a solution which is uniformly extended over all f sites. In Figure 3 we plot the relationship between the nonlinearity parameter, γ, and the localization measure, α, for some stationary solutions. Full curves represent stable solution; dotted curves represent unstable solutions. Some of the solutions are localized for γ = 0 owing to the linear localization caused by the random distribution of the site frequencies (Anderson localization). These solutions become delocalized as soon as γ is increased. Also most of these solutions become unstable. Other solutions which do become very localized as γ is increased are those with lowest energy. These results thus illustrate linear and nonlinear localization effects in an harmonic model of the dynamics in a globular protein molecule.

Acknowledgements

Financial support from the NATO Special Programme Panel on Chaos, Order and Patterns under grant no. 35-0948/88, the EEC Science Programme under grant no. 89 100079/JU1 and the Danish Technical Research Council under grant no. 16-4307.E is gratefully acknowledged.

References

[1] J.C. Eilbeck, P.S. Lomdahl, and A.C. Scott, Physica D16, 318 (1985).
[2] A.C. Scott, P.S. Lomdahl, and J.C. Eilbeck, Chem.Phys.Lett. 113, 29 (1985).
[3] J.C. Eilbeck, P.S. Lomdahl, and A.C. Scott, Phys.Rev. B30 , 4703 (1984).
[4] A.C. Scott and J.C. Eilbeck, Phys.Lett. A119, 60 (1986)

[5] A.C. Scott, L. Bernstein, and J.C. Eilbeck, J.Biol.Phys. 17, 1 (1989).
[6] L. Bernstein, J.C. Eilbeck, and A.C. Scott, "The Quantum Theory of Local Modes in a Coupled System of Nonlinear Oscillators", Nonlinearity (in press).
[7] P.L. Christiansen, J.Molec.Liquids 41, 113 (1989).
[8] A.C. Scott and P.L. Christiansen, Physica Scripta 42, 257 (1990).
[9] L. Cruzeiro-Hansson, H. Feddersen, R. Flesch, P.L. Christiansen, M. Salerno, and A.C. Scott, Phys.Rev. B42, 522 (1990).
[10] H.K. Feddersen, R. Flesch, M. Salerno, and P.L. Christiansen, J.Chem. Phys. 92, 2117 (1990).
[11] H. Feddersen, P.L. Christiansen, and M. Salerno, "Quantum Chaology in the Discrete Self-Trapping Equation in the Presence of Arnold Diffusion", Physica Scripta (in press).
[12] H. Feddersen, "Localization of Vibrational Energy in Globular Protein", Preprint, Laboratory of Applied Mathematical Physics, The Technical University of Denmark, DK-2800 Lyngby, Denmark (1990).
[13] R.P. Feynman, F.L. Vernon, and R.W. Hellworth, J.Appl.Phys. 28, 49 (1957).
[14] P.F. Byrd and M.D. Friedman, "Handbook of Elliptic Integrals for Engineers and Physicists", Springer, Berlin (1971).
[15] M.V. Berry and M. Robnik, J.Phys. A17, 2413 (1984).
[16] M.V. Berry and M. Tabor, Proc.R.Soc.(London) A356, 375 (1977).
[17] O. Bohigas, M.J. Giannoni, and C. Schmit, Phys.Rev.Lett. 54, 1 (1984).
[18] B.V. Chirikov, F.M. Izrailev, and O.P. Shepelyansky, Physica D33, 77 (1988).

The Nonlinear Schrödinger Equation on a Disordered Chain

R. Scharf and A.R. Bishop

Theoretical Division and Center for Nonlinear Studies,
Los Alamos National Laboratory, Los Alamos, NM 87545, USA

The integrable lattice nonlinear Schrödinger equation is a unique model with which to investigate the effects of disorder on a discrete integrable dynamics, and its interplay with nonlinearity. We first review some features of the lattice nonlinear Schrödinger equation in the absence of disorder and introduce a 1- and 2-soliton collective variable approximation. Then we describe the effect of different types of disorder: attractive and repulsive isolated impurities, spatially periodic potentials, random potentials, and time dependent (kicked) long wavelength perturbations.

1. Introduction

Recent years have seen an enormous progress in understanding the effects of disorder in linear systems like Anderson localization and the transition from insulating to conducting behavior in $d = 3$ dimensions at zero temperature upon decreasing the disorder. Completely integrable nonlinear systems in $d = 1$ are also fairly well understood by now. The understanding of the interplay between disorder and nonlinearity on the other hand is still in its infancy [1].

Both, nonlinearity and disorder, may give rise to self-localized excitations (solitons or Anderson localized wave packets, respectively). Therefore it is natural to ask, how these effects might reinforce, complement or frustrate each other. Transport properties in disordered, nonlinear materials for example strongly depend on whether solitons behave as "particles" in the presence of disorder, or interact very strongly with other degrees of freedom. Does the nonlinearity lead to adaptive behavior of excitations in disordered materials which preserves coherence? How do nonlinear excitations interact with each other in the presence of disorder? These and other issues are of great experimental concern in fields from nonlinear optics [2], to polaron formation in solid state materials [3,4], to vibron localization in natural and synthetic biomolecules [5,6,7].

In all the cases mentioned so far perturbation theory allows us to calculate which of the unperturbed structures becomes unstable and how they start to decay [8]. But it does not tell us, what excitations in fully developed disorder look like. Collective variable approximations together with numerical experiments allow us to investigate these properties in greater detail.

As a first step, in this rich program, we consider the *discrete* nonlinear Schrödinger equation (NLS eq.) in $(1 + 1)$ dimensions and investigate some elementary forms of parametric spatial and time periodic disorder – namely constant bias, isolated impurities, periodic and random spatial variations, and time periodic (kicked) perturbations. Several novel features have been found, some of them being a result of the spatial discreteness of the dynamics.

2. The completely integrable lattice NLS equation

Ablowitz and Ladik [9] have introduce a completely integrable discretization of the NLS equation. We generalize it by including an onsite potential energy term:

$$i\dot{\psi}_n = -(\psi_{n+1} + \psi_{n-1})(1 - \kappa|\psi_n|^2) + V_n\psi_n, \tag{1}$$

with $\kappa = -1$ or $+1$ (attractive or repulsive self-interaction, respectively). This equation can be derived from the Hamiltonian

$$H = -\sum_n (\psi_n\psi^*_{n+1} + \psi^*_n\psi_{n+1}) - \frac{1}{\kappa}\sum_n V_n \log(1 - \kappa|\psi_n|^2), \tag{2}$$

with nonstandard Poisson brackets

$$\{\psi_m, \psi^*_n\} = i(1 - \kappa|\psi_n|^2)\delta_{mn}, \qquad \{\psi_m, \psi_n\} = \{\psi^*_m, \psi^*_n\} = 0, \tag{3}$$

and $\kappa = \pm 1$. For $\kappa = 1$ we assume $|\psi_n| < 1$. One can easily verify that the following "norm" is conserved:

$$N = -\frac{1}{\kappa}\sum_n \log(1 - \kappa|\psi_n|^2). \tag{4}$$

We now show that eq. (1) is completely integrable, not only for constant $V_n = V$, but also in the case $V_n = an + b$ with $a \neq 0$. To accomplish this we cast eq. (1) in the form af a zero curvature condition with time-dependent spectral parameter λ. Following ref. 10, the compatibility condition for a vector $F_n(t, \lambda(t))$ obeying the equations

$$\begin{aligned} F_{n+1} &= L_n(t,\lambda)\, F_n, \\ \frac{dF_n}{dt} &= W_n(t,\lambda)\, F_n, \end{aligned} \tag{5}$$

is of the following form:

$$\frac{d}{dt}L_n + L_n W_n - W_{n+1} L_n = 0. \tag{6}$$

With the choice

$$\begin{aligned} L_n(t,\lambda) &= \begin{pmatrix} \lambda & \sqrt{\kappa}\psi^*_n \\ \sqrt{\kappa}\psi_n & \lambda^{-1} \end{pmatrix}, \\ W_n(t,\lambda) &= i\begin{pmatrix} 1 + \kappa\psi^*_n\psi_{n-1} - \lambda^2 + f_n & \sqrt{\kappa}(\lambda^{-1}\psi^*_{n-1} - \lambda\psi^*_n) \\ \sqrt{\kappa}(\lambda^{-1}\psi_n - \lambda\psi_{n-1}) & -1 - \kappa\psi_n\psi^*_{n-1} + \lambda^{-2} - f_n \end{pmatrix}, \end{aligned} \tag{7}$$

the zero curvature condition (6) leads to eq. (1) with

$$\begin{aligned} f_n &= \gamma n + \delta, \\ \lambda(t) &= \lambda_0 e^{i\gamma t}, \\ V_n &= 2 + f_n + f_{n+1} = 2\gamma n + 2(\delta + 1) + \gamma = an + b. \end{aligned} \tag{8}$$

In the case of periodic boundary conditions, for example, a hierarchy of (time dependent for $V_n = an + b$) integrals of motion may be constructed from the trace of powers of the (2×2)-matrix

$$L = L_M L_{M-1} \dots L_2 L_1, \tag{9}$$

with $L_{M+1} \equiv L_1$. For example, $\log |\det L|$ leads to the norm N, whereas $\operatorname{tr} L$ lead to the kinetic part of the Hamiltonian H.

3. Constant and linear onsite potentials

In what follows we will always assume attractive self-interaction and set $\kappa = -1$. For $V_n \equiv V$ or $V_n = an + b$ travelling solutions exist and can be expressed in terms of elliptic functions. Of special interest are the 1-soliton solutions. For $V_n \equiv V$ they are of the form

$$\begin{aligned} \psi_n(t) &= \sinh \beta \, \operatorname{sech}\big(\beta(n - ut - x_0)\big) \, e^{-i(\omega t + \alpha n + \phi_0)}, \\ \omega &= -2 \cosh \beta \, \cos \alpha + V, \\ u &= -\frac{2 \sinh \beta \, \sin \alpha}{\beta}. \end{aligned} \tag{10}$$

The existence of a maximum velocity $u_{\max} = 2 \sinh \beta / \beta$ is clearly a discreteness effect. Energy and norm for the 1-soliton solution (10) are found to be

$$\begin{aligned} E &= -4 \sinh \beta \, \cos \alpha, \\ N &= 2\beta. \end{aligned} \tag{11}$$

For $V_n = an + b$ the solitons have the same shape but move in a different manner (with $\alpha_0 = \phi_0 = 0$ for simplicity):

$$\begin{aligned} \psi_n(t) &= \sinh \beta \, \operatorname{sech}\big(\beta(n - x(t))\big) \, e^{-i(\phi(t) + V_n t)}, \\ \phi(t) &= -\frac{2}{a} \cosh \beta \, \sin(at) + bt, \\ x(t) &= \frac{2}{a\beta} \sinh \beta \, \big(\cos(at) - 1\big) + x_0. \end{aligned} \tag{12}$$

The solitons move exactly harmonically in a linear potential (see Fig. 1). Their norm is unchanged, $N = 2\beta$, but their energy is given by

$$\begin{aligned} E &= -4 \sinh \beta \cos(at) + 2\beta\big(ax(t) + b\big) \\ &= -4 \sinh \beta + 2\beta(ax_0 + b). \end{aligned} \tag{13}$$

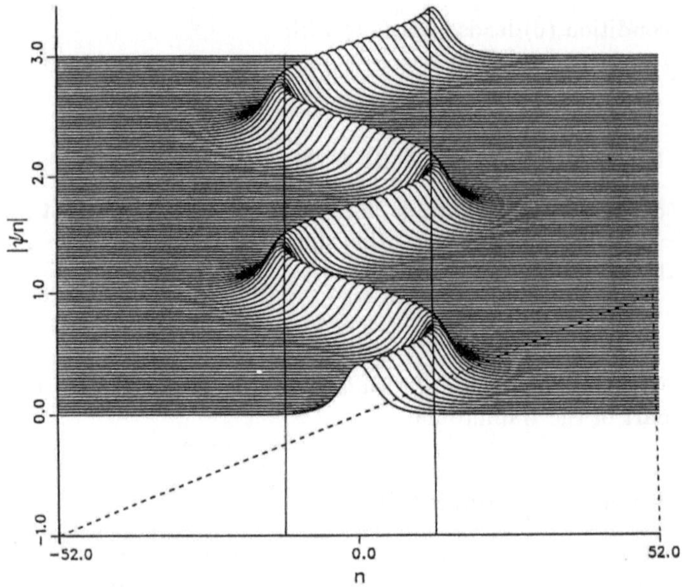

Fig. 1. *Soliton trapped on a tilted plane:* Solution of eq. (1), for $l_{\text{chain}} = 104$; zero boundary conditions; $V_n = 16n/l_{\text{chain}}$; initial soliton parameters: $\alpha = -1.571$, $\beta = 0.4$; integration time $T = 90$. The vertical marks indicate the turning points given by eq. (16).

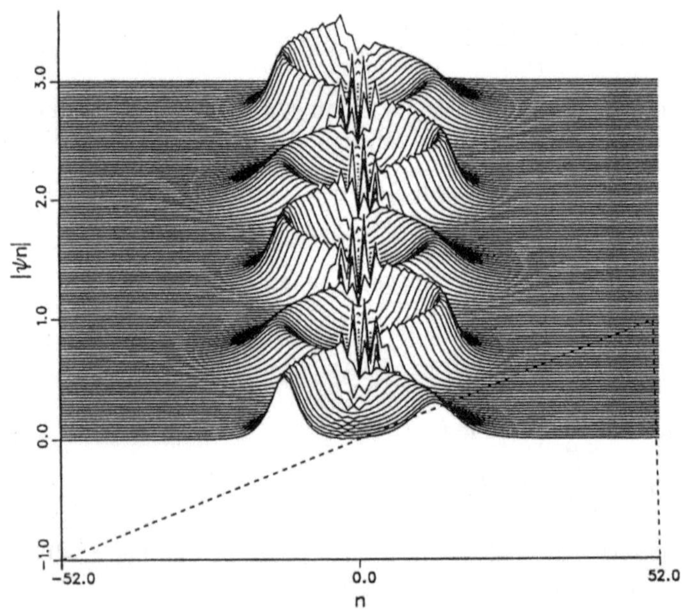

Fig. 2. *Two soliton collision on a tilted plane:* Solution of eq. (1), for $l_{\text{chain}} = 104$; zero boundary conditions; $V_n = 16n/l_{\text{chain}}$; initial soliton parameters: $\alpha_1 = \pi$, $\alpha_2 = 0.$, $\beta_1 = 0.5$ and $\beta_2 = 0.3$; integration time $T = 90$.

The explicit 2-soliton solutions are, as usual, quite involved. Therefore we only show the collision of two different solitons for $V_n = an + b$ in Fig. 2, thereby illustrating the complete integrability in this case. In the continuum limit the NLS eq. is also integrable for harmonic potentials [11]. Whether this is also true in the discrete case seems to be unknown.

4. The 1-soliton collective variable approximation

As noted above, in the case of a linear potential solitons move in a harmonic way without changing their shape. This is no longer true for a more general form of the potential. But as long as the curvature of the potential felt by the soliton is small one might assume that the soliton emits only a negligeable amount of radiation and adjusts its parameters to comply with the constancy of energy and norm. Introducing $V(x)$ for $V_n(n = x)$, this lead to $N = 2\beta$ being independent of $V(x)$ and

$$E = -4 \sinh \beta \, \cos \alpha(x) + 2\beta V(x) = \text{const.} \tag{14}$$

Now x is interpreted as the position of the soliton, the relevant collective variable. The soliton might be regarded as a particle moving in an effective potential with a velocity $u(x) = -2 \sinh \beta \sin \alpha / \beta$:

$$\frac{u^2}{2} + \frac{V(x)(\beta V(x) - E)}{2\beta} = \frac{16 \sinh^2 \beta - E^2}{8\beta^2}, \tag{15}$$

where $V_{\text{eff}}(x) = \frac{1}{2\beta} V(x)(\beta V(x) - E)$ plays the role of the effective potential and the r. h. s. of eq. (15) plays the role of the particle energy, both depending on the initial conditions. A constant potential V leads to a constant effective potential $V_{\text{eff}}(x)$, a linear potential $V(x)$ lead to a harmonic $V_{\text{eff}}(x)$.

That the effective potential $V_{\text{eff}}(x)$ is the relevant one and not the original potential $V(x)$ can be seen in Fig. 3, which shows the trapping of a soliton on a *maximum* of $V(x)$!

Solitons trying to percolate on a onedimensional lattice in a smooth random potential with small curvature have to observe not only a lower bound for their energy, but also an upper one. There exist lower and upper bounds for the potential that the soliton is allowed to enter without being reflected, depending on the initial condition ($E = -4 \sinh \beta \, \cos \alpha_0 + 2\beta V_0$):

$$\begin{aligned} V_{\text{max}} - V_0 &= \frac{2 \sinh \beta}{\beta} (1 - \cos \alpha_0), \\ V_0 - V_{\text{min}} &= \frac{2 \sinh \beta}{\beta} (1 + \cos \alpha_0). \end{aligned} \tag{16}$$

This window of allowed potential values acts as a filter for solitons having certain initial values for α and β, as long as the collective variable approximation holds. In the continuum limit the unusual lower bound disappears.

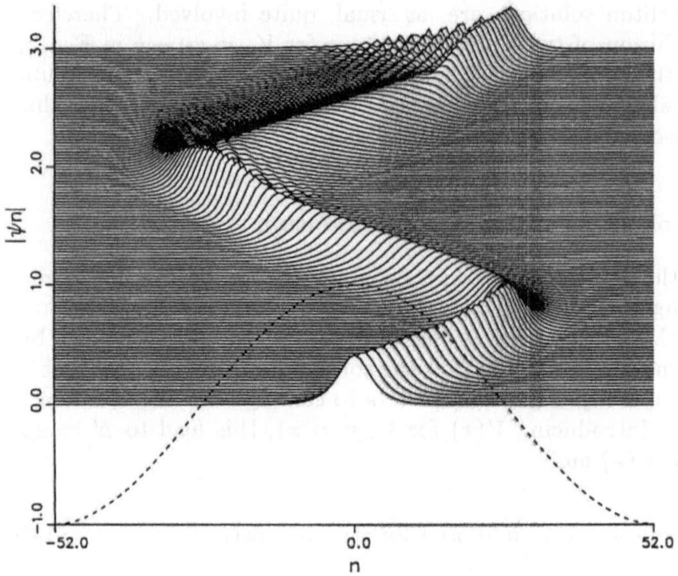

Fig. 3. *Soliton trapped on potential maximum:* Solution of eq. (1), for $l_{\text{chain}} = 104$; periodic boundary conditions; $V_n = 2\cos(2\pi n/l_{\text{chain}})$; initial soliton parameters: $\alpha = -1.571$, $\beta = 0.4$; integration time $T = 90$.

5. Trapping on an impurity

The case of an isolated single impurity of the form $V_n = V\,\delta_{n,0}$ is important in two respects. First it illustrates that the stationary solutions do *not* tell the full story of what is going on in the system, as opposed to the case of the linear Schrödinger equation; and second it shows that a collective variable approach can be useful even in the case when one soliton breaks in two.

Stationary solutions with the frequency $\omega = -2\cosh\beta$ are readily found to be of the form

$$\psi_n(t) = \sinh\beta\,\operatorname{sech}\big(\beta(|n| - \operatorname{sgn}(V)\xi)\big)\,e^{-i\omega t}, \tag{17}$$

where $\xi > 0$ and $\cosh\beta\xi = 2\sinh\beta/\sqrt{\omega^2 - 4 - V^2}$. For an attractive impurity ($V < 0$) the solution ψ_n has a maximum at $n = 0$; for a repulsive impurity ($V > 0$) it has a minimum at $n = 0$ between maxima at $\pm n_0 \approx \xi$. For $V = 0$ the original soliton with $\xi = 0$ is recovered. In all cases solutions (17) exist only for sufficiently large β: $\sinh\beta \geq |V|/2$.

What really happens dynamically in the vicinity of an impurity might look quite different from the trapped oscillatory solution (17). There may be scattering and emission of phonon-like excitations as well as multi-soliton processes. To simplify matters we choose a soliton at rest ($\alpha = 0$) on top of the impurity and assume that the main decay channel is through generation of two solitons.

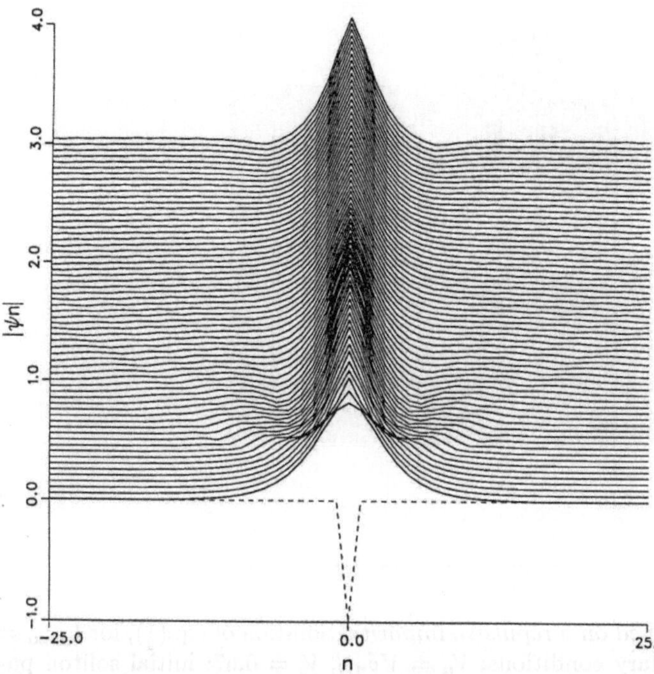

Fig. 4. *Soliton trapped on an attractive impurity:* Solution of eq. (1), for $l_{\text{chain}} = 314$; periodic boundary conditions; $V_n = V\delta_{n,0}$, $V = -0.3$; initial parameters: $\alpha = 0.$, $\beta = 0.4$; integration time $T = 30.$; $|\psi|$-magnification: 2.

As the initial condition is symmetric with respect to spatial inversion, we expect the two solitons to have the same shape but opposite velocity. Initially we have for the norm: $N = 2\beta$, and for the energy: $E = -4\sinh\beta + 2V\log(\cosh\beta)$. Well after the decay into 2 travelling solitons (and hopefully negligeable radiation) with exponentially small overlap with the impurity we find for the norm $N' = 4\beta'$, and for the energy $E' = -8\sinh\beta' \cos\alpha'$. If we assume that the amount of radiation emitted during the decay is negligeable ($N = N'$, $E = E'$) we find:

$$\cos\alpha' = \cosh(\beta/2) - \frac{V\log(\cosh\beta)}{4\sinh\frac{\beta}{2}}. \tag{18}$$

This equation has solutions for α' only if V falls into the following interval:

$$0 < \frac{4\sinh\frac{\beta}{2}(\cosh\frac{\beta}{2} - 1)}{\log(\cosh\beta)} < V < \frac{4\sinh\frac{\beta}{2}(\cosh\frac{\beta}{2} + 1)}{\log(\cosh\beta)}. \tag{19}$$

In the continuum limit these bounds got to 0 and $+\infty$. The upper bound in eq. (19) corresponds to a relatively strong potential for which more complicated things are expected to happen than only 2-soliton decay without radiation. The lower bound on the other hand is small for small β and our ansatz is reasonable

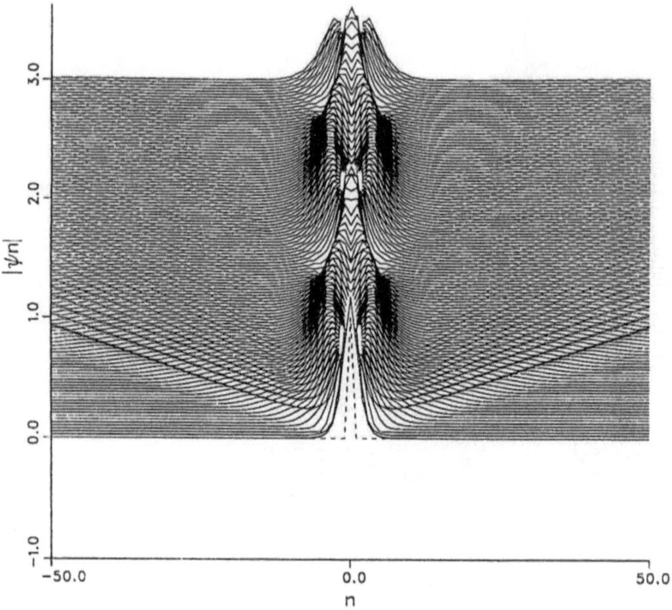

Fig. 5. *Soliton trapped on a repulsive impurity:* Solution of eq. (1), for $l_{\text{chain}} = 314$; periodic boundary conditions; $V_n = V\delta_{n,0}$, $V = 0.67$; initial soliton parameters: $\alpha = 0.$, $\beta = 1.0$; integration time $T = 90$.

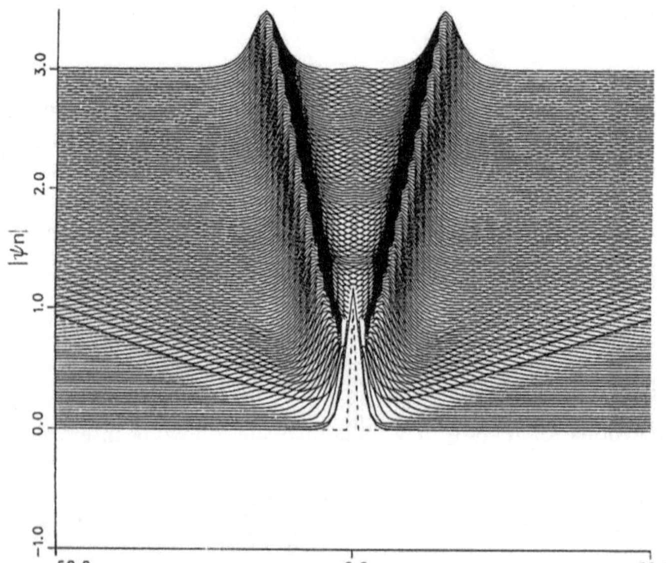

Fig. 6. *Soliton on a strong, repulsive impurity:* Solution of eq. (1), for $l_{\text{chain}} = 314$; periodic boundary conditions; $V_n = V\delta_{n,0}$, $V = 0.70$; initial soliton parameters: $\alpha = 0.$, $\beta = 1.0$; integration time $T = 90$.

in this case. We therefore expect bound states for potentials V smaller than the lower bound in eq. (19) — even if V is repulsive! Fig. 4 shows how a soliton at rest centered at an isolated attractive impurity adjusts and stays trapped. Fig. 5 illustrates the same but for a repulsive impurity below the critical value V_c for soliton breakup (the lower bound in eq. (19)). Fig. 6 shows the breakup into two unbound solitons. The actual value for V_c is about 10% higher than the one given by eq. (19) because of radiative losses, reducing the kinetic energy available for the emerging solitons.

It is interesting to compare V_c here with the crossover suggested in disordered "polaron" problems (with electron-phonon, exciton-phonon, magnon-phonon coupling) discussed by Anderson [12] and others [13]. These authors have considered the competition of disorder- and nonlinearity-induced localization within a stationary approximation ($\psi_n(t) = \exp(-i\omega t)\chi_n$), whereas here we have explored the full time-dependence.

6. Spatially periodic onsite potentials

If a soliton experiences a spatially periodic onsite potential one crucial parameter that governs its behavior is the ratio between the width of the soliton and the modulation length of the potential. The other is the ratio between the soliton amplitude $\max_n |\psi_n|^2$ and the amplitude of the potential. Let's start with the latter. If this ratio is large then the nonlinearity effects dominate. In the opposite case the soliton is expected to decay into different coherent structures centered around the minima of the potential. But what is going to happen exactly will be governed by the lengthscale ratio mentioned first. If the soliton is narrow compared to the wavelength of the potential then it will essentially move like a particle in an effective potential, as described in section 4. If the soliton is broad compared to the wavelength of the potential and it moves sufficiently fast then a smoothing over the potential is likely to occur and the soliton will only feel a reduced effective potential. A situation of major physical interest is when nonlinearity and onsite perturbation or width of the soliton and wavelength of the perturbation are comparable: then a breakdown of a simple particle (soliton) picture can be anticipated, corresponding to strong scattering. Thus the periodic potential is very important for understanding disorder of general color.

In Fig. 7 we illustrate the case of a broad soliton ($\beta = 0.1$) slowly moving ($\alpha = 0.5$) in a periodic potential. The period of the potential is comparable to the width of the soliton, the amplitude of the potential is comparable to the height of the soliton. It is clearly visible that the soliton steepens up each time it encounters a maximum of the potential. It tries to adjust as it travels over the wiggles but is able to do so only in an imperfect manner: it leaves behind some excitations which get trapped in the minima of the potential. If the soliton moves even more slowly initially ($\alpha = 0.3$) it decays rapidly into a couple of humps which stay trapped in the potential minima. On the other hand if the soliton moves very fast ($\alpha = 1.57$) it smoothes over the potential and is not visibly affected by it. One way to interprete these results is in terms of

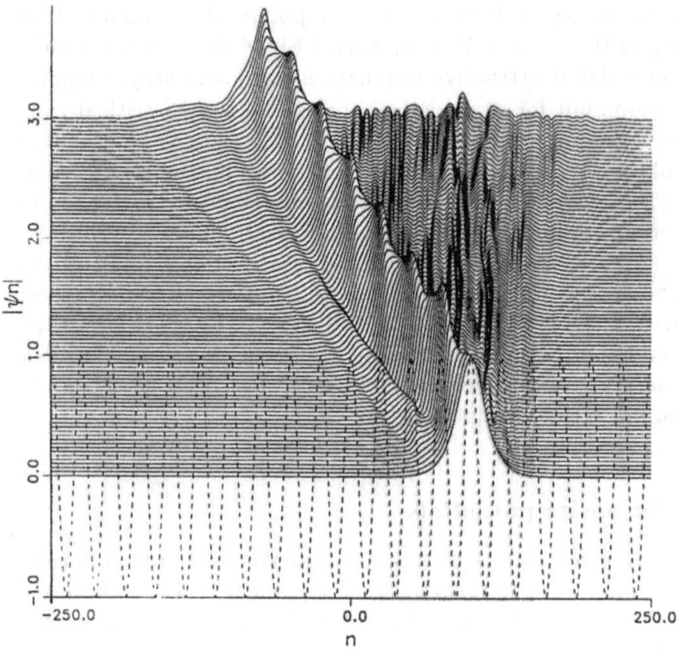

Fig. 7. *Soliton in a periodic potential:* Solution of eq. (1), for $l_{\text{chain}} = 500$; periodic boundary conditions; $V_n = \frac{1}{10}\cos(2\pi n/25)$; initial soliton parameters: $\alpha = 0.5$, $\beta = 0.1$; integration time $T = 180.$; $|\psi|$-magnification: 10.

excitation of shape modes. If the soliton moves slowly enough over the wiggler potential there is strong excitation of shape modes, besides other excitations which are left behind and get trapped. But for sufficiently large velocity of the soliton and correspondingly higher wiggler frequency the excitation of shape modes of that frequency is strongly reduced. A detailed stability analysis of the shape modes is in progress. The interplay between the characteristic frequency ω of the unperturbed soliton, the frequency of the shape mode oscillation, and the wiggler frequency might give rise to phenomena well known from the circle map, as, for example, mode locking, and chaos.

7. Spatially random onsite potentials

Of major interest is the case of a single soliton travelling along a disordered chain. Several questions arise in this case: What are the radiative losses the soliton suffers? How far does the soliton travel before it decays? How does the mean velocity of the soliton depend on the strength of the disordered potential? Will the emitted radiation become Anderson localized ? Will it self-focus into new solitons?

We have investigated how one or two broad solitons ($\beta \ll 1$) travel in a random onsite potential V_n equally distributed in the interval $[-V, V]$. Slowly travelling solitons in weak potentials decay rapidly and are soon overtaken by the radiation they generate. Fast solitons on the other hand stay ahead of their radiation for a longer time and — which seems to be more important — they dynamically smooth out the random potential. This is quite similar to the case of periodic potentials as shown in sect. 6. Strong fluctuations in the potential give rise to modulations in the shape of the soliton, which become more pronounced the longer it takes for the soliton to pass over the fluctuations. The modulations of the soliton shape are finally left behind as excitations travelling in the opposite direction with about the same velocity as the soliton. The velocity of a broad soliton travelling in a small random potential is essentially unchanged in comparison to its value without disorder: $u \approx -2\sin\alpha$. The radiation front which the soliton generates travels with a maximum velocity $u_0 \approx 2$. Therefore only solitons with $\alpha \approx \pi/2$ stay ahead of their own radiation for a sufficiently long time.

We also investigated the case of two identical solitons, travelling in the same direction with a separation of a few soliton widths. The soliton that travelled in the wake of the leading one was only negligeably influenced by the wake. The superposition of the wakes of both solitons contained more pronounced structures than each of the wakes alone, as is to be expected because of the self-focussing nonlinearity. The ultimate fate of the structures and the radiation that is left behind is still unclear — whether the structures decay into radiation which in turn then Anderson localizes, or whether the structures that are to small for nonlinearity to have any effect are "recycled" by larger structures. Whatever happens will be interesting to understand as it will involve the full timedependence of the nonlinear dynamics with disorder and not only properties of noninteracting stationary solutions [14].

Recently Newell and coworkers gave an account of their work on the transmission properties of the nonlinear Schrödinger chain with disorder [15]. They show that the transmission properties are only improved by the nonlinearity if it contributes to self-focussing into soliton pulses.

8. The periodically kicked NLS chain

The different kinds of disorder we introduced so far were time independent. Now we allow for a periodical driving of a NLS chain with periodic boundary conditions ($\psi_{M+n} = \psi_n$). We choose a driving which facilitates the analysis in terms of a 1-soliton collective variable approximation, namely time periodic δ-kick perturbation:

$$i\dot{\psi}_n = -(\psi_{n+1} + \psi_{n-1})(1 + |\psi_n|^2) + KV_n\psi_n \sum_{m=-\infty}^{+\infty} \delta(t - mT) \qquad (20)$$

with $V_{n+M} = V_n$. The continuum limit of this equation for $V_n = \cos(2\pi n/M)$ was recently investigated by Casati and coworkers [16]. They found sensitive dependence of the time evolution on the initial conditions not only for single solitons, but also for more general initial conditions. For vanishing nonlinearity their model is equivalent to the quantum kicked rotor which shows dynamical localization in momentum representation — in striking contrast to the deterministic diffusion observed in the classical limit for sufficiently large kicking constant K. The connection between this dynamical localization and Anderson localization in tight binding models was pointed out by Fishman, Grempel and Prange [17]. Casati and coworkers claim that the dynamical localization persists even in the case with nonlinearity, which certainly deserves further investigation.

We propose to use the discrete model (20) to investigate these questions in greater detail. Here we present some first results. We assume that the state of the chain is given at time $mT-0$, i. e. immediately before the kick at time mT. Upon integrating eq. (20) from $mT-0$ to $mT+0$ one finds the following map

$$\psi_n(mT+0) = \psi_n(mT-0)\, e^{-iKV_n}. \tag{21}$$

This map is succeeded by the completely integrable evolution from $mT+0$ to $(m+1)T-0$, completing one iteration step. For the special case of a single, narrow soliton at time $mT-0$ the effect of the kick can be calculated in 1-soliton collective variable approximation. Neglecting the unimportant overall phase factor ϕ_0 (see eq. (10)) the important soliton parameters are $x_0(t)$, $\alpha(t)$, and $\beta(t)$ with $t = mT$. Neglecting the curvature of V_n over the soliton hump we can approximate the kick by

$$\psi_n(mT+0) = \psi_n(mT-0)\ \exp^{-iK\left(V(x_0)+V'(x_0)(n-x_0)\right)}. \tag{22}$$

Neglecting again overall phases this shows that x_0 and β are left unchanged by the kick: $x_0(mT+0) = x_0(mT-0)$ and $\beta(mT+0) = \beta(mT-0)$. The phase modulation α on the other hand does change:

$$\alpha(mT+0) = \alpha(mT-0) + KV'\big(x_0(mT-0)\big). \tag{23}$$

The integrable evolution between the kicks leaves β unchanged, too. Therefore β is an adiabatic invariant for the dynamics (20). The integrable evolution also leaves α, the velocity parameter, unchanged: $\alpha\big((m+1)T-0\big) = \alpha(mT+0)$. But the position of the soliton changes:

$$x_0\big((m+1)T-0\big) = x_0(mT+0) - \frac{2T\sinh\beta}{\beta}\sin\alpha(mT+0). \tag{24}$$

Now we introduce position and momentum variables, both ranging between 0 and 2π:

$$\begin{aligned} q_m &= \frac{2\pi}{M}x_0(mT-0), \\ p_m &= -\alpha(mT-0). \end{aligned} \tag{25}$$

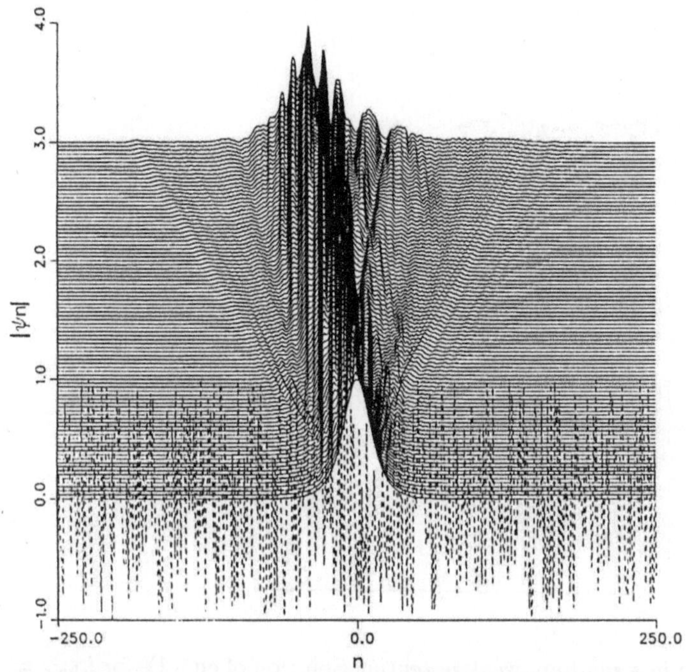

Fig. 8. *Slow soliton in a random, weak potential:* Solution of eq. (1), for $l_{\text{chain}} = 500$; periodic boundary conditions; $V_{\max} = 0.1$; initial soliton parameters: $\alpha = 0.3$, $\beta = 0.1$; integration time $T = 90.$; $|\psi|$-magnification: 10.

With these new variables the soliton motion in 1-soliton collective variable approximation is described by a symplectic map:

$$\begin{aligned} p_{m+1} &= p_m - K\tilde{V}'(q_m), \\ q_{m+1} &= q_m + \tilde{T}(\beta)\sin(p_{m+1}), \end{aligned} \tag{26}$$

where $\tilde{V}(q_m) = V(x_0(mT - 0))$ and $\tilde{T}(\beta) = 4\pi T \sinh\beta / M\beta$. For $\tilde{V}(q) = \cos q$ and in the region $|p| \ll 1$ the standard map is recovered. The full map then takes the doubly periodic form

$$\begin{aligned} p_{m+1} &= p_m + K\sin(q_m), \\ q_{m+1} &= q_m + T(\beta)\sin(p_{m+1}), \end{aligned} \tag{27}$$

with $T(\beta) = T(0)\sinh\beta/\beta$, upon skipping the wiggle. $T(\beta)$ takes into account that narrow solitons move faster than broad ones.

Eq. (27) is non-integrable for $KT(\beta) \neq 0$. For $K = T = 2$, for example (see Fig. 10), regions of regular and chaotic motion coexist. We place a single soliton in a regular region near an elliptic fixed point (Fig. 11, with q shifted by π in comparison with Fig. 10), and in a chaotic region near a hyperbolic

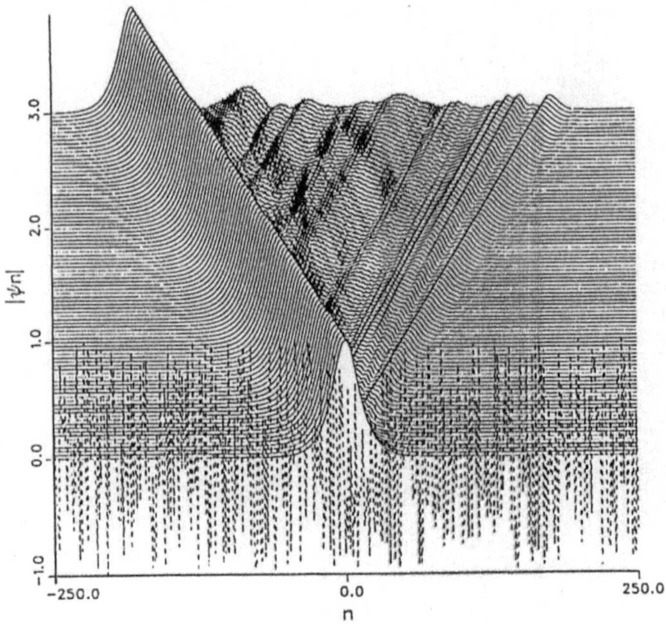

Fig. 9. *Fast soliton in a random, weak potential:* Solution of eq. (1), for $l_{\text{chain}} = 500$; periodic boundary conditions; $V_{\max} = 0.1$; initial soliton parameters: $\alpha = 1.57$, $\beta = 0.1$; integration time $T = 90.$; $|\psi|$-magnification: 10.

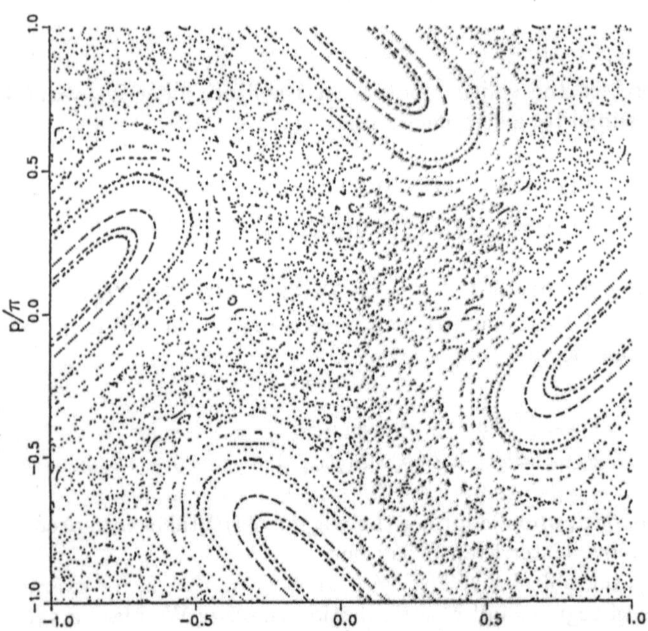

Fig. 10. *The "double-sine-map":* phase plane for the map (27) with $K = T = 2$.

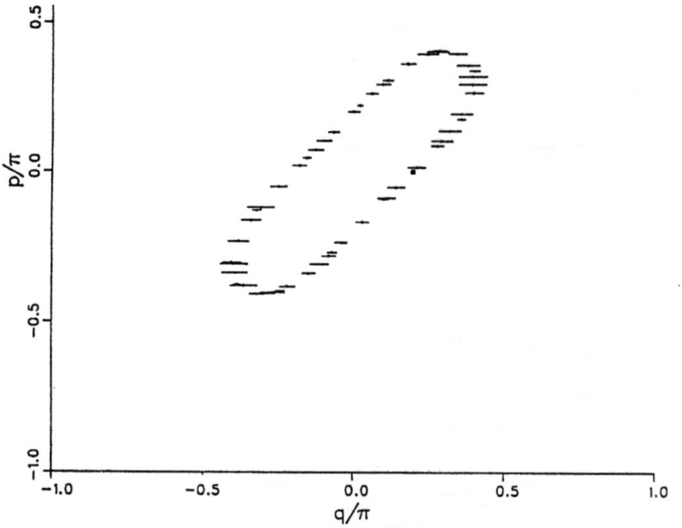

Fig. 11. *Soliton moving in a regular region:* $K = -T = -2.$, $q_0 = 2\pi/10$, $p_0 = 0$; 50 iterations forward and backward; length of the bars corresponds to δq_n, the spread of the solution in q-direction at time n.

fixed point (Fig. 12). The solitons display clearly different kinds of behavior. Fig. 13 shows that the motion of the soliton in the second case closely follows the chaotic classical iteration. Surprisingly it does so without breaking apart. The amount of radiation that is emitted is small and the shape of the soliton is practically unchanged.

Finally we address the question whether the dynamics of the chain is reversible and how this depends on the motion of the soliton. We monitor the spread of the soliton plus emitted radiation ($\sqrt{< q^2 > - < q >^2}$) over a number of time steps and then we time-reverse the dynamics. Because of inevitable numerical errors maps like the one given by eq. 27 are not reversible numerically over say 100 iterations for initial conditions with positive Lyapunov exponets. For the full dynamics of the chain (eq. (20)) integrated numerically with sufficiently large accuracy it turns out that the system returns to its initial condition upon time-reversal if the soliton is moving regularly (see Fig. 14). If the soliton is moving in a chaotic manner no available numerical accuracy was found to be sufficient to restore the initial conditions upon time-reversal. This illustrates the sensitive dependence of the dynamics on the initial condition if the soliton is started in a chaotic region. Our results clearly show that solitons can be forced to move chaotically with only negligeable excitation of other degrees of freedom.

Further interesting questions are for example: (i) how fast do solitons decay when the curvature of V_n is not negligeable; (ii) does the dynamics (20) show localization in p-direction when the corresponding map (26) shows deterministic diffusion (on the torus); (iii) how do two solitons interact in the

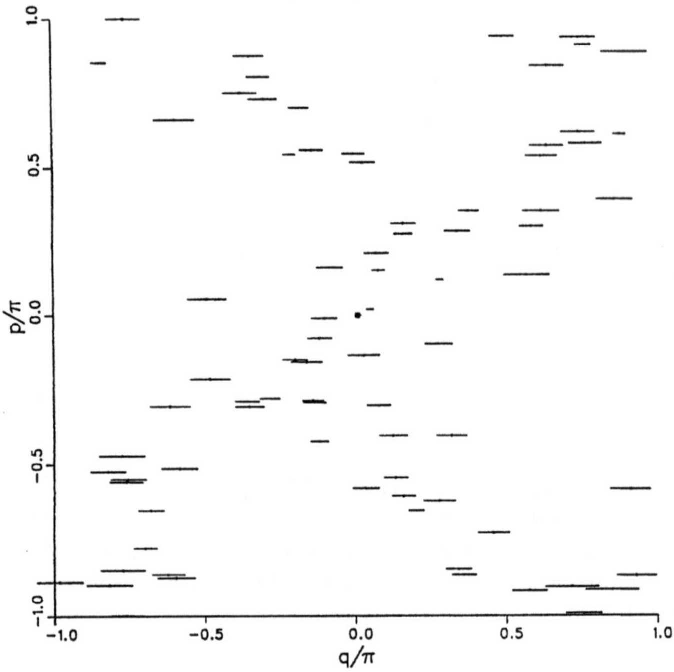

Fig. 12. *Soliton moving in a chaotic region:* $K = T = 2.$, $q_0 = \pi/100$, $p_0 = 0$; 50 iterations forward and backward.

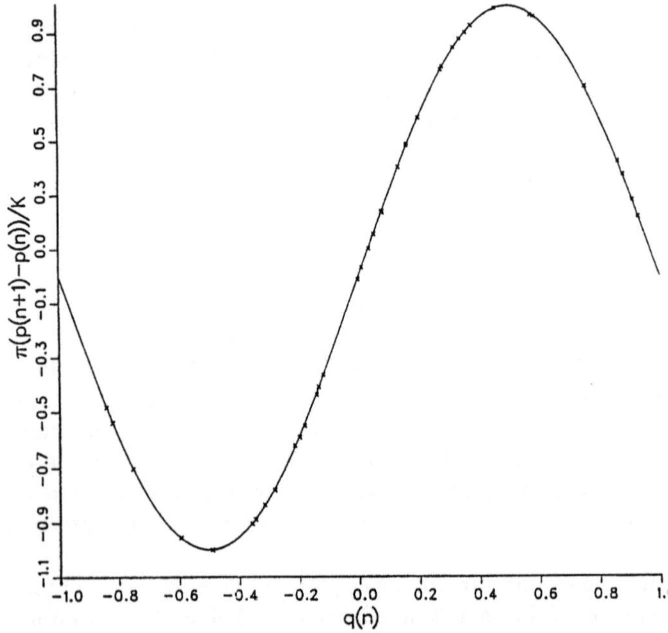

Fig. 13. *Full dynamics versus collective variable approximation:* $\pi(p_{n+1} - p_n)/K$ versus q_n compared with $\sin q_n$ for the chaotic case shown in Fig. 12.

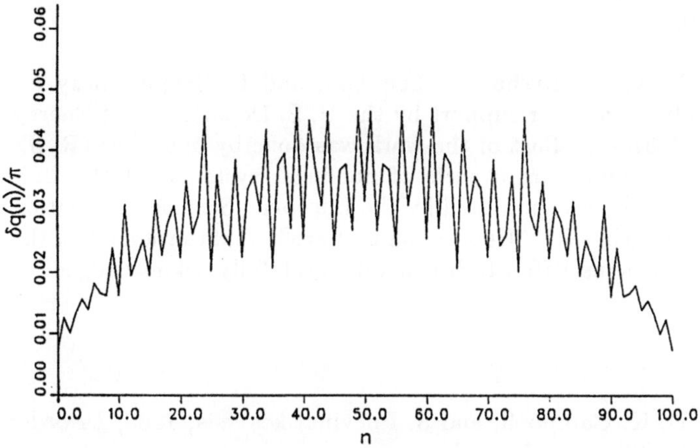

Fig. 14. *Time-reversal for "regular" soliton:* Spread $\delta q_n/\pi$ versus n; time reversal at $n = 50$; initial condition same as in Fig. 11.

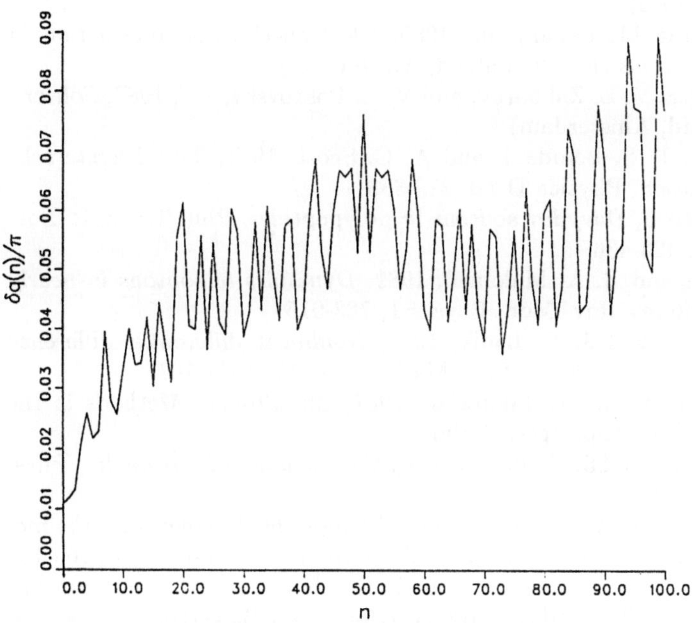

Fig. 15. *Time-reversal for "chaotic" soliton:* Spread $\delta q_n/\pi$ versus n; time reversal at $n = 50$; initial condition same as in Fig. 12.

presence of the kicking. These and further questions will be addressed in a forthcoming publication [18].

Acknowledgments

We thank F. Abdullaev, Y. Kivshar, P. Lomdahl, and D. Shepelyansky for valuable help and advice, and for support by the U. S. Department of Energy and NATO Grant RG674-88. Part of the work was done by one of us (R. S.) during a workshop on "Quantum aspects of nonlinear systems" at the Universität-GHS Essen which was sponsored by the European Science Foundation and the Sonderforschungsbereich "Unordnung und große Fluktuationen" of the Deutsche Forschungsgemeinschaft which is hereby gratefully acknowledged.

References

[1] A. R. Bishop, D. K. Campbell, and S. Pnevmatikos, eds, 1989, *Disorder and Nonlinearity* (Springer, Berlin)

[2] A. Hasegawa, 1989, *Optical Solitons in Fibers* (Springer, Berlin)

[3] N. Mott, 1987, *Conduction in Non-Crystalline Materials* (Oxford University Press, Oxford)

[4] A. L. Efros and M. Pollak, eds, 1985, *Electron-Electron Interaction in Disordered Systems* (North-Holland, Amsterdam)

[5] S. E. Trullinger, V. E. Zakharov, and V. L. Pokrovsky, eds, 1986, *Solitons* (North-Holland, Amsterdam)

[6] J. C. Eilbeck, P. S. Lomdahl, and A. C. Scott, 1985, *The discrete self-trapping equation*, Physica D **16**, 318-338

[7] A. C. Scott, 1985, *Davydov solitons in polypeptides*, Phil. Trans. R. Soc. Lond. A **315**, 423-436

[8] Y. S. Kivshar and B. A. Malomed, 1989, *Dynamics of solitons in nearly integrable systems*, Rev. Mod. Phys. **61**, 763-915

[9] M. J. Ablowitz and J. F. Ladik, 1976, *Nonlinear differential-difference equations and Fourier analysis*, J. Math. Phys. **17**, 1011-1018

[10] L. D. Faddeev and L. A. Takhtajan, 1987, *Hamiltonian Methods in the Theory of Solitons* (Springer, Berlin)

[11] R. Balakrishnan, 1985, *Soliton propagation in nonuniform media*, Phys. Rev. A **32**, 1144-1149

[12] P. W. Anderson, 1972, *Effect of Franck-Condon displacements on the mobility edge and the energy gap in disordered materials*, Nature **235**, 163-165

[13] M. H. Cohen, E. N. Economou, and C. M. Soukoulis, 1983, *Polaron formation near a mobility edge*, Phys. Rev. Lett. **51**, 1202-1205

[14] R. Scharf and A. R. Bishop, to be published

[15] J. G. Caputo, A. C. Newell, and M. Shelley, 1990, *Nonlinear wave propagation through a random medium and soliton tunneling*, preprint

[16] F. Benvenuto, G. Casati, A. S. Pikovsky, and D. L. Shepelyansky, 1990, *Manifestation of classical and quantum chaos in nonlinear wave propagation*, preprint

[17] S. Fishman, D. R. Grempel, and R. E. Prange, 1982, *Chaos, quantum recurrences, and Anderson localization*, Phys. Rev. Lett. **49**, 509-512

[18] R. Scharf, to be published

Dynamics of Sine-Gordon Soliton Interactions with Impurities

W.C. Kerr[1], *A.R. Bishop*[2], *and S.N. Kerr*[3]

[1]Olin Physical Laboratory, Wake Forest University, Winston-Salem, NC 27109-7507, USA

[2]Theoretical Division and Center for Nonlinear Studies, Los Alamos National Laboratory, Los Alamos, NM 87545, USA

[3]Department of Mathematics and Computer Science, Winston-Salem State University, Winston-Salem, NC 27110, USA

Abstract. Molecular dynamics simulations of the sine-Gordon system containing static impurities are described. A pictorial device which serves as a "soliton detector" shows that the impurities can reflect solitons and breathers, dynamically trap and release them, and cause sufficiently large local distortions to produce trapped solitons. These effects of the impurities are exhibited in the dynamical scattering functions as a very sharp static "central peak" superimposed on the dynamical central peaks produced by soliton and breather propagation.

One purpose of the conference recorded in these proceedings was to advance understanding of *coexisting* influences of strong nonlinearity and disorder when both are present in a particular physical system. Nonlinearity by itself generally leads to self-focussing into soliton-like states, while disorder alone leads to localized modes and Anderson localization. If both are present with comparable length scales for focussing and for localization, then competition arises leading to instabilities and dynamical switching. If both are present, but the soliton size is much larger than the scale of the disorder, then soliton structures can smooth out disorder. Conversely, if the soliton size is much less than the scale of the disorder, then solitons can be trapped on pinning sites within the disordered regions. Finally, if thermal fluctuations are present, then transitions among all of these possibilities can be expected, at least for certain ranges of temperature, nonlinearity, and disorder.

The system studied here is intended to illustrate some of these possibilities in a simple but physically interesting context. The system is the discrete sine-Gordon (sG) system, with disorder introduced through static, short-range impurity potentials added to the periodic potential in which the particles move. We use the molecular dynamics (MD) technique, that is, computer integration of the equations of motion for a large number of particles, to explore the range of phenomena which are possible in this system. The pure sG system has been analyzed previously by MD methods [1],[2]; the results presented here should be compared with those earlier studies.

The equations of motion used for the MD calculations are

$$M\frac{d^2u_n}{dt^2} = -\frac{2\pi}{a}A\sin\left(\frac{2\pi}{a}u_n\right) + C\,(u_{n+1} - 2u_n + u_{n-1})$$

Editors: F. Abdullaev · A.R. Bishop · S. Pnevmatikos

$$+ \frac{2V_0}{\sigma^3\pi}\sum_{j=1}^{N_I}[(n-1)a+u_n-R_j]\,e^{-[(n-1)a+u_n-R_j]^2/\sigma^2},$$

for $n = 1,...,N$. Here N is the number of particles, M is the mass of the particles, A is the amplitude of the periodic potential, and C is the force constant of the linear coupling forces. The period of the periodic potential *and* the equilibrium length of the coupling forces are both a. The displacement of the n-th particle from its equilibrium position at $(n-1)a$ is u_n. The N_I impurities are located at the positions R_j, which are chosen randomly. The impurity potential is a Gaussian with strength parameter V_0 and width σ. Periodic boundary conditions are used, for a system of $N = 1000$ particles. In solving these equations, the system of units defined by choosing $M = 1, A = 1$, and $a = 2\pi$ is used. If the continuum limit of (1) is taken, which is justified if the displacement u_n varies sufficiently slowly, the first two terms on the right-hand-side give the standard sG equation which has well known soliton, breather, and phonon solutions [3]. In order to make this situation obtain, we chose the value of the harmonic coupling constant to be C = 29.22 (in the units described above). Then the width of a stationary soliton is about 11 lattice constants and its energy is 43.2. For the strength of the impurity potential V_0, we chose two different values: *strong* impurities with $V_0 = 5112$ and *weak* impurities with $V_0 = 1022$. Finally, the computer simulations were done at a non-zero temperature $k_BT \approx 12.4$, corresponding to $0.29\, E_K$ where E_K is the bare sG soliton energy . The simulations are microcanonical with conserved total energy, and the temperature is obtained from the average kinetic energy per particle. The simulations are started from random initial conditions, and are "aged" to remove any artifacts of the particular initial conditions. The calculations described in the rest of this paper are done after thermal equilibrium was achieved by this procedure.

To understand how the nonlinear excitations of the sG equation interact with the impurities, we have found a pictorial device called a "soliton detector" to be very useful. The same device has been used previously to show soliton and breather motions for the pure sG system [1],[2]. A soliton is the solution of (the continuum limit of) the equations of motion which interpolates smoothly between two adjacent minima of the periodic potential. When a soliton propagates through the system, it moves particles from their initial potential well to an adjacent well (the well to the left for a soliton, the well to the right for an anti-soliton). Thus the motion of a soliton can be followed by marking on a space-time plot (site n versus time t) which well each particle is in as a function of time. If it is in its initial well (centered at site $(n-1)a$ for particle n), we mark the n-th position with white, and when it moves to an adjacent well in either direction, we change the mark to black. If it moves over two wells, the color is changed back to white, and so on alternately. With this scheme, the center of a soliton appears as the boundary between large black and white regions on the picture. Breathers whose amplitude is large enough to take them over the maxima of the periodic potential, *i.e.* larger than $a/2$, appear as small bubbles of white (*resp.* black) within a black (*resp.* white) background. Small amplitude breathers and phonons do not appear in the soliton detector.

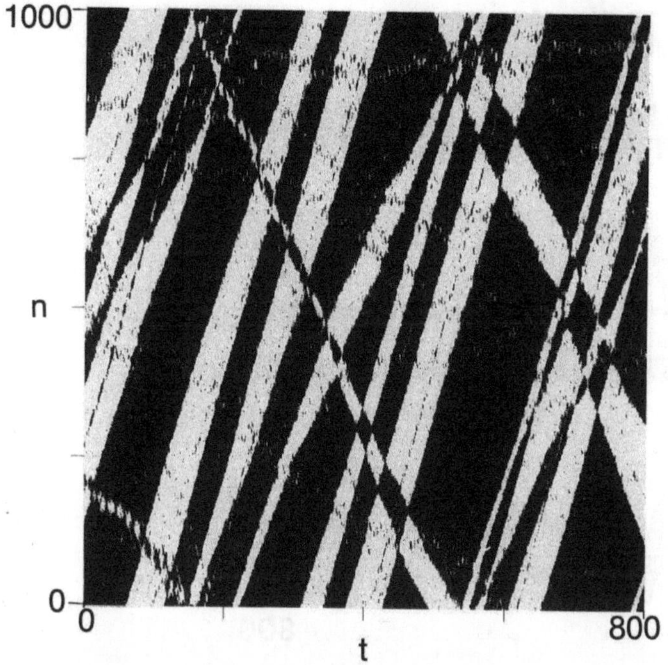

1. Soliton detector for the pure sG system.

Figure 1 shows the soliton detector for the pure sG system (a different but identically constructed detector picture is shown in Ref. [1]) and Figs. 2 and 3 are soliton detectors for specific configurations of strong and weak impurities, respectively. Time is plotted horizontally, lattice site index vertically; the positions of 10 randomly placed impurities are shown by the arrows along the right-hand side of Figs. 2 and 3.

For the pure system one sees unhindered soliton and breather propagation with phase shifts occurring at collisions. The phase shifts are easily visible near the bottom left corner where a large, slow breather collides with several solitons in succession. This breather subsequently decays into a soliton - anti-soliton pair. (Recall that the periodic boundary conditions mean that entities moving off the top of the picture reappear at the bottom and *vice versa*). This decay results from having a discrete system and would not occur for a continuum system.

For the system with strong impurities, the following features should be noticed. (1) An impurity can produce static displacements in its neighborhood which are sufficiently large to show on the detector; see e.g. the first, second, third and fifth impurities (counting from the bottom). These are appropriately called "trapped" solitons [4]. (2) Solitons can be reflected from impurities with their velocities reversed, but with no change in speed, or they can be reflected with decreased speed; both of these effects occur between the sixth and seventh impurities. (3) Similar reflections occur for breathers as seen between the fifth and sixth

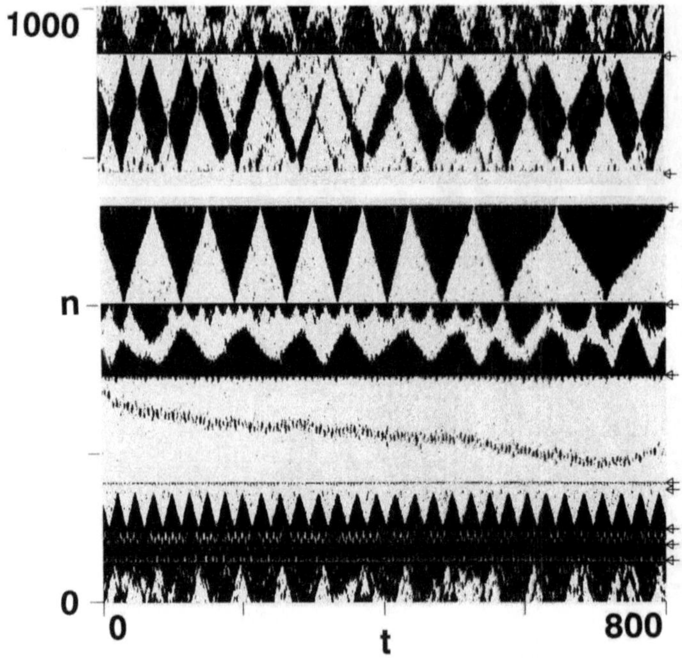

2. Soliton detector for the sG system with strong impurities.

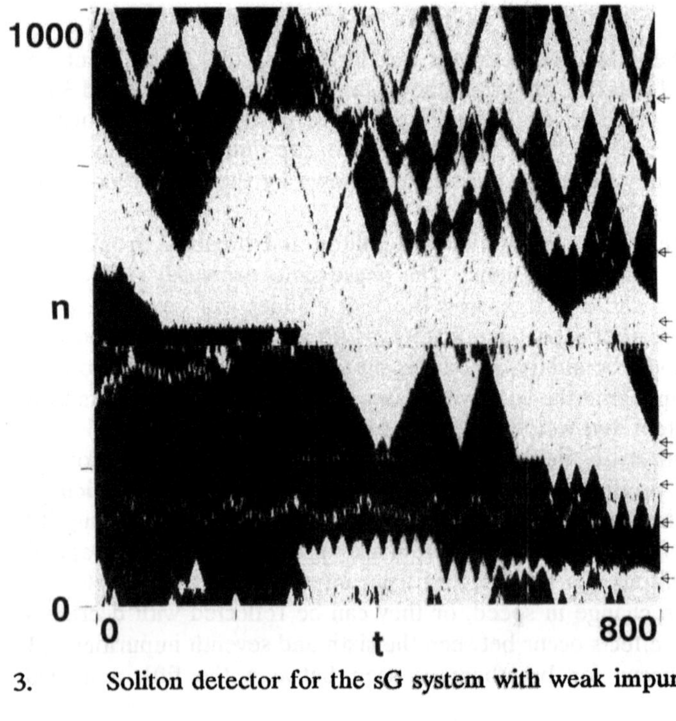

3. Soliton detector for the sG system with weak impurities.

impurities, and in the center of the figure between the ninth and tenth. (4) Features which occur for the pure system are also evident here, although affected by the presence of the impurities. These include solitons and breathers passing through each other between reflections from the impurities, essentially unmodified except for phase shifts; this occurs between the ninth and tenth impurities and also between the tenth and first (again recall the periodic boundary conditions).

For the system with weak impurities, the following additional features should be noted. (1) A soliton can approach an impurity, be trapped for some time interval and subsequently be released. This occurs on the fifth, seventh, and tenth impurities at various times in this figure. This effect is "dynamical" trapping, to be contrasted with the static trapping described in the previous paragraph. (2) Solitons can also pass directly over impurities, and then be trapped on the next impurity they encounter. This effect can be seen on the third, fifth and ninth impurities at different times.

Quantitative measurement of the impurity effects can be done by calculation of power spectra of different functions of the displacement variable. Two of these, which can be measured directly by inelastic neutron scattering on appropriate systems (e.g. magnetic chain materials [5]), are the sine and cosine of the phase variable, i.e.

$$s_n(t) = \sin\left[\frac{2\pi}{a}u_n(t)\right], \qquad c_n(t) = \cos\left[\frac{2\pi}{a}u_n(t)\right].$$

The time dependence of these functions at each site is obtained directly from the solution of the equations of motion; Fourier transforms with respect to site index and time are performed on these functions to give the spectral functions $S(q,\omega)$, according to standard methods (see e.g. [1],[2]). The sine and cosine functions give quite different information because they are, respectively, even and odd functions of the displacement. The sine function contains information on the linear phonons (since its lowest approximation is just the displacement) and for symmetry reasons is insensitive to breathers. The cosine function contains information on breathers and on multiphonon processes. Both functions show features due to solitons.

An important point to make here is that the spectra for the systems with impurities are averaged over an impurity ensemble with 10 different impurity configurations; each member of the ensemble has the same strength for the impurities but has different (random) impurity locations.

Figures 4-6 show the spectral function of the sine function for the pure system, the strong impurities and the weak impurities, respectively. The major feature common to all three systems is the peak due to the phonons. The position of this peak follows the temperature-renormalized dispersion relation for the linear Klein-Gordon equation

$$\omega(q) = \sqrt{\omega_0^2\langle\cos\varphi\rangle + \frac{2C}{M}\left[1-\cos(qa)\right]}\,,$$

where the pointed brackets denote thermal average and $\omega_0^2 = (2\pi)^2A/Ma^2$. This

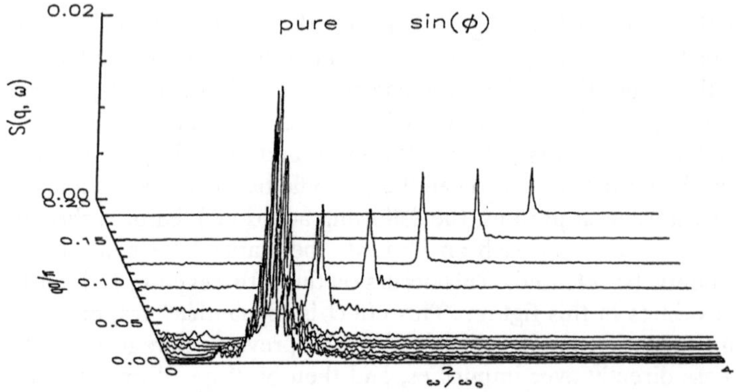

4. Wavevector and frequency dependent spectral function for the correlation function of the sine of the displacements for the pure system.

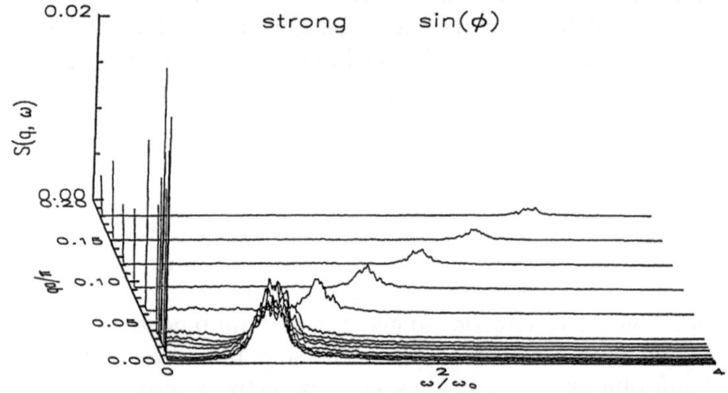

5. Wavevector and frequency dependent spectral function for the correlation function of the sine of the displacements for the system with strong impurities.

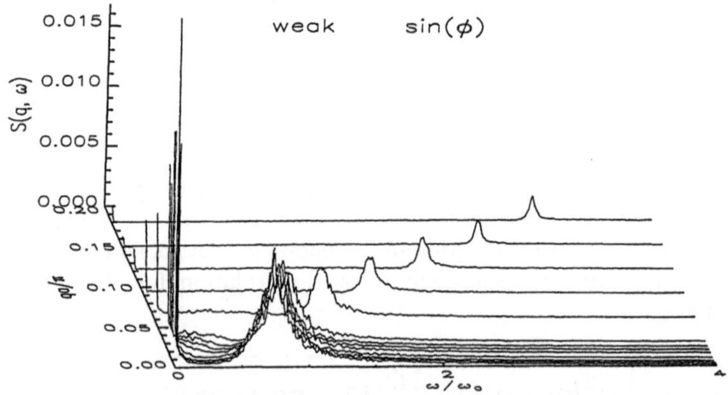

6. Wavevector and frequency dependent spectral function for the correlation function of the sine of the displacements for the system with weak impurities.

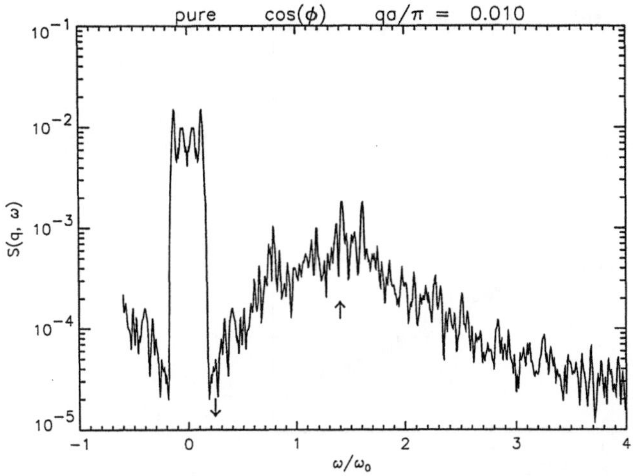

7. Frequency dependence of the spectral function for a particular wavevector for the correlation function of the cosine of the displacements for the pure system.

phonon peak is made broader and weaker by the impurities, but for each case it follows this same dispersion relation. A second feature in the spectrum which is common to all three systems is quasi-elastic scattering due to soliton propagation. In the spectral function for the sine function this feature has much less intensity than is in the linear phonon peak. However it is recognizable because in the pure system it has a sharp cutoff close to $\omega = c_0 q$ [where $c_0 = \sqrt{(Ca^2/M)}$ is the linearized sound velocity]; this cutoff is predicted by ideal soliton gas phenomenology [3] and is confirmed by the computer simulations (see also Ref. [1]. The impurities make the cutoff less sharp, so it is not so noticeable in those spectra. The third feature in the spectra, which occurs only in the systems with impurities, is the very intense and very narrow peak located essentially at $\omega = 0$, a so-called "central peak". The intensity of this peak falls slowly with increasing q; the scale is determined by the spatial extent of the distortion of the displacement field around the impurities, and this extent is rather small. The impurity-induced central peak is present for both strong and weak impurities, and is somewhat more intense for the stronger impurities.

Figures 7 and 8 show the spectral function for the cosine function of the displacements for the pure system and for the strong impurities; the corresponding function for the weak impurities is very similar to the one shown for the strong impurities. These two figures show the spectrum for one wavevector q, on a semi-logarithmic scale to bring out the contrasting features; the spectrum shape is similar for other wavevectors. Again there are three features. The weight at high frequency, which is present for both pure and impure systems, is due to the internal oscillation of the breathers. The upward pointing arrow locates the internal oscillation frequency of the "most-favored breather" [3] which arises from a form factor describing breather shape. Around

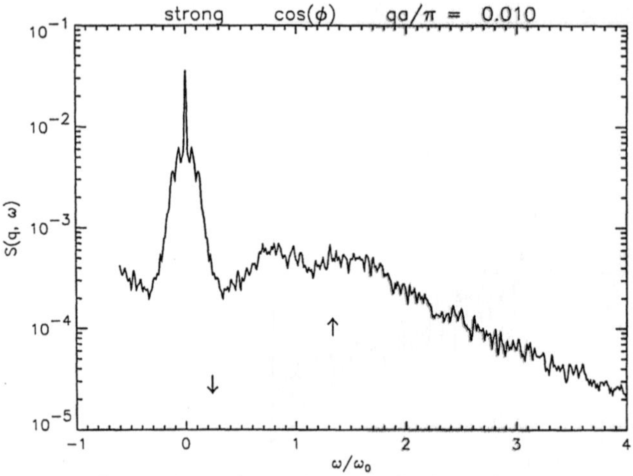

8. Frequency dependence of the spectral function for a particular wavevector for the correlation function of the cosine of the displacements for the system with strong impurities.

$\omega = 0$, there is again quasi-elastic scattering, which is more prominent in these cosine spectra than in the sine function spectra. For both the pure and the impure systems, the wavevector-dependent part of the quasi-elastic scattering is due to soliton propagation. It has the same cutoff as for the sine spectra, moving with wavevector q according to $\omega = c_0 q$ (marked with the downward pointing arrow). These two features have nearly the same intensity for the pure and impure systems. Finally, the distinctive feature for the impure system is an additional impurity-induced central peak.

The question of the origin(s) of quasi-elastic scattering ("central peaks") has been of recurring interest since it was first observed at second-order structural phase transitions [6]. These calculations show definitely that quasi-elastic scattering can arise from several mechanisms; both soliton propagation and impurity induced mechanisms are illustrated here.

The work at Wake Forest University was partially supported by a grant from the Research Corporation. The work at Los Alamos National Laboratory was supported by the US Department of Energy.

References

1. W. C. Kerr, D. Baeriswyl, and A. R. Bishop, Phys. Rev. B **24**, 6566-6578 (1981).

2. T. Schneider and E. Stoll, Phys. Rev. B **23**, 4631 (1981).

3. See e.g. A. R. Bishop, J. Phys. A: Math. Gen **14**, 1417-1430 (1981) and references therein.

4. D. Baeriswyl and A. R. Bishop, J. Phys. C: Solid St. Phys. **13**, 1403 (1980).

5. M. Steiner, J. Villain, and C. G. Windsor, Adv. Phys. **25**, 87 (1976); M. Steiner and A. R. Bishop, in *Solitons*, edited by S. E. Trullinger, V. E. Zakharov and V. L. Pokrovsky (North Holland, Amsterdam, 1986), p. 783.

6. T. Riste, E. J. Samuelson, K. Otnes, and J. Feder, Solid State Commun. **9**, 1455 (1971); S. M. Shapiro, J. D. Axe, G. Shirane, and T. Riste, Phys. Rev. B **6**, 4332 (1971); see also the review by A. D. Bruce and R. A. Cowley, Adv. in Phys. **29**, 219 (1980).

Chaotic Dynamics of Nonlinear Schrödinger Soliton Interaction with an Oscillating Impurity

F.Kh. Abdullaev[1] *and A.A. Abdumalikov*[2]

[1]Thermal Physics Department, Uzbek Academy of Sciences, 700 135 Tashkent, Chilanzar, Katartal Str. 28, Uzbekistan
[2]Physics Department, Tashkent State University, 700 095 Tashkent, Uzbekistan

Abstract. The problem of nonlinear Schrodinger soliton interaction with an oscillating impurity has been considered. The regimes of soliton motion for cases of periodic and random oscillations of the impurity near an equilibrium point have been investigated. As shown in the first case there exist regions of parameters under which a soliton dynamics becomes chaotic. The comparison of statistic characteristics of chaos in the region of dynamic and standard stochasticity has been carried out. The coefficient of soliton diffusion is found.

1. Introduction

The investigation of local oscillation spectrum and structure on impurities is of great interest in the solid state theory. A linear regime of oscillations is studied quite well(see e.g. [1]). But under intensive external actions nonlinear localized excitations (solitons) can be generated and in this case local oscillation dynamics will be defined by soliton interaction with impurities. To describe these problems it is necessary to make use of other methods. As shown in [2] that in onedimensional system where nonlinear localized excitations of crystals are described by solitons of nonlinear Schrodinger (NLS) equation, local oscillations can be described as NLS soliton interacting with point (δ -like) static impurity. In general, in this paper the untrapped regimes are studied. The investigation of NLS soliton interaction with dynamic impurity is of special interest. The thing is that the impurity can interact with other types of excitations in crystal (in the case of magnetic one,e.g.with phonons).and may be nonstationary. As shown in this paper the soliton dynamics in the field of such an impurity appears to be rather complicated.

In particular, there exist regimes of nonlinear resonances in soliton motion with a field of oscillations and at strong resonance overlapping a solitons moves by a random way and there is a dynamic chaos. We can carry out an analysis in comparison with a case of randomly oscillating impurity. In the first case the energy increases by a law $t^{1/3}$ and in the second one by ordinary diffusive law ~ t.

2. Formulation of problem

Let us consider NLS soliton interaction with localized (δ-like) inhomogeneity. It can be for example an impurity in crystal, that locally changes a characteristic frequency defining a spectrum of linear excitations [2]. In usual dimensionless variables NLS equation takes the form

$$iu_t + u_{xx} + 2|u|^2u = \varepsilon\delta(x)u \ . \tag{1}$$

In the case of dynamic impurity we have instead of (1) an equation

$$iu_t + u_{xx} + 2|u|^2u = \varepsilon\delta[x + \varphi(t)]u \ . \tag{2}$$

Here $\varphi(t)$ describes an impurity motion and can be both a periodic function and a random one. We get

1. $\varphi(t) = \varphi_o cos(\Omega t)$ (3)

2. $\varphi(t)$ is the gaussian δ-correlated random process with properties

$$\langle\varphi\rangle = 0 \ , \quad \langle\varphi(t)\varphi(t')\rangle = 2\sigma_f^2\,\delta(t - t') \ . \tag{4}$$

Let us analyze the NLS soliton dynamics, interacting with impurity dynamics in the framework of the equation (2). The soliton solution (2) at $\varepsilon = 0$ takes the form [3]

$$u_s(x,t)=2i\eta\,\frac{exp[-2i\xi - 4i(\xi^2 - \eta^2)t]}{cos[2\eta(x - 4\xi t)]} \ . \tag{5}$$

Here η is the soliton amplitude $v_s = -4\xi$ is the soliton velocity.

3. Soliton dynamics in a field of moving impurity

Let us consider the NLS soliton motion in a field of nonstationary local impurity. If the intensity of linear wave radiation by a soliton during the interaction is small, one can use an adiabatic approximation to investigate the soliton evolution. Let us derive a system of equations defining parameter changes of soliton with the help of the perturbation theory [4,5]. Introducing $x_o= -4\int_o^t \xi(t)dt$ one obtains a system

$$\frac{d\xi}{dt} = -2\varepsilon\eta^2 sech^2 z_o tanh z_o \ , \tag{6}$$

$$\frac{dx_o}{dt} = -4\xi \ , \tag{7}$$

$$\frac{d\delta}{dt} = 4(\eta^2 - \xi^2) + \ldots\ , \tag{8}$$

$$\eta = \eta_o \ , \tag{9}$$

where $z_o=-2\eta[x_o+\varphi(t)]$. Hence as follows the soliton amplitude is not changed while interacting with impurity. Only a coordinate center, velocity and soliton phase are changed. As follows from (6)-(7) a motion of soliton center is described by the equation

$$\frac{d^2x_o}{dt^2} = 8\varepsilon\eta^2 sech^2 z_o \tanh z_o \,. \tag{10}$$

Introducing new variables $y=-z_o=2\eta[\varphi(t)+x_o]$ we obtain, after substituting in (10)

$$\frac{d^2y}{dt^2} = 2\eta\varphi''(t) + 16\varepsilon\eta^3 sech^2 y \tanh y \,, \tag{11}$$

i.e

$$\frac{d^2y}{dt^2} = f(t) - \frac{\partial V}{\partial y} \,, \tag{12}$$

where

$$f(t) = 2\eta\varphi''(t) \,, \quad V(y) = \frac{8\varepsilon\eta^3}{\cosh^2 y} \,. \tag{13}$$

As follows from (12) the investigation of soliton dynamics in the framework of equation (2) is reduced to the investigation of a particle of single mass in a field of potential V(y) and nonstationary field f(t). The same problem arises on investigation of beam propagation in periodically inhomogeneous waveguide [6]. At $\varepsilon > 0$ a soliton reflection (passage) from impurity is realized, at $\varepsilon < 0$ the oscillating motion of soliton near inhomogeneity takes place. In the oscillating regime an interaction with periodic field f(t) can lead to resonant motions and deterministic chaos in soliton motion. At $\varepsilon < 0$ and $f = 0$ a solution of equation (12) takes the form

$$y(t) = \sinh^{-1}\left\{\sqrt{\frac{|v_o|}{|E|} - 1}\, \sin\left[2|E|(t - t_o)\right]\right\} \,, \tag{14}$$

$$v_o = 8\varepsilon\eta^3 \,.$$

Here E is the total energy of particle. Further we will make use of action-angle variables I-θ. For a potential (13) one can find them in explicit form. For action we have:

$$I = \frac{1}{2\pi}\oint p\,dq = \left(\sqrt{2|v_o|} - \sqrt{2|E|}\right) . \tag{15}$$

Here we find E through I

$$E(I) = -\frac{1}{2}\left[-I + \sqrt{2|v_o|}\right]^2 \,. \tag{16}$$

An action value upon the separatrix equals

$$I_s = \sqrt{2|v_o|} \,. \tag{17}$$

Hence $I = I_s - \sqrt{2|E|}$. Applying (16) we find a cyclic frequency ω

$$\omega = \frac{\partial E}{\partial I} = \sqrt{2|E|} = (I_s - I) . \quad (18)$$

At $I \to I_s$, $\omega(I) \to 0$.

Applying action-angle variables one can write an equation (12) in the system of two equations of the first order

$$\frac{dI}{dt} = - \Phi_o \frac{\partial H_{int}}{\partial \theta} , \quad \frac{d\theta}{dt} = \omega(I) + \Phi_o \frac{\partial H_{int}}{\partial \theta} , \quad (19)$$

where $\Phi_o = 2\eta\varphi_o\Omega^2$.

Here H_{int} is the part of total Hamiltonian

$$H = H_o + H_{int} , \quad (20)$$

$$H_o(I) = \frac{1}{2}(I - I_s)^2 ,$$

$$H_{int} = y(I,\theta)f(t) = \sinh^{-1}\{\alpha \sin\theta\} f(t) , \quad (21)$$

where $\alpha = \sqrt{|V_o/E| - 1}$.

4.Analysis of the motion types

Nonlinear resonances must appear in the particle motion under the action of nonlinear potential V(y) and variable field f(t). To find nonlinear resonances let us expand H_{int} into Fourier series.

$$H_{int}(I,\theta;t) = \Phi_o \sum_{n=-\infty}^{\infty} H^{(n)}(I) \exp(in\theta + i\Omega t) . \quad (22)$$

As seen there exist only even terms i.e. n=2m+1 in the Fourier series.Then the condition of nonlinear resonance takes the form

$$(2m + 1) \omega(I) = \Omega \quad . \quad (23)$$

Hence for large m we have

$$m \approx \Omega/2\omega . \quad (24)$$

Let us find a distance between nonlinear resonances

$$\delta\omega = |\omega_{m+1} - \omega_m| \approx \Omega/2m^2 = 2\omega^2/\Omega , \quad (25)$$

$$m >> 1.$$

A nonlinear resonance width equals

$$\Delta\omega = \max\ |\omega(I) - \omega(I_s)| = 4|\phi_o H_{int}^{(m)}\ d\omega/dI\ |^{1/2} . \qquad (26)$$

A calculation demonstrates that $H_{int}^{(m)}$ has the following behavior

At $\alpha >> m >> 1$,

$$H_{int}^{(m)} \approx 4\sqrt{\pi}/m . \qquad (27)$$

At $\alpha \simeq m >> 1$,

$$H_{int}^{(m)} \approx \frac{\sqrt{\pi}}{m} \frac{exp\left\{-\sqrt{4m^2/\alpha^2} + 1/2\ (1 - \sqrt{1 + 16m^2/\alpha^2}\right\}}{\sqrt{\sqrt{1 + 16\ m^2/\alpha^2} - 1}} . \qquad (28)$$

At $m >> \alpha >> 1$ (that corresponds to $\Omega >> 1$,i.e.a high frequency limit), we have

$$H_{int}^{(m)} \approx \frac{2\sqrt{\alpha\pi}}{m^{3/2}}\ exp(-2m/\alpha) . \qquad (29)$$

Let us first study the case ,when $\alpha >> m >> 1$. Then the nonlinear resonance width equals:

$$(\Delta\omega)^2 = \frac{32\sqrt{\pi}\ \phi_o\omega(I)}{\Omega} . \qquad (30)$$

In the case when the nonlinear resonance width is larger or the same as the distance between them, then one can expect the appearance of chaotic regime (Chirikov criterion [7])

$$K = (\Delta\omega/\delta\omega)^2 \geqslant 1 . \qquad (31)$$

Applying (25), (30) to (31) we find

$$K = \frac{32\phi_o\ \Omega\ \sqrt{\pi}}{(I - I_s)^3} \geqslant 1 , \qquad (32)$$

Hence as follows under any ϕ_o , Ω there exists a region in the separatrix vicinity, where (32) is satisfied and the NLS soliton motion is chaotic.

By similar way, applying (25) and (26), (29) one obtains chaoticity criterion in high-frequency limit

$$K = 16 \frac{\sqrt{2\phi_o^2\alpha\Omega\pi}}{\omega^{5/3}(I)}\ exp\left[-\Omega/\alpha\omega(I)\right] . \qquad (33)$$

A physical pattern looks like by the following way. Soliton absorbs resonantly energy from the impurity oscillation field. If the initial values of soliton parameters correspond to the stochastic region near separatrix, the boundary frequency of which is

$$\bar{\omega} = (16\varphi_0\ \Omega)^{1/3},$$

then the soliton motion becomes chaotic. In result soliton is stochastically accelerated and soliton depinning occurs. The depinning time is equal to

$$t_d \sim 1/\ \bar{\omega} = (16\varphi_0\ \Omega)^{-1/3}. \tag{34}$$

Appearance of random motion leads also to diffusion in the space of action. One can calculate a diffusion coefficient D in the region of dynamical chaos. Diffusion occurs in the action space and is equal to [8,9]

$$D = \frac{1}{4\pi} \int_0^{2\pi} (\Delta I)^2 d\theta\ . \tag{35}$$

In the Ω low-frequency region we have

$$D \simeq \frac{8\phi_0^2\ \Omega^2}{|E|^3}\ . \tag{36}$$

Here out we obtain for the energy the relation

$$E \simeq E_0 + 2\phi_0^{2/3}\ \Omega^{2/3}\ t^{1/3}\ . \tag{37}$$

Let us compare this result with the diffusion in the case of initial randomly oscillating impurity (see eqs.(4)). The standard estimates [10] give

$$E \simeq E_0 + \sigma_f^2\ t\ . \tag{38}$$

Though the rate of energy growth being described by the equation (37) is less in comparison with ordinary diffusion rate (38) by the parameter choice one can get rapid chaotic depinning of a soliton from the dynamic impurity due to dynamic stochasticity mechanism.

References

1. A.M.Kosevich. Elements of crystallic lattice theory. Moscow, Nauka, 1978 (In Russian).
2. Yu.S.Kivshar, A.M.Kosevich and O.A.Chubykalo. JETP, **93**, p.968 (1987) (In Russian).

3. V.E.Zakharov, S.V.Manakov, S.P.Novikov and L.P.Pitaevsky. Theory of Solitons. Moscow, Nauka Publ. 1980. (Transl. by Consultant Bureau, New York).
4. V.I.Karpman, E.M.Maslov. Sov.Phys.-JETP , **46**, 281 (1977).
5. D.J.Kaup and A.C.Newell. Proc.Roy.Soc. **A361**, 413 (1978).
6. S.S.Abdullaev, G.M.Zaslavsky. JETP, **80**, 524, (1981) (In Russian).
7. B.V.Chirikov. Phys.Rep., **52** , 263 (1979).
8. A.J.Lichtenberg and M.A.Lieberman. Regular and Stochastic motion. Springer-Verlag, Berlin, 1983.
9. G.M.Zaslavsky. Chaos in dynamic systems. New York, Harwood (1985).
10. R.L.Stratonovich. Selected topics in noise. Pergamon Press, (1967) London.

Modulated Dark Soliton: Features of Creation and Propagation

V.V. Konotop and V.E. Vekslerchik

Institute for Radiophysics and Electronics, Academy of Sciences, Proscura Str. 12, 310 085 Kharkov, Ukraine

Abstract. We study the "stable" nonlinear Schrodinger equation with inititial conditions being close to the dark soliton one (modulated dark soliton). It is considered in the framework of the perturbation theory based on the inverse scattering technique. Modulations result in both the variation of soliton parameters and creation of new solitons. Corresponding problems are treated analytically. The criterion for the perturbation theory to be valid is stated.

1. Introduction

The "stable" nonlinear Schrodinger equation (NSE):

$$iq_t + q_{xx} - 2|q|^2q = 0 \tag{1}$$

is one of the well-studied nonlinear evolution systems. The Cauchy problem for this equation can be solved in the framework of the inverse scattering technique (IST) [1]. The growing today attention to the NSE is stipulated by its practical significance. So, the NSE is successfully used to describe the propagation of the electromagnetic pulses along single-mode optical fibers [2] and nonlinear planar waveguides [3]. A series of experiments confirming that has been carried out recently by various groups of scientists [4-6].

Nowadays theoretical investigations are aimed to bring mathematical consideration nearer to experementators'needs, i.e. to develope more realistic models in order to describe the propagation of dark optical pulses. That includes, in particular, studying of dynamics of dark pulses against a finite background [7,8] (while most of the mathematical works devoted to the dark pulses deal with the NSE under nonzero boundary conditions which corresponds to the case of an infinite background), taking into account the dissipation [8], Raman effect [9-11], high order dispersion [11], etc. A problem of considerable interest is that of generation of dark solitons. The main question here is to determine parameters of the solitons forming from the given initial pulse. This problem was a subject of interest of a number of recent theoretical papers [8,12,13]. It also will be discussed in the present paper.

Creation of dark solitons is a thresholdless process [8]. That is so because the dark soliton energy is negative: the energy of the system 'background + soliton' is less than that of the background alone. As a result , however weak the initial pulse is , solitons can, on certain conditions, appear in the system.

Similarly, in the case of the initially modulated dark soliton, which is the case when an initial pulse slightly differs from the pure soliton one (for instance, due to fluctuations), we can observe more than one soliton, even if the initial modulation is vanishingly small. Obviously, one of them has to be close to the unperturbed dark soliton (we shall call it 'prime'). Additional, or 'new-born', solitons (or a soliton) are small as far as the initial modulation is small. Nevertheless, in order to describe the prime soliton one has to take them into account, since in the case under consideration the energy of the intial pulse is divided among the prime soliton, the new-born solitons (if there are any) and quasilinear modes.

In the present paper we consider changing of the soliton parameters caused by a initial modulation. To this end we use the perturbation method described in Sec.2. In Sec.3 we obtain statistical characteristics of the parameters of the prime soliton for the case of a random initial modulation . Sec.4 is devoted to the creation of additional solitons. In conclusion we summarize the main features of the dark soliton dynamics in comparison with the bright soliton one.

2. Mathematical statement of the problem

The more covenient form of the stable NSE can be obtained by the substitution $q \to q\exp(-2iq_0^2 t)$ into the Eq.(1). Thus, in what follows we consider the equation

$$iq_t + q_{xx} + 2(q_0^2 - |q|^2)q = 0 \tag{2}$$

The corresponding boundary conditions can be written in the form

$$q(x=-\infty) = q_0; \quad q(x=+\infty) = q_0\exp(i\theta) \tag{3}$$

Here q_0 is a positive real constatnt which plays a part of the background pulse amplitude and θ is a total phase shift.

2.1. Inverse scattering technique.

According to the IST (here and below we follow Takhtajan and Faddeev [14]) one can associate with the NSE the linear scattering problem, the so-called Zakharov - Shabat system, such that the initial condition $q(x) = q(x,t=0)$ forms the scattering potential.

$$dF(x;\lambda)/dx = U(x;\lambda)F(x;\lambda) \qquad (4)$$

Here F is a 2x2 matrix,

$$U(x;\lambda) = \begin{pmatrix} \lambda/2i & q^*(x) \\ q(x) & -\lambda/2i \end{pmatrix} \qquad (5)$$

λ is a spectral parameter and an asterisc means the complex conjugation.

The spectral problem has the continuous spectrum R_w at real λ: $\lambda^2 > \omega^2$ where ω is defined as $\omega = 2q_o$. Besides, the system (4) may have eigenvalues, constituting the discrete spectrum, in the lacuna $(-\omega, \omega)$.

The scattering data can be obtained from the scattering matrix $T(\lambda)$ linking the normalized solutions of Eq.(4)

$$T_-(x;\lambda) = T_+(x;\lambda)T(\lambda) \qquad (6)$$

Matrix Jost functions $T_\pm(x;\lambda)$ are defined by their asymptotics

$$T_\pm(x;\lambda) \to \exp[(1\pm1)\theta\sigma_3/4i]E(x;\lambda) \quad \text{as} \quad x \to \pm\infty \qquad (7)$$

Here

$$E(x;\lambda) = \begin{pmatrix} 1 & -i\omega/z\lambda) \\ i\omega/z(\lambda) & 1 \end{pmatrix} \exp[k(\lambda)x/2i] \qquad (8)$$

σ_3 = diag(1,-1) is the Pauli matrix, and parameters z and k are introduced according to the relations

$$k^2(\lambda) = \lambda^2 - \omega^2, \quad \text{sign } k = \text{sign } \lambda \quad (\lambda \in R_\omega) \qquad (9)$$

$$z(\lambda) = \lambda + k(\lambda) \qquad (10)$$

For $\lambda \in R_\omega$ the scattering matrix $T(\lambda)$ can be written in the form

$$T(\lambda) = \begin{pmatrix} a(\lambda) & b^*(\lambda) \\ b(\lambda) & a^*(\lambda) \end{pmatrix} \qquad (11)$$

If we cosider $a(\lambda)$ as a function of z , defined by (10), the set of zeroes of $a(\lambda(z))$ in the upper halfplane of complex z,

$$z_n = \lambda_n + i\nu_n \qquad (n = 1, \ldots, N) \qquad (12)$$

where

$$\nu_n = (\omega^2 - \lambda_n^2)^{1/2} \tag{13}$$

and N is the number of eigenvalues of the scattering problem (i.e. the number of solitons) is in the one-to-one correspondence with the discrete spectrum of the scattering problem.

Since for all n $z_n = \omega$ the discrete spectrum may be characterized by the set of angles $\{\theta_n\}$ defined by

$$z_n = -\omega \exp(-i\theta_n/2) \tag{14}$$

It follows from the analysis of analytical properties of the scattering matrix that

$$\exp(-i\theta) = \prod_{n=1}^{N} \exp(-i\theta_n) \exp\left\{ \frac{1}{\pi i} \int_{-\infty}^{+\infty} \frac{d\lambda}{k(\lambda)} \ln|a(\lambda)|^2 \right\} \tag{15}$$

(the so-called θ-relation [14]).

The NSE as an completely integrable system possesses an infinite number of conservation laws. Further we shall make use of one of them which may be written in the form

$$\frac{1}{2} \sum \lambda_n \nu_n = \frac{1}{2\pi} \int_{-\infty}^{+\infty} \frac{d\lambda}{k(\lambda)} (\lambda^2 - \omega^2/2) \ln|a(\lambda)|^2 - J_2[q] \tag{16}$$

where

$$J_2 = \frac{1}{2i} \int_{-\infty}^{+\infty} dx\, (q_x q^* - q q_x^*) \tag{17}$$

2.2 Perturbation theory

To study the features of the dynamics of an initially modulated dark soliton consider the following initial condition:

$$q(x,t=0) = q_s(x) + \delta q(x) \tag{18}$$

where q_s is the "soliton" one:

$$q_s(x) = \frac{1 + \exp(\nu x + i\theta)}{1 + \exp(\nu x)} \tag{19}$$

The Zakharov-Shabat system with the scattering potential given by (18) cannot be solved analytically for arbitrary δq. That is why we regard δq as small and make use of a perturbation method.

The problem under consideration differs from the analogous one for the bright soliton case. First, a dark soliton is one-parametrical while a bright pulse is described by two parameters. Secondly, one has to take into account the fact that the scattering data have singularities at the edges of the continuous spectrum,

$$a(\lambda \rightarrow \pm\omega) = a_{\pm}/k(\lambda) + O(1) \tag{20a}$$

$$b(\lambda \rightarrow \pm\omega) = b_{\pm}/k(\lambda) + O(1) \tag{20b}$$

which are absent if the unperturbed dark soliton is considered. Therefore, δT cannot be regarded as small when δU is small in spite of the fact that $\delta T=0$ when $\delta U=0$. Then, an initial perturbation, even a small one, can change the number of solitons, in contrast to the bright soliton case. Earlier we have developed a perturbation method that takes into account these features of the dark soliton dynamics [13]. Here we outline its main points.

It is convenient to develop the perturbation theory for the angle θ_1 which characterizes the amplitude and velocity of the prime soliton, rather than for λ_1 and ν_1 directly.

To overcome the abovementioned singularities of the Jost coefficient $a(\lambda)$ consider the function $c(\lambda)$ defined by

$$a(\lambda) = a_s(\lambda)[1 + ic(\lambda)/k(\lambda)] \tag{21}$$

The quantity $c(\lambda)$ is regular for all λ from R_ω, and is small as far as δq is small.

If the perturbation δq would not change the number of solitons, then one could use one of the conservation laws, say, given by (16), which, for $N=1$, expresses soliton amplitude ν_1 in terms of $q(x,t=0)$ and $|a(\lambda)|$. This is a way that has been used for the description of the dynamics of the modulated bright solitons [15]. The possibility of creation of new solitons poses a nontrivial problem: to "extract" the contribution of the prime soliton to the sum in the l.h.s. of (16). To this end, instead of using (16), one can use the proper combination of (16) and the θ-relation (15). That leads to the following result:

$$(\theta_1 - \sin\theta_1) - (\theta - \sin\theta) = \frac{2}{\pi\omega^2}\int_{-\infty}^{+\infty} d\lambda\; k(\lambda)\,\ln|a(\lambda)|^2 - \frac{4}{\omega^2}\,\delta J_2 \tag{22}$$

where $\delta J_2 = J_2[q_s+\delta q]-J_2[q_s]$.

From (22) one can get after some algebra that up to the second order in δq

$$\delta\theta_1 = \frac{2}{\nu^2}\left[-\delta J_2 + \frac{\lambda}{\nu^3}(\delta J_2)^2 + \frac{1}{2\pi}\int_{R_\omega}\frac{d\lambda}{k(\lambda)}c^2(\lambda)\right] \qquad (23)$$

where $\delta\theta_1 = \theta_1 - \theta$. Formula (23) is the key equation of the proposed perturbation method. By the use of (12) and (14) one can express addenda to the velocity $\delta\lambda_1 = \lambda_1 - \lambda_s$ and to the amplitude $\delta\nu_1 = \nu_1 - \nu$ in terms of $\delta\theta_1$ while the r.h.s. of (23) may be expressed in terms of δq. To evaluate $c(\lambda)$ one can use the following relation linking the variation δT of the scattering matrix with the variation δU of the matrix U from (4) which is caused by the variation δq of the initial condition

$$\delta T(\lambda) = \int_{-\infty}^{+\infty} dx\; T_+^{-1}(x;\lambda)\; \delta U(x)\; T_-(x;\lambda) \qquad (24)$$

where $T_\pm$ are the one-soliton Jost functions, i.e. the Jost functions of the scattering problem (4) with q_s as the potential. This yields

$$c(\lambda) = 2\,\mathrm{Re}\int_{-\infty}^{+\infty} dx\; [q_s^* + i(\lambda-\lambda_s)^{-1}(q_s^*)_x]\;\delta q \qquad (25)$$

To obtain the expression for δJ_2 one has to use straightforwardly the definition of J_2 (17) together with (18). We do not present here the resulting formulae since they are rather cumbersome . To demonstrate how this method can be used we next consider few examples. All of them deal with stochastic initial perturbations of the dark soliton.

3. Characteristics of soliton parameters

As was mentioned above, the most interesting application of the NSE is that for the description of the optical pulse propagation. The NSE initial condition $q(x,0)$ in this sense corresponds to the form of a pulse applied to the fiber input. Real laser pulses have not only a regular soliton component, but a random one, as well. Now we obtain the statistical characteristics of the prime soliton in such a realistic situation. Since different types of initial modulations can be taken into account additively [15] as far as the intensities are small enough, we consider them separately.

3.1. Phase fluctuations.

First consider the case of phase modulations

$$q(x,t=0) = q_s(x)\exp[i\phi(x)] \qquad (26)$$

Here $\phi(x)$ is a real random process of zero mean value and the correlator

$$\langle\phi(x)\phi(x')\rangle = \Phi(x-x') = \int_{-\infty}^{+\infty} dy\ \hat{\Phi}(y)\ \exp[iy(x-x')/2] \qquad (27)$$

vanishing at the infinities: $\phi(\pm\infty) = 0$.

After insertion (26) into both the definition of J_2 and expression for $c(\lambda)$ a simple algebra invoking the direct averaging of (23) and evaluating of the integral over R_ω in (23) yields the mean value of $\delta\theta_1$

$$\langle\delta\theta_1\rangle_{ph} = 3\lambda_s\Delta_{ph}/v \qquad (28)$$

Together with (23) it also gives

$$\langle(\delta\theta_1)^2\rangle = 4\Delta^2_{ph} \qquad (29)$$

where

$$\Delta^2_{ph} = (\pi^2/16v^4)\int_{-\infty}^{+\infty} dy\ \hat{\Phi}(y)\ y^4\operatorname{cosec}^2(\pi y/2v) \qquad (30)$$

The suffix "ph" indicates the case of phase fluctuations.

In the case of sufficiently weak Gaussian initial fluctuations the random quantity $\delta\theta_1$, and therefore $\delta\lambda_1$ and δv_1 may be regarded as distributed according to the normal law. Then formulae (28)-(30) completely describe statistics of $\delta\theta_1$, and, hence, statistics of the addenda $\delta\lambda_1$ and δv_1. So, mean values of the shifts of velocity and amplitude due to phase modulation (26) may be written as follows

$$\langle\delta\lambda_1\rangle_{ph} = \lambda_s\Delta^2_{ph} \qquad (31)$$

$$\langle\delta v_1\rangle_{ph} = -v(3\lambda_s^2/2v^2 + 1/2)\Delta^2_{ph} \qquad (32)$$

3.2. Amplitude fluctuations.

Let us obtain the statistical characteristics of a dark soliton generated from initial random pulse of the form (18) with

$$\delta q(x) = g(x)f(x) \qquad (33)$$

where $g(x)$ is a regular function vanishing at the infinities and $f(x)$ is a random process with the zero mean value and correlators

$$\langle f(x)f(x')\rangle = A(x-x') = \int_{-\infty}^{+\infty} dy\, \hat{A}(y)\, \exp[iy(x-x')/2] \qquad (34)$$

$$\langle f(x)f^{*}(x')\rangle = B(x-x') = \int_{-\infty}^{+\infty} dy\, \hat{B}(y)\, \exp[iy(x-x')/2] \qquad (35)$$

(we make use of the term 'amplitude' for the fluctuations defined by (33)). Note that such a choice of δq is rather natural. So in the experiments by Krokel *et. al.* [5] the "ultrafast light-controlled optical fiber modulator" has been used. The work of this device is based on the interaction of a long signal pulse with a driving one [16]. Thus the addendum δq may be considered as a random modulation of the driving pulse. Correspondingly, below we use the traditional form of a laser signal envelope

$$g(x) = \mathrm{sech}[\upsilon x/2] \qquad (36)$$

If a scale of fluctuations R (or, in other words, a correlation radius) is much less than the dark soliton width $(\upsilon^{-1} \gg R)$ it is natural to use the approximation of the delta-correlated random process. Correspondingly assume

$$\begin{pmatrix} A(x) \\ B(x) \end{pmatrix} = \begin{pmatrix} \exp(2i\alpha+i\theta) \\ 1 \end{pmatrix} A_0 \delta(x) \qquad (37)$$

Then, it can be shown that

$$\langle \delta\lambda_1 \rangle_0 = -3\lambda_s \Delta_0^2/2 \qquad (38)$$

$$\langle \delta\upsilon_1 \rangle_0 = \upsilon(\lambda_s^2/\upsilon^2 - 1/2)\Delta_0^2 \qquad (39)$$

where

$$\Delta_0^2 = 8(A_0 \cos\alpha)^2/15\upsilon \qquad (40)$$

Similar results may be obtained in the opposite limiting case

$$\begin{pmatrix} A(x) \\ B(x) \end{pmatrix} = \begin{pmatrix} \exp(2i\alpha+i\theta) \\ 1 \end{pmatrix} A_\infty \qquad (37)$$

which corresponds to fluctuations with large correlation radius $(R \gg \upsilon^{-1})$. In this case

$$\langle \delta\lambda_1 \rangle_\infty = -3\lambda_s \Delta_\infty^2 \qquad (42)$$

$$\langle \delta\upsilon_1 \rangle_\infty = \upsilon(5\lambda_s^2/2\upsilon^2 - 1/2)\Delta_\infty^2 \qquad (43)$$

with

$$\Delta_{\infty}^{2} = (\pi A_{\infty}\cos\alpha/2v)^{2} \tag{44}$$

As for the dispersion of the soliton parameters the corresponding formulae for the all cases considered above (initial phase modulations (26), delta-correlated fluctuations (37) and large scale initial amplitude modulations (41)) can be unified as follows

$$\langle(\delta\lambda_{1})^{2}\rangle = v^{2}\Delta^{2} \tag{45}$$

$$\langle(\delta v_{1})^{2}\rangle = \lambda_{s}^{2}\Delta^{2} \tag{46}$$

where Δ is one of the Δ_{ph}, Δ_{o} and Δ_{∞}.

3.3. Discussion

Formulae (31), (38) and (42) imply that the averaged velocity of a dark soliton increases under initial phase modulations and decreases in the case of amplitude perturbations. A mean velocity is not chahged if its unperturbed value is equal to zero, i.e. a rest soltion remains to be rest in all above cases. The dispersion of velocity fluctuations rises under the modulations of the both types. The dependence of the average amplitude of a dark soliton on the initial conditions is more complicated. If phase fluctuations result in the decreasing of the mean deep against a background carrier wave, the effect of amplitude fluctuations depends on their correlation radius and parameters of the unperturbed soliton. It can be seen from (39) and (43) that there is a critical velocity λ_{c} which depends on correlation radius R, such that a slow soliton ($|\lambda_{s}| < \lambda_{c}$) becomes more shallow, while a fast one ($|\lambda_{s}| > \lambda_{c}$) becomes more deep. So, for delta-correlated fluctuations $\lambda_{c}(R=0)=\omega_{o}/\sqrt{3}$ and for large correlation radii $\lambda_{c}(R=\infty)=\omega/\sqrt{6}$.

The effect of initial amplitude perturbations essentially dependes on the phase α. In particular, fluctuations and shifts of the prime soliton parameters decreases to zero with α changing from 0 to $\pm\pi/2$. If $\alpha=\pm\pi/2$ soliton and noise are phase-contrast, while they are phase-coherent when $\alpha=0$.

In conclusion of this section we consider the dependence of our results on the unperturbed soliton amplitude v. If $v \to 0$ ($\lambda_{s} \to \pm\omega$) all mean values go to the infinity in the case of amplitude noise and to a nonzero values in the case of phase fluctuations. Mathematically it is stipulated by the singularities at the edges of the continuous spectrum (see Sec. 2.2). It means, that the developed perturbation theory fails near the points $\pm\omega$. To specify this region and, hence, to give the estimates for the perturbation theory to be valid, we require

$$\Delta < const\ \frac{v}{\lambda_s} \tag{47}$$

In the limit $v \to 0$ $(\lambda_s \to \pm\omega)$ this condition provides (in the statistical meaning) the location of the perturbed root λ_1 inside the interval $(-\omega, \omega)$. The constant in (47) is of unity order and has to be determined in each case separately. If the requirement (47) fails , the results for mean values are unphysical. In any case, small and fast dark solitons are affected by the initial noise most effectively (*cf.* [8],[10]).

4. On creation of new solitons

It was mentioned above that creation of new solitons due to an initial weak perturbation is a distinguishing feature of the dark soliton dynamics. Considering the dynamics of an weakly modulated bright soliton one can neglect the possibility of a new soliton to arise (as well as initial one to be destroyed).

The situation is different in the dark soliton case. The representation (21) shows that in spite of smallness of $c(\lambda)$ which is of order of fluctuations, the addendum to the Jost coefficient $a(\lambda)$ takes all values because of the factor $k(\lambda)^{-1}$. Thus we have to take into account the possibility of new eigenvalues to appear near the singular points $\pm\omega$. Note that eigenvalues close to $\pm\omega$ correspond to the solitons of small amplitudes. Namely in this region a pair of solitons is born from a pure stochastic pulse against a nonsoliton background [8].

4.1 The dispersion relation

Using (24) together with (18) one can write $a(\lambda)$ as follows:

$$a(\lambda) \doteq a_s(\lambda) \left[1 + \frac{2i\omega}{k(\lambda)} \left[\Delta_r + \frac{2\omega}{\lambda - \lambda_s} \Delta_i \right] \right] \tag{48}$$

where to the first order

$$\Delta_r = \omega^{-1}\, Re \int_{-\infty}^{+\infty} dx\ q_s^* \,\delta q \tag{49}$$

$$\Delta_i = \omega^{-2}\, Im \int_{-\infty}^{+\infty} dx\ q_s^* \,(\delta q)_x \tag{50}$$

Equating $a(\lambda)$ with zero one can obtain the dispersion relation for the discrete spectrum of the scattering problem (4) corresponding to the initial condition (18):

$$(\lambda-\lambda_s)(\omega^2-\lambda^2)^{1/2} + 2\omega\ [(\lambda-\lambda_s)\Delta_r + \omega\Delta_i) = 0 \tag{51}$$

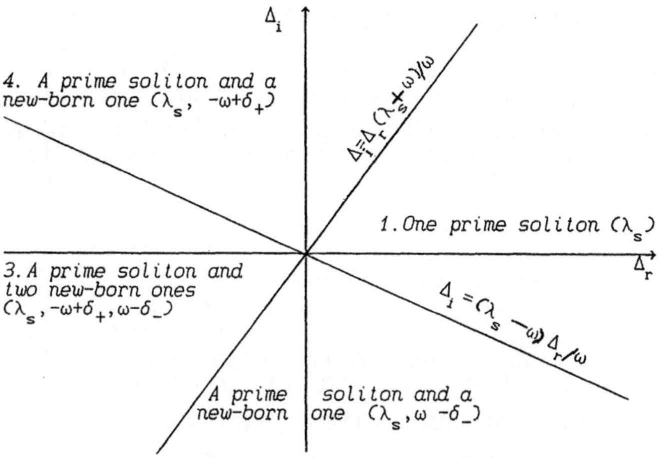

Fig.1. The complex plane of $\Delta=\Delta_r+i\Delta_i$ can be divided by two lines, crossing zero, into four parts which correspond to different number of solitons in the system. Note that areas 2 and 4, corresponding to the existence of two solitons, differs from each other with opposite directions of moving of new-born solitons

In accordance with the above discussion, it is naturally to look for solutions of this equation in two regions close to the banding points $\pm\omega$ and in the neibourhood of λ_s. Thus, the roots can be written as follows: $\lambda_1=\lambda_s+\delta$; $\lambda_\pm=\pm(\omega-\delta_\pm)$, with δ, $\delta_\pm \ll \omega$. Inserting these values into the dispersion relation and analysing corresponding equations for δ, $\delta_\pm$, one can picture the diagram for the existence of soltions in the system (see Fig.1). Following this way, one can also obtain

$$\delta=-2\omega^2\Delta_i/v \tag{52}$$

i.e. the sign of the addendum to the soliton velocity does not depend on the sign of the velocity itself. It is determined only by Δ_i.

In order to obtain $\delta_\pm$ one has to insert $\lambda_\pm$ into (51). Having made this one can find

$$\delta_\pm = 2\left[\frac{\omega\Delta_i \mp (\omega\pm\lambda_s)\Delta_r}{\omega\pm\lambda_s}\right] \tag{53}$$

Thus, $\delta_\pm$ are of the second order in the small parameter. It seems to be in contradiction with that the perturbation expansion was provided in this section only to the first order. To explain this, note that all conservation laws for the stable NSE are expressed in terms of υ_n (see e.g. [14] and formula (16)). It means that using the definition of a small parameter, as, for example, a relation between noise and the prime soliton energies, one has first to expand the parameter υ_n and consider the expansion for λ_n produced from the series for υ_n. It manifests itself in the dispersion relation (51) if one rewrite it in terms of υ_n. It can be easely found that amplitudes of new-born solitons are of the first order of the small parameter. On the other hand, if we take into account the next order in the expansion of $c(\lambda)$ near the banding points it gives terms of order of $\Delta^2_{r,i}$. Than the direct solution of (51) gives that addenda to $\delta_\pm$ are of order of $\Delta^3_{r,i}$, i.e. the may be neglected.

4.2. Examples

Let us give some illustrations of the developed perturbation theory. Begining with the case of initial phase modulation (26) we have

$$\Delta_r = 0; \quad \Delta_i = -\frac{\sin(\theta/2)}{\omega}\int_{-\omega}^{+\infty} \frac{dx\ \phi''(x)}{1+\exp(\upsilon x)} \tag{54}$$

where a prime means the derivation with respect to the argument. All obtained formulae can be applied to both regular and stochastic problems. However, as far as we consider the last ones, we have to treat the results in the statistical meaning. In particular, that means that phase modulations always lead to the creation of a new-born soliton (see Fig.1). However, its velocity (and correspondingly the direction of moving) is a random value. Both the initially modulated dark soliton and the new-born one move in the same direction. If $\phi(x)$ is a Gaussian process, the velocity of the prime soliton is destributed according to the normal law with mean value being equal to λ_s. But the velocity of a new-born soltion is not a Gaussian quantity (see Eq.(53)).

In the case of real odd modulations $\delta q(x)=-\delta q(-x)$ we have

$$\Delta_i = 0; \quad \Delta_r = (\cos\theta-1)\int_{-\infty}^{+\infty} dx\ \delta q(x)\ \tanh(\upsilon x/2) \tag{55}$$

Thus, a number of solitons is odd. In order to state how many (one or three) solitons are there in the system it is necessary to evaluate the integral in (55). As it follows from (52) (and from the symmetry of the problem under consideration) the velocity of the prime soliton does not change.

At last, in the case of real even modulations

$$\Delta_r = (\cos\theta - 1) \int_{-\infty}^{+\infty} dx\, \delta q(x) \tag{56}$$

$$\Delta_i = \frac{\upsilon \sin\theta}{8\omega} \int_{-\infty}^{+\infty} dx\, \delta q(x)\, \mathrm{sech}(\upsilon x/2) \tag{57}$$

i.e. all variants of new soliton creation are possible. The shift of the prime soliton velocity depends on both the form of the initial modulation and the velocity of the unperturbed soliton.

Conclusion

To conclude we summarize the main points that distinguish the dynamics of a dark soliton from the dynamics of a bright one. These differences have as a beginning, from the mathematical point of view, two features of the scattering problem for the NSE with nonzero boundary conditions: the eigenvalues of the scattering problem are real and the scattering data have singularities in the general case.

Since the discrete spectrum is real, a dark soliton is one-parametrical, while the bright one is two-parametrical. That is why it was convenient to develop the perturbation theory for the parameter θ rather than for the amplitude and velocity of the soliton directly. As a result a perturbation changes simultaneously both the velocity and amplitude of a dark soliton, while in the bright soliton case the phase fluctuations do not result in fluctuations of a soliton amplitude and initial amplitude modulations do not effect the velocity [15,17].

The main cosequence of the fact that the scattering data are, in general, singular is the possibility of the creation of new solitons due to small perturbations. Though this effect is not of much importance from the point of view of applications (since amplitudes of new-born solitons are as small as noise generated in the system), it is the main feature distinguishing the dark soliton dynamics from the bright soltion one. The developed perturbation theory allows to take this process into account. The creation of new solitons is described in the framework of the dispersion relation. Moreover, solving this equation one can find up to the first order not only the parameters of new-born soltions but the displacement of the prime root also. It allows in "regular" problems to restrict the consideration only by this dispersion relation and not to carry out the direct expansion of the Sec.3. However, in stochastic problems, in which all mean values are of the second order in fluctuations, the accurate description of the prime soliton can be done only with help of the direct expansion up to the second order.

It turns out that pure soliton solutions are unstable: a number of solitons can be changed by a vanishingly small perturbations. Physical origing of this phenomenon is in the fact that creation of solitons is energetically profitable. Mathematical explanation of this instability may be given as follows. The pure soliton case is an exceptional case, when singularities are absent: $|a_s(\pm\omega)|=1$, $a_{\pm}=0$. A perturbation, even being small, "returns" us to the general case. In this sense quasilinear constituent may play stabilizing part, since its presence implies $a_{\pm}\neq 0$. So, if "unperturbed" initial condition is not pure soliton one, a small initial perturbation leads to a small change of $a_{\pm}$, which is not crucial provided that unperturbed values of $a_{\pm}$ are nonzero.

References.

1. V.E.Zakharov and A.B.Shabat, Zh.Exp.Teor.Fiz. **64**, 1627 (1973) [Sov.Phys.JETP 37, 823 (1973)].
2. A.Hasegawa and F.Tappert, Appl.Phys.Lett. **23**, 171 (1973).
3. P.-A.Belander and P.Mathieu, Appl.Opt. **26**, 111 (1977).
4. P.Emplit, J.P.Hamaide, F.Reynand, G.Froehly, and A.Barthelemy, Opt.Comm. **62**, 29 (1987).
5. D.Krokel, N.J.Halas, G.Giuliani, and D.Grischkovsky, Phys.Rev.Lett. **60**, 29 (1988).
6. A.M.Weiner, J.P.Heritage, R.J.Hawkins, R.N.Thurson, E.M.Krischner, D.E.eaird, and W.J.Tomlinson, Phys.Rev. Lett. **61**, 2445 (1988).
7. W.J.Tomlinson, R.G.Hawkins, A..Weiner, J.P.Heritage, and R.N.Thurston. J.Opt.Soc.Am. B **6**, 329 (1989).
8. S.A.Gredescul, Yu.S.Kivshar, and .M.V.Yanovskaya, Phys. Rev.A **41**, 3994 (1990).
9. W..Tomlinson, R..Stolen, R.J.Hawkins, and A.M.Weiner, in Nonlinear Guided-Wave Phenomena: Physics and Applications, 2, 132 (1989).
10. Yu.S.Kivshar, Phys Rev.A, **42**, 1757 (1990).
11. F.G.Bass, V.V.Konotop, S.A.Puzenko, and V.E.Vekslerchik (submitted to press).
12. S.A.Gredescul and Yu.S.Kivshar, Opt. Commun. (in press).
13. V.V.Konotop and V.E.Vekslerchik, J.Phys.A. (in press).
14. L.A.Takhtajan and L.D.Faddeev. Hamiltonian Approach in Soliton Theory (Moscow, Nauka, 1986)(In Russian)
15. V.V.Konotop, Kvant.Elektronika **16**, 1032 (1989) [Sov.J. Quant.Electron. **19**, 669 (1989)].
16. N.J.Halas, D.Krokel, and D.Grischkowsky, Appl.Phys.Lett. **50**, 886 (1987).
17. F.G.Bass, Yu.S.Kivshar, V.V.Konotop, and S.A.Puzenko, Opt. Commun. **68**, 385 (1988)

Dynamics of a Stochastically Perturbed ϕ^4 Model

A. Sánchez and L. Vázquez

Departamento de Física Teórica I, Facultad de Ciencias Físicas,
Universidad Complutense, E-28040 Madrid, Spain

Abstract. A stochastic perturbation of the ϕ^4 model is studied. Computer simulations are carried out to determine the validity regime of the usual adiabatic approach. The threshold noise strength is found to depend on the initial kink speed. The main effects induced by strong perturbations are subsequently described. Possible explanations of our results are briefly discussed.

1 Introduction

The ϕ^4 model was introduced in 1975 by Aubry [1] and Krumhansl and Schrieffer [2] to study structural phase transitions in ferrodistortive and perovskite-like compounds, characterized by a mode becoming soft at some critical temperature. Since then, the range of applications of the model has rapidly grown; for instance, it reproduces fairly well the phenomenology of topological excitations in macromolecules [3], such as those of polymers [4] or hydrogen-bonded chains [5]. However, these systems, as well as the generality of real physical ones, are never ideal, and factors like thermal fluctuations, impurities, imperfections, external fields and so on, should play an essential role in any model intended to be realistic. This motivates the increasing amount of effort devoted to study perturbations of nonlinear systems (see [6] for very recent, thorough reviews). The work we report here belongs to that exciting field. It is aimed, on the one hand, to complete previous related research [7,8,9], and, on the other hand, to study a model that not only might be useful for actual practical purposes, but also allows to draw some conclusions of wider applicability concerning nonlinear phenomena.

2 Survey of the model and its main properties

The standard version of the one-dimensional discrete ϕ^4 model [1,2] consists of a chain of particles of identical mass. Each one of them interacts with its two nearest neighbours through harmonic coupling, and undergoes the influence of a double well potential. The particles positions are referred to a lattice of points na, a being the lattice spacing. This definition corresponds to the following dimensionless Hamiltonian:

Springer Proceedings in Physics, Vol. 67 Nonlinearity with Disorder
Editors: F. Abdullaev · A.R. Bishop · S. Pnevmatikos

$$H = \sum_{n=1}^{N} \left[\frac{1}{2} \left(\frac{d\phi_n}{dt} \right)^2 + \frac{\epsilon_0}{4} (1 - \phi_n^2)^2 \right] + \sum_{n=1}^{N-1} \frac{1}{2} (\phi_{n+1} - \phi_n)^2. \tag{1}$$

In (1) ϕ_n stands for the nth-particle displacement from its reference position na. The system equations of motion as derived from (1) are

$$\frac{d^2\phi_n}{dt^2} - (\phi_{n+1} - 2\phi_n + \phi_{n-1}) - \epsilon_0(\phi_n - \phi_n^3) = 0. \tag{2}$$

In the continuum limit, i.e., when the lattice spacing is much less than the wavelength of the excitations propagating along the chain, the set of equations (2) becomes

$$\frac{\partial^2 \phi}{\partial t^2} - \frac{\partial^2 \phi}{\partial x^2} - \epsilon_0 (\phi - \phi^3) = 0. \tag{3}$$

Notice that we have replaced the N variables ϕ_n by the field $\phi(x,t)$. Equation (3) happens to be the celebrated ϕ^4 field equation. Its main properties are well known (see e.g. [10] for a review). The most important one is that it possesses a solitary-wave solution, that it is usually termed "soliton" or "kink", though it does not fulfill the conditions of their strict mathematical definition. These objects and their stochastic stability will be our main interest.

The stochastic ϕ^4 model which we deal with here is obtained from the deterministic one (1) by setting

$$\epsilon_0 \equiv 1 + \xi(t); \tag{4}$$
$$< \xi(t) > = 0, < \xi(t)\xi(t') > = 2D\, \delta(t - t'), \tag{5}$$

the brackets $< \dots >$ standing for averages over the random process realizations. So, ϵ_0 is transformed into a gaussian white noise with unit mean and variance $2D$, and hence we are left with a ϕ^4 model in which the potential varies stochastically in time. This kind of perturbation represents large-spatial-scale changes (large compared to soliton width) of the lattice of on-site potentials, possibly due to interaction with noisy external fields, fluctuations of the underlying structure responsible for the nonlinear potential (electrons in polyacetylene for example [4]), etc. Our work on this model will center on single kink dynamics, trying to establish what the effects of disorder on soliton propagation are. In this context, we must remark that, at the early stages of the evolution, the random perturbation affects only the kink center or kernel (the zone where ϕ is distinctly different from ± 1) due to the form of the Hamiltonian (1), a fact that will be important in order to understand our findings.

It would be desirable to have analytical approaches to treat the stochastic ϕ^4 model. However, the most powerful techniques in nonlinear problems, like inverse scattering perturbation theory, do not apply to it because the unperturbed model is not integrable. Then, series expansions are the only available way to study the stochastic chain; in fact, they prove very useful and can be shown to explain most of the results we will mention below [11]. Calculations are rather involved and we will go into details about them here. For our present purposes, it is enough to recall the so called adiabatic equations, that are

$$\frac{dv}{dt} = 0, \tag{6}$$

$$\frac{dz}{dt} = v(t) - \frac{1}{2}v(t)\,[1 - v^2(t)]\,\xi(t), \tag{7}$$

where v and z stand for the kink center speed and position, respectively. These formulae can be obtained [11] by a nowadays standard procedure first proposed by McLaughlin and Scott [12], though there exist other derivations [7,8,9]. The hypothesis needed to arrive to (6), (7) can be expressed in several equivalent ways. Among them, the most physical one is perhaps to assume that the kink distortion is negligible and that the noise influence concentrates on the evolution of the kink center and speed in time.

Standard stochastic calculus yields the averages of the center position, the only magnitude affected by noise in the adiabatic approach:

$$< z(t) > = v_0 t, \tag{8}$$

$$\sigma_z^2(t) \equiv < z^2(t) > - < z(t) >^2 = \frac{D}{2} v_0^2\,(1 - v_0^2)\,t. \tag{9}$$

In the remaining of this report, we will compare these predictions to numerical simulations of the model for different noise intensities and initial kink speeds. These simulations were carried out by means of the generalized Strauss-Vázquez numerical scheme also used in [8,9,11]. In these previous works it was established that this procedure showed very nice consistency features, reproducing exact predictions [13] concerning the energy evolution. Besides, the scheme was rather unsensitive to changes of the number of trajectories entering the ensemble over which we average the relevant magnitudes. These verifications (see details in [9,11]) allow us to trust our numerical computations, and, hence, to achieve definite conclusions about soliton dynamics in our model. We have already given a complete account of the parameters we used in our simulations [9,11], and we will not come back to it here.

3 Results

Firstly, we concern ourselves with the weak noise regime. We define weak noises as perturbations under which solitons verify the adiabatic predictions (8), (9), if not exactly, at least as a good approximation. The weak noise-strong noise threshold must be computed from the simulations, because we can not know *a priori* the value of $2D$ for which is valid the perturbative theory. We have found that this threshold was neither a sharp value, nor the same value for all the initial speeds of solitons. Rather well, we discovered that speaking in a proper way of weak noise, requires that the soliton speed we are referring to should be specified. Nevertheless, attending to noise effects, we can make a rough classification of solitons into two main kinds, namely slow ones (initial speed $0 < v_0 < 0.5$) and fast ones ($0.8 < v_0$), having always in mind that there is a continuous transition from one class to another. If we consider slow kinks, noises under or around $2D = 0.01$ are weak. On the opposite, fast (sometimes

Table 1. Parameters characterizing the kink behavior as obtained from the simulations. All the entries are defined in the text. The noise scaled exponent is defined as the energy exponent t_e^{-1} divided by the noise strength, a quantity that it is around 0.2 except for relativistic kinks.

Noise Strength	Initial Speed	Energy fit $\xi \equiv t_e^{-1}$	Energy fit E_0	Noise Scaled Exponent	Dispersion fit δ	Dispersion fit σ_0
1.0×10^{-3}	0.00	1.678×10^{-4}	0.9432	0.1678	—†	—†
	0.05	1.885×10^{-4}	0.9438	0.1885	1.212	3.718×10^{-7}
	0.10	2.299×10^{-4}	0.9475	0.2299	1.023	2.396×10^{-6}
	0.20	1.953×10^{-4}	0.9621	0.1953	1.165	6.564×10^{-6}
5.0×10^{-3}	0.20	1.059×10^{-3}	0.9615	0.2118	1.161	3.770×10^{-5}
1.0×10^{-2}	0.00	1.741×10^{-3}	0.9468	0.1741	—†	—†
	0.05	2.547×10^{-3}	0.9418	0.2547	1.534	2.746×10^{-6}
	0.10	2.143×10^{-3}	0.9458	0.2143	1.087	2.410×10^{-5}
	0.20	1.932×10^{-3}	0.9611	0.1932	1.144	7.389×10^{-5}
2.5×10^{-2}	0.20	4.606×10^{-3}	0.9663	0.1842	1.483	1.169×10^{-4}
5.0×10^{-2}	0.20	1.042×10^{-2}	0.9599	0.2084	1.880	2.179×10^{-4}
1.0×10^{-1}	0.00	1.936×10^{-2}	0.9596	0.1936	—†	—†
	0.025	1.462×10^{-2}	0.9601	0.1462	1.757	1.144×10^{-5}
	0.05	1.718×10^{-2}	0.9387	0.1718	2.380	1.815×10^{-5}
	0.10	2.436×10^{-2}	0.9278	0.2436	2.065	1.080×10^{-4}
	0.20	1.883×10^{-2}	0.9656	0.1883	—‡	—‡
	0.50	1.845×10^{-2}	1.0711	0.1845	1.975	1.527×10^{-3}
	0.80	7.704×10^{-3}	1.5651	0.0770	1.446	1.364×10^{-3}
	0.99	3.790×10^{-4}	6.6438	0.0038	1.161	7.845×10^{-6}
2.5×10^{-1}	0.20	8.327×10^{-2}	0.8617	0.3331	2.213	1.368×10^{-3}

† The center of the kink remained at rest within an error of 10^{-11}, and therefore the dispersion results were meaningless.
‡ Not available data.

we will also name them relativistic) kinks nearly obey the adiabatic equations if noise is $2D \leq 0.1$. This conclusion can be drawn from table 1, where we summarize a number of our experiments. In the rightmost columm we show the values of two parameters, δ and σ_0, for several runnings. They were obtained by fitting power law behaviors to the dispersion $\sigma_z^2(t)$ computed directly from the simulations, that is, we supposed

$$\sigma_z^2(t) \sim \sigma_0 t^\delta \tag{10}$$

and so we got σ_0 and δ. In table 1 it can be seen that if noise is under $2D = 0.025$, almost all of the values (notice that they correspond to slow kinks) of δ are around 1, the exponent predicted in (9). We must add that the values found for σ_0 were in fair agreement with (9) as well.

It is important to stress that for each noise strength, the fastest the kink was, the closest we get to the adiabatic value for δ. This trend can be appreciated in the values we obtained when $2D = 0.1$, where we carried out the greatest number of our simulations, because it turned out to be a crucial intensity. As we mentioned above, such noise is not weak anymore for slow kinks while it does remain weak for fast ones; notice, e.g., the value $\delta = 1.161$ for a $V_0 = 0.99$ kink reported in the table. Instead, slow kinks are described by exponents δ around 2. See also that there is not a sharp transition between these different situations. Moreover, these two different behaviors are correlated with emission of radiation by the moving kink. Remember that if the equations (6) and (7) are to hold, it must be supposed that linear waves emitted by the soliton are negligible. This is not so when noise becomes strong and we could clearly see it in the simulations; at $2D = 0.1$, the amplitude of phonons emitted by slow kinks was sensitively nonzero, and they were propagating mainly backwards. That is then the reason why the adiabatic approach ceases to be valid.

Let us now briefly describe the effects of strong noises on the solitons. From the δ exponents, it is evident that, instead of the brownian-like motion of kinks around their center, characteristic of both the adiabatic approach and the experiments on weak noise, it suffers an anomalous diffusion process, characterized by $\delta \approx 2$. Besides, its speed, that it was kept practically unchanged by weak perturbations, gradually diminished. At the end of the runnings, the kink center has not travelled so far as if its velocity remained constant. Finally, large amounts of radiation appeared along the kink tails, mainly in the back one. When we further increased the noise strength, up to $2D = 1$, slow kinks were destroyed before the end of the running, while fast ones were preserved though severely distorted. By distroyed we mean that a blow up occurred, not due to intrinsic effects of the scheme but as a consequence of the dramatic deformations of the kink.

To finish this section, we must devote some explanations to the energy evolution. At first, we monitored how this quantity changed in time to check the predictions of [13], but later on we discovered that it always followed an exponential law, namely

$$E(t) \sim E_0 e^{\xi t} \equiv E_0 e^{t/t_e}. \tag{11}$$

The ξ exponents, as obtained from the simulations, are shown in the middle of table 1. A very remarkable fact is that the quantity $t_e/2D$ is almost independent of both $2D$ and v_0 (for slow kinks) and its value fluctuates around 0.2 (see corresponding column in the table). Indeed, we found that this magnitude t_e, that has the sense of a characteristic time, the time for the energy to become e times its initial value, grew essentially linearly with $2D$ for slow kinks. This is an important point, because of the following feature of the model: the blow up, i.e., the destruction of the kink, happened to take place soon after the energy reached some point around twice or three times $E(0)$. Then, had we a slow kink initially unperturbed, we could approximately predict the time it would maintain its integrity under each value of noise. As far as we know, this is the first time that such a predictive result is obtained in stochastic, nonlinear systems.

4 Discussion and conclusions

So far, we have expounded the main properties of the stochastic ϕ^4 model. A comment is now in order about possible physical reasons for the behavior of the kinks under strong noises; of course, it is obvious from the definition of weak noises that we can easily account for the phenomena related to them through the adiabatic approach. In the strong perturbation regime, the principal properties we should understand are the following:

- Anomalous diffusion ($\sigma_z(t) = \sigma_0 t^\delta$, with $\delta > 1$) happens when strong noise acts on kinks. More precisely, up to times $t \approx D^{-1}$ it can be described by the adiabatic approach, but at later times it can not: it crosses over to the anomalous diffusive regime.
- The energy growth (it was predicted in Ref.[13], although its precise form was left undetermined) is actually exponential. The exponent governing this evolution depends approximately linearly on the noise strength for slow kinks.
- For each noise strength, all of the noise effects become weaker as the initial kink speed approaches unity.

Probably, the clue to understand soliton dynamics in the stochastic ϕ^4 model is to realize that only the particles lying outside the bottom of the two potential wells are affected by noise. Consequently, at the beginning of each simulation, fluctuations act on the kink kernel and do not influence the kink tails. It is a while later, when the soliton has been propagating along the chain for some time, that the radiation developed (it can be shown [11] that the power of the process of phonon generation is constant in time) becomes visible. It seems quite evident that the time interval necessary for the radiation to be important will depend on the noise strength. This is the first part of the kink evolution, and it is the one that can be described by the adiabatic approach.

Now, let us recall that the chain particles constituting the radiation are oscillating around the bottom of a potential well. Hence, they are indeed driven by the noise, that is to say, once the kink passed over a place and excited certain amount of linear waves, they evolve undergoing the perturbation. It can be proven by linearizing (3) around the soliton [11] that the radiation evolution has to do with the so called stochastic parametrical resonance phenomenon, induced by the random term that appears in the frequency of waves. This leads to an exponential growing of the initial waves, as well as to changes in their spectrum: actually, the amplitude of phonons grows in time with an exponent that depends on their frequency, that asymptotically decreases as its inverse square. Thus, due to this stochastic parametrical resonance, the spectrum of linear noise initially generated by the kink is modified, enriching itself with long waves more effectively than with short ones. As, of course, these phonons contribute to the energy density in the chain, the above reasoning qualitatively explains the soliton dynamics at not so short times: notice that the diffusion process of the kink center is accelerated by phonons, and that energy increases with the amplitude of waves. At last, the perturbative series expansion predicts [11] that all of the changes in the kink shape (the seed from which waves begin to grow) are given by expressions whose leading factor is $(1 - v^2)$ or powers of it. Then, the fastest the initial kink is, the smaller the amplitude of the radiation induced is, and the smaller the noise effects become. This is clearly in good agreement with the outcome of our simulations and the dependence of the results on the speed.

We finish with a summary of our findings. First, using a reliable [8,9,11] numerical scheme, we have found that the features of the stochastic ϕ^4 model crucially depend on the speed the kink has initially. In spite of the fact that there is not a sharp transition between them, we can loosely speak of slow ($v_0 \leq 0.5$) and fast ($v_0 \geq 0.8$) solitons. These two classes of kinks exhibit the same evolution for different range of noises: while the adiabatic approach holds approximately for slow kinks under noises of strength $2D \leq 0.01$ and fails for larger values, fast ones verify the simple theoretical prediction up to $2D \leq 0.1$. Indeed, the validity limit of the approach is continuously related to the kink speed. Second, we have described soliton dynamics in the strong noise regime. Its main characteristics are the kink slowing down and the anomalous diffusion, that behave as t^δ where δ grows with noise strength up to values around 2. Third, we have found an empirical law for energy behavior, that is exponentially increasing, with an exponent dependent mainly on the noise strength and besides being greater for slow kinks than for fast ones. As the kink stability is correlated with the energy evolution, this gives some predictive power to our study, allowing to estimate a mean stability time for each initial kink. Finally, we have indicated some qualitatively satisfactory explanations for all these facts, based on perturbative series expansions valid for short times and on the process of stochastic parametrical resonance for later ones. These theoretical considerations, as well as other simulations concerning generalizations of the model, will be reported in full detail elsewhere [11].

Acknowledgments.

We are grateful to V.V.Konotop for very useful conversations on this work. We thank the kind permission of the Centro de Investigaciones Energéticas, Medio Ambientales y Tecnológicas (C.I.E.M.A.T, Spain) to carry out our simulations in their IBM 3090. This work has been supported in part by the Instituto de Cultura y Ciencia Soviéticas de la Universidad Complutense. A.S. was supported by a fellowship from the program Formación de Personal Investigador of the Ministerio de Educación y Ciencia of Spain.

References

1. Aubry, S.: J.Chem.Phys **62**, 3217 (1975); **64**, 3392 (1976).
2. Krumhansl, J.A., and Schrieffer, J.R.: Phys.Rev.B **11**, 3535 (1975).
3. Davydov, A.S.: *Solitons in Molecular Systems*. Reidel, Dordrecht (1985).

4. Campbell, D.K., Schonfeld, J.F., and Wingate, C.A.: Physica **9D**, 1 (1983); Rice, M.J.: Phys.Lett. **71A**, 153 (1979).
5. Pnevmatikos, St.: Phys.Lett.A **122**, 249 (1987).
6. Bass, F.G., Kivshar, Yu.S., Konotop, V.V., and Sinitsyn, Yu.A.: Phys.Rep. **157**, 63 (1988); Kivshar, Yu.S., and Malomed, B.A.: Rev.Mod.Phys. **61**, 763 (1989).
7. Bass, F.G., Konotop, V.V., and Sinitsyn, Yu.A.: Izv. Vyssh. Uchebn. Zaved., Radiofiz. **27**, 718 (1984) [Sov.J. Radiophys. Elec. **27**, 491 (1984)]; Bass, F.G., Konotop, V.V., and Sinitsyn, Yu.A.: Izv. Vyssh. Uchebn. Zaved., Radiofiz. **29**, 921 (1986) [Sov.J. Radiophys. Elec. **29**, 708 (1987)].
8. Rodríguez-Plaza, M.J., and Vázquez, L.: Phys.Rev.B **41**, 11437 (1990).
9. Sánchez, A., and Vázquez, L.: in *Lecture Notes in Physics no. 353: Nonlinear Coherent Structures*, eds. M.Barthes and J.Léon. Springer, Berlin-Heidelberg (1990).
10. Rajaraman, R.: Phys.Rep. **21**, 229 (1975).
11. Sánchez, A., Vázquez, L. and Konotop, V.V.: Phys.Rev.B (submitted); Konotop, V.V., Sánchez, A. and Vázquez, L.: Phys.Rev.B (submitted).
12. McLaughlin, D.W., and Scott, A.C.: Phys.Rev.A **18**, 1652 (1978); in *Solitons in Action*, eds. K.Longrenn and A.C.Scott. Academic, New York (1978).
13. Parrondo, J.M.R., Mañas, M., and de la Rubia, F.J.: J.Phys.A **23**, 2363 (1990).

On the Threshold of KdV Soliton Production

Kh.T. Kholmurodov

Tadjik Polytechnical Institute, 10 Kujbishev Avenue,
734041 Dushanbe, Tadjikistan

In this paper the existence of a threshold of the soliton formation for the Korteweg–de Vries (KdV) equation, $U_t+6UU_x+U_{xxx}=0$, in decaying of "zero area" pulses, $\int_{-\infty}^{+\infty} U(x,0)\,dx=0$, is established. The following zero area pulses were considered: (1) the rectangular pulse (expression (1)); (2) the "sinusoidal" pulse, $U(x,0)=U_0\cdot\sin kx$, $x\in(-\pi/k,\pi/k)$; (3) the Rayleigh pulse, $U(x,0)=U_0\cdot x\exp(-k^2x^2)$.

It is note that an approximate method for analytical study of soliton formation as a result of the decay of initial pulses has been proposed by Karpman and Sokolov [1] for the KdV equation. However their method is relevant only when the nonsoliton sector contains a small part of excitation. The decay of zero area pulses is known to be such that the non-soliton tail takes away a considerable part of excitation energy and under some conditions solitons are not formed [2,3].

In Ref. [2], by using linear representation of the KdV equation, the problem reduced to finding the discrete energy levels of the Schrödinger equation,

$$-\Psi_{xx}+U(x)\Psi=E\Psi,$$

with the zero area potential,

$$U(x)=\begin{cases}-U_0, & -a<x<0,\\ 0, & x<-a,\ a<x<b-a,\ x>b,\\ +U_0, & b-a<x<b.\end{cases}\qquad(1)$$

The following transcendent equation for the discrete energy levels E was obtained [2]:

$$|E|=(U_0-|E|)\cdot\left\{\frac{f(E)\,\mathrm{ctg}[(U_0-|E|)^{1/2}\cdot a]+1}{\mathrm{ctg}[(U_0-|E|)^{1/2}\cdot a]-f(E)}\right\}^2,\qquad(2)$$

where f is complicated function of E [2].

Table 1. The discrete energy levels E for potential (1) versus the well-barrier distance d.

$U_o = 1.$					
a					
0.5		3.5		4.0	
N					
1		2		3	
d	|E|	d	|E|	d	|E|
4.0	0.05	4.0	0.69	4.0	0.73
			0.19		0.38
					0.10
3.0	0.04	3.0	0.69	3.0	0.73
			0.18		0.38
					0.09
2.0	0.03	2.0	0.69	2.0	0.73
			0.16		0.37
					0.08
1.0	0.01	1.0	0.68	1.0	0.73
			0.0		0.34
					0.06
0.5	0.0	0.5	0.67	0.5	0.72
			0.0		0.30
					0.0
0.0	0.0	0.0	0.65	0.0	0.70
			0.0		0.21
					0.0

We have investigated equation (2) numerically. The Table 1 shows the dependences of the discrete energy levels of the well width parameter a and the well-barrier distance d=b-a (N denotes the number of the discrete energy levels).

From Table 1 for values a#Ø we find the threshold for a discrete energy level to exist.This means that the discrete energy levels do not exist sometimes depending on the well-barrier distance d.At the same time adding a second,symmetrically located barrier is well known to make the energy of the levels lower.This quite unexpected behaviour of quantum mechanical particles leads to the threshold for KdV solitons to be formed through the decay of the zero are pulses [2].

References

1.Karpman,V.I. and Sokolov,V.P. Sov.Phys.JETP 54,1568(1958);
Berezin,Y.A. and Karpman,V.I. Sov.Phys.JETP 51,1557(1966).
2.Kholmurodov,Kh.T.,et al.,Physica Scripta,v.39,pp.9-12,1989.
3.Lamb,C.L. Jr.,Elements of Solitons Theory,Wiley,New York (1980).

One-Dimensional Localization and Wave Propagation in Linear and Nonlinear Media

S.A. Gredeskul and L.A. Pastur

Institute for Low Temperature Physics and Engineering,
Lenin Avenue 47, 310 164 Kharkov, Ukraine

Abstract. The paper is a brief survey of some works on the wave propagation in the media with one-dimensional disorder. It is shown that the Anderson localization gives rise to the existence of quasi-stationary states in open disordered systems. As a result 1)the random stratified layer has effective waveguide properties, and 2) in nonlinear optical fibers the random input pulse generates soliton-like excitations.

1. Introduction

It is well-known that due to Anderson localization [1] the transmission coefficient t and transparency $T = |t|^2$ of the one-dimensional disordered segment are exponentially small (see e.g. Ref.[2]) when the segment size L tends to infinity:

$$T \sim \exp(-L/l) \tag{1}$$

(l is the localization length). This asymptotic equation and all other such formulas in this text are written with logarithmic accuracy, i.e. they mean that e.g. in the case (1) we have

$$-L \ln T = l^{-1}(1 + o(1))$$

when L tends to infinity. It follows from Eq.(1) that the difference between unity and the reflection coefficient r modulus is also exponentially small on a typical realization:

$$1 - |r| \sim \exp(-L/l). \tag{2}$$

Asymptotic equations (1) and (2) lead to interesting peculiarities in wave propagation in linear and nonlinear media. In Section 2 of the report we'll show that due to Eq.(2) in the open one-dimensional disordered system the quasi-stationary states exist [3]. They play the leading role in wave propagation in the randomly stratified layer, which is an effective fluctuation waveguide [4] (Section 3). In the last Section 4 the wave propagation in nonlinear optical fibers is considered. It is shown that in the case of the random input pulse the existence of quasi-stationary states in associated linear Dirac equation leads to creation of solitonlike excitations [5].

2. Quasi-stationary states

Let's consider the one-dimensional Schroedinger equation

$$-\frac{d^2\Psi}{dz^2} + U(z)\,\Psi = E\,\Psi \tag{3}$$

where the potential $U(z)$ differs from zero only on a segment $[0, L]$:

Springer Proceedings in Physics, Vol. 67
Editors: F. Abdullaev · A.R. Bishop · S. Pnevmatikos
Nonlinearity with Disorder

$$U(z) = \begin{cases} v(z) & z \in [0, L], \\ 0 & \text{otherwise} \end{cases} \tag{4}$$

and $v(z)$ is a spatially homogeneous random function with the correlations disappearing at the infinity. The continuous spectrum of the problem coincides with the positive semiaxis $0 < E < \infty$. Every positive value $E > 0$ is two-fold degenerated. One of the states corresponding to it has at the positive infinity only outgoing wave

$$\Psi(z, E) \sim \begin{cases} [1 - r_-(E) r_+(E)] \exp(ikz) + \\ + \frac{r_1}{r_-^*} [r_-^*(E) - r_+(E)] \exp(-ikz), & z < 0 \\ t_1(E)\, t_2(E) \exp[+ik(z-L)], & L < z \end{cases} \tag{5}$$

Here $r_+(E)$ ($r_-(E)$) are the reflection coefficients of the plane wave incident from the left (right) with energy E from semisegments $[z_0, L]$ ($[0, z_0]$), z_0 is arbitrary point within the interval $[0, L]$, $r_1(E)$ is the reflection coefficient of a plane wave incident from the left with energy E from semisegment $[0, z_0]$ and $t_{1,2}(E)$ are the transmission coefficients of the semisegments.

It is seen from Eq.(5) that on a typical realization when moduli of reflection coefficients $r(E)$ are exponentially close to unity, the amplitude $[1 - r_+(E)\, r_-(E)]$ of the initial wave becomes exponentially small in all points E_n for which the product $r_+(E)\, r_-(E)$ becomes real and positive. If the segment length L is sufficiently large in comparison with the localization length $l(E_n)$, $L \gg l(E_n)$ then using the equation for the reflection coefficient phase (Ref.[6], see also Ref.[2], p.394) we obtain

$$E_n = \left(\frac{\pi n}{L}\right)^2 [1 + o(1)]\,. \tag{6}$$

The second term in the r.h.s. of this equation essentially depends on a realization and is small in comparison with the universal first one, but not small in comparison with the distance $(2\pi^2 n/L^2)$ between the neighboring energy levels. In the ε- complex plane in the vicinities of the points E_n of the real axis there exist points ε_n, $\mathrm{Im}\, \varepsilon_n < 0$, $|\varepsilon_n - E_n| \sim \exp(-L/l)$ for which the equation

$$1 - r_-(\varepsilon_n) r_+(\varepsilon_n) = 0 \tag{7}$$

satisfies and in the expression (5) for the wave function only. outgoing waves are present. So $\Psi(z, \varepsilon_n)$ corresponds to quasi- stationary states in the quantum mechanical sense.

Some simplifications may be done in the case of semi axis when the solution of the problem (3),(4) satisfies the boundary condition

$$\left.\frac{\Psi'}{\Psi}\right|_{z=0} = -ik \frac{1 - r_-}{1 + r_-} \tag{8}$$

and the continuous spectrum is non-degenerated. In particular when $r_- = 1$

(full reflection) it may be shown that for arbitrary positive E the envelope of the state Ψ is localized within the segment $[0,L]$ (but not normalized because of oscillating tails).

3. Waveguide properties of the random stratified layer

Let's consider the random stratified layer $0 < z < L$ of a medium with the dielectric constant

$$\varepsilon(z) = \begin{cases} \varepsilon_0 + \delta\varepsilon(z) & 0 < z < L \\ \varepsilon_0 & L < z \end{cases}$$

where $\delta\varepsilon(z)$ is a random function of coordinate z with a zero mean value, correlation radius r_c and variance $\sigma_\varepsilon^2 = \langle\delta\varepsilon^2(z)\rangle$. Such a layer in the case of the infinite thickness $L \Rightarrow \infty$ is an ideal reflector for the plane wave. So if we place inside the layer the point source then its energy will be more or less canalized along the layer. We may hope that in the case of finite but large thickness this property in a certain sense will be conserved.

Let's consider now a field $G(\mathbf{R},\mathbf{R}_0)$, $\mathbf{R} = (\rho,z)$, which is produced in the layer by a monochromatic source placed at the point $\mathbf{R}_0=(0,z_0)$ above the ideally reflecting plane $z = 0$. The field G satisfies the following equation

$$\Delta G(\mathbf{R},\mathbf{R}_0) + \frac{\omega^2}{c^2}\varepsilon(z)\, G(\mathbf{R},\mathbf{R}_0) = \delta(\mathbf{R}-\mathbf{R}_0),$$

the self-adjoint boundary condition of the type of Eq.(8) with $|r_-| = 1$ and the radiation condition at the infinity. The Fourier transform $G(\mathbf{k},z)$ of the field $G(\mathbf{R},\mathbf{R}_0)$ with respect to the longitudinal coordinate ρ as a function of z coincides with the Green function of Eq.(3) satisfying the same boundary condition as $G(\mathbf{R},\mathbf{R}_0)$. The potential $U(z)$ and the energy parameter E
in this case are

$$U(z) = -E_0\,\frac{\delta\varepsilon(z)}{\varepsilon_0}, \quad E_0 = \varepsilon_0\,\frac{\omega^2}{c^2}, \quad E = E_0 - k^2. \tag{9}$$

A characteristic feature of the waveguide propagation is cylindrical divergence of the energy-flux density

$$\mathbf{S}(\mathbf{R}) = 2\,\mathrm{Im}\,[G^*(\mathbf{R})\,\nabla G(\mathbf{R})]$$

at a large distance from the source. Therefore we shall investigate two energy fluxes:

$$\Phi_d(z) = \lim_{\rho\Rightarrow\infty} \rho \int_0^z dz' \int_0^{2\pi} d\psi\, S_\rho(\rho,z'),$$

i.e. the flux through the side surface of the cylinder with the infinite radius bounded by two planes, $z' = 0$ and $z' = z$, and the flux

$$\Phi_c = \lim_{z\Rightarrow\infty}\lim_{z\Rightarrow\infty} \int S_z(\rho,z)\, d^2\rho$$

through the upper base of such cylinder with the infinite height. We obtained the exact dynamic (i.e. valid at an individual realization) expressions for the fluxes

$$\Phi_d(z) = \frac{1}{2}\sum_j \psi_j^2(z_0)\int_0^z \psi_j^2(\zeta)d\zeta, \tag{10}$$

$$\Phi_d = \lim \Phi_d(z) = \frac{1}{2}\sum_j \psi_j^2(z_0), \tag{11}$$

$$\Phi_c = \frac{1}{2}\int_0^{E_0} \psi_E^2(z_0)\, dE .$$

Here $\psi_j(z)$ are the eigenfunctions of the discrete spectrum of Eqs.(3), (9) normalized to unity and $\psi_E(z)$ are the eigenfunctions of the continuous spectrum normalized to the delta function of energy.

A measure of realizations for which the discrete spectrum is absent is exponentially small in the parameter L/r_c, therefore at the almost all realizations the pure waveguide effect ($\Phi_d > 0$) takes place connected with the existence of the discrete spectrum. But the height distribution of the energy flux $\Phi_d(z)$ [Eq.(10)] has some peculiarities. First of all the flux $\Phi_d(z)$ along the layer is not a self-averaging quantity and therefore it has the mesoscopic fine structure depending on the individual properties of a realization. The existence of such a structure gives rise a possibility of resonant spectroscopy of every realization (an analogous procedure was proposed earlier in Ref.[7]). As suggested by Eq.(10) the j-th state of the discrete spectrum contributes to the flux $\Phi_d(z)$ only when $z - z_j \geq l_j$ where l_j is the localization length. This contribution $\frac{1}{2}\psi_j^2(z_0)$ considerably differs from zero only if the localization center z_j of the state is vertically separated from the source by no more than the localization length l_j and has in this case the order of l_j^{-1}. Therefore the total energy flux along the layer Φ_d [Eq.(11)] in the thick layer with $L >> l$ consists only of a small group of waves for which $|z_0 - z_j| \leq l_j$ and is of the order of $(\Delta z)^{-1}$ – inverse distance between the neighboring localization centers.

But more interesting is a role of the continuous spectrum whose states are delocalized. The field associated with the spectrum does not stay within the layer but propagates to the outside. However we shall show that the property of open disordered systems which is responsible for the formation of the continuous spectrum results in the radical rearrangement of the structure of that part of the field which is associated with this spectrum and greatly enhances (as compared with regular structures) the waveguide effect.

Indeed, let us analyze the flux Φ_c emerging from the layer and rewrite it in the most simple case $z = 0$ as

$$\Phi_c = \frac{1}{2\pi}\int_0^{E_0} \frac{dE}{\sqrt{E}}\,\rho(E)$$

where the density $\rho(E)$ equals

$$\rho(E) = \frac{1 - |r_+(E)|^2}{|1 - r_-(E)r_+(E)|^2} . \tag{12}$$

It is seen from this relation and estimate (2) that at every realization the main contribution to the flux density Φ_c is connected with the points E_n (6). At these points the density $\rho(E)$ is exponentially large $\sim \exp(L/l)$ being however exponentially small $\sim \exp(-L/l)$ outside their vicinities with the widths of the order of $\exp(-L/l)$. As a result the total flux Φ_c is of the order of its average $\langle \Phi_c \rangle$ or (what's the same) the value $\Phi_c^{(0)}$ in a homogeneous space with $\varepsilon(z) = \varepsilon_0$. On the other hand we may use the well-known representation [8] for the field of the source G as an integral on a path in a complex plane with integrand containing in denominator the multiplier $1-r_-r_+$. Analysis of this expression shows that it reduces to the sum of the residues at the denominator zeroes ε_n, corresponding in this case to the quasi-stationary states, and to the integrals on the branching-on sides [8]. The contribution from one of these poles contains factor $\exp(i\text{æ}_n\rho)$ where $\text{æ}_n = \sqrt{E_0 - \varepsilon_n}$ has the positive imaginary part of the order of $\exp(-L/l_n)$. It means that the essential part of the field is contained in quasihomogeneous waves corresponding to the quasistationary states and attenuates at L/l exponentially large distances

$$D_n \sim L \exp(L/l_n)$$

from the source. This attenuation is connected not with the dissipation but with the upward emergence of the field. However when $\rho < D_n$ quasi-homogeneous waves are locked within the layer and therefore the last represents an effective fluctuation waveguide.

4. Soliton-like excitations created in nonlinear optical fibers by the random input pulse.

In this part we'll discuss the possibility of the soliton's creation in optical fibers described by the non-linear Schroedinger equation

$$iu_z - u_{tt} + 2|u|^2u = 0 . \tag{13}$$

Associated linear equation has the Dirac-type form

$$-i\sigma_z \frac{d\vec{\psi}}{dt} + (v(t)\sigma_x - w(t)\sigma_y)\vec{\psi} = E\vec{\psi}, \qquad \vec{\psi} = \begin{pmatrix} \psi_1 \\ \psi_2 \end{pmatrix} , \tag{14}$$

where $v(t)$ and $w(t)$ are respectively the real and imaginary parts of input signal $u(0,t)$ from Eq.(13). The values of spectral parameter E belonging to the discrete spectrum correspond to the solitons developing from the initial condition $u(0,t)$. It is well known that in this case only so-called dark solitons may exist and the necessary condition for theirs existence has the form

$$\lim_{|t|\to\infty} |u(O,t)| = u_0 \qquad (15)$$

(nonlinear excitations on a cw background). But this condition is never satisfied at the real experimental situation because the limit in l.h.s. of Eq. (15) always equals to zero. Nevertheless dark solitons (soliton-like excitations to be precise) were experimentally observed recently (see Refs. [9-11]). In particular in Ref.[11] the interpretation of the experiment was based on an assumption about the creation of dark solitons by the random input signal.

The possible mechanism of this effect may be following. All states of Eq. (14) with the random potential $v + iw$ on the whole axis are localized. So the estimate (2) for the modulus of the reflection coefficient from the disordered segment $[O,T]$ occupied by the random potential is also valid in this case. Then Eq.(7) has solutions with real parts described by the equation similar to (6) and imaginary parts which are exponentially small in parameter L/l. Such solutions relate to quasistationary states corresponding soliton-like excitations The base moment in the conclusions is the proof of the first of this statements.
Let's prove at the beginning the positiveness of the Lyapunov exponent of Eq.(14). The state $\vec{\psi}$ may be parameterized by the formulae

$$\vec{\psi} = e^{\xi - i\psi} \begin{pmatrix} 1 \\ e^{i\phi} \end{pmatrix}$$

with real-valued ξ, ψ, ϕ. It may be shown that the Lyapunov exponent

$$\gamma(E) = \lim_{|t|\to\infty} \frac{\langle\xi(t)\rangle}{t} \qquad (16)$$

equals to

$$\gamma(E) = \langle v \sin\phi - w \cos\phi \rangle_{st}$$

where the symbol $\langle\ldots\rangle_{st}$ denotes an averaging over the stationary distribution $p(\phi, v, w)$ of random functions $v(t)$, $w(t)$, $\phi(t)$.

The phase $\phi(t)$ satisfies the equation

$$2^{-1}\dot{\phi} = v \cos\phi + w \sin\phi - E.$$

If the random potentials $v(t)$ and $w(t)$ are Markovian then triad (ϕ, v, w) forms three-dimensional Markov process. Its probability density $p(\phi, v, w)$ satisfies the Fokker-Plank equation

$$\frac{\partial p}{\partial t} + \frac{\partial}{\partial\phi}\Big[2(v\cos\phi + w\sin\phi - E\Big] + \mathcal{L}_u p = 0$$

where the operator $\mathcal{L}_u$ $(u = (v,w))$ acts on the arguments v, w only. So for the Lyapunov exponent (16) we obtain the expression

$$\gamma(E) = 4\sum_{m=1}^{\infty}\int \frac{p_m^*(u)\,\mathcal{L}_u\,p_m(u)}{p_0(u)}\,d^2u$$

where we introduced the Fourier-transforms of the stationary distribution $p(\phi, u)$:

$$p_m(u) = \int_0^{2\pi} e^{-im\phi} p(\phi, u)\, d\phi.$$

Following the proof scheme used in Ref.[12] (see also Ref.[2], P.P.136-139) it is easy to show that the Lyapunov exponent $\gamma(E)$ is positive at all energies with the probability 1. Then in accordance with the result of Ref.[13] for the pure point character of the spectrum of the Eq.(14) e.g. with the white noise Gaussian potential on the whole axis it is sufficient to prove that the phase probability density $p(\phi)$ is a continuous function. But the last statement immediately follows from the formula (A6) for $p(\phi)$ in Ref.[12].

The next step is to obtain an explicit expressions for the real parts E_n of energies $\mathcal{E}_n$ of quasi-stationary states . As the results (6) these expressions may be found in quasi-classical region from the equation for the reflection coefficient phase and have the form (compare with Eq.(6))

$$E_n = \frac{\pi n}{T}(1 + o(1))$$

where T is the "length" of the disordered segment. Imaginary parts of $\mathcal{E}_n$ are of the order of $\exp(-T/\tau)$, where τ is the characteristic localization time analogous to the localization length l for the Schroedinger equation (3). In the case of real ($w = 0$) white noise potential when

$$\langle v(t)\rangle = 0, \qquad \langle v(t)\, v(t')\rangle = 2D\,\delta(t-t') \tag{17}$$

the localization time τ in quasi-classical region $D \ll E$ equals to $\tau(E) = 1/2D$ and in the main approximation in the relation D/E doesn't depend on the spectral parameter E. This result is also valid for the random potential with the short but finite correlation time τ_c when $\tau_c \ll \omega_c^{-1}$ (where ω_c and τ_c are characteristic values of the amplitude of the correlation function $\langle v(t)v(t')\rangle$ and the scale of its changing. But the region of its validity in this case is $D \ll |E| \ll \tau^{-1}$.

The spectral density

$$|a(E)|^{-2} = |t(E)|^2 \tag{18}$$

which describes the pulse intensity (we use standard notations - see e.g. Refs.[5, 14]) may be rewritten in the form

$$|a(E)|^{-2} = \frac{|t_1 t_2|^2}{|1 - r_+ r_-|^2}$$

where $t_{1,2}$ and $t_{1,2}$ have the similar sense as in Eq. (5). But this spectral density essentially differs from the spectral density (12) used in the section 3. Indeed as we saw above the density (12) represented the system of exponentially high peaks with the exponentially narrow widths concentrated near *all* points E_n. On the contrary the density (18) doesn't exceed unity and receive the values of the order of unity only for the special E_n. Introducing the quantity $\Delta = -\ln|r|$ describing the reflection coefficient modulus let's write the expression (18) in the form

$$|a(E)|^{-2} = \frac{4\Delta_+\Delta_-}{(\Delta_+ + \Delta_-)^2}$$

Using this formula we see that $|a(E)|^{-2}$ has the order of unity only if two exponentially small (on the typical realizations) values coincide with the exponential accuracy. Therefore only solitons excitate which correspond *not all* quasi-stationary states but *only resonant transparent ones* with $|t|^2 \sim 1$.

References

1. P.W.Anderson, Phys. Rev. 109, 1492 (1958).
2. I.M.Lifshits, S.A.Gredeskul and L.A.Pastur, *Introduction to the Theory of Disordered Systems* (Wiley, New-York, 1988).
3. S.A.Gredeskul and V.D.Freylikher, Izv. Vysch. Uchebn. Zaved. "Radiofiz." 33, 61 (1990).
4. V.D.Freylikher and S.A.Gredeskul, J. Opt. Soc. Am. A7, 868 (1990).
5. S.A.Gredeskul, Kivshar Yu.S. and Yanovskaya M.V., Phys. Rev. A41, 3394 (1990).
6. G.Papanicolaou, SIAM J. Appl. Math. 21, 13 (1971).
7. M.Ya.Azbel, Solid State Commun. 45, 527 (1983).
8. L.M.Brekhovskikh, *Waves in Layered Media* (Nauka, Moscow, 1957).
9. D.Krökel, N.J.Halas, G.Guliani and D.Grishkowski, Phys, Rev. Lett. 60, 29 (1988).
10. A.M.Weiner, J.P.Heritage, R.J.Hawkins, R.N.Thurson, E.M.Kirschner, D.E.Leaird and W.J.Tomlinson, Phys. Rev. Lett. 61, 2445 (1988).
11. A.E.Grudinin, E.M.Dianov, A.M.Prokhorov and D.V.Khaidarov, Pis'ma Zh. Tech. Fiz. 14, 1010 (1988).
12. E.N.Bratus', S.A.Gredeskul, L.A.Pastur and V.S.Schumeiko, Teor. i Mat. Fiz. 76, 401 (1988).
13. S.Kotani, Contemp. Math. 50, 277 (1986).
14. V.E.Zakharov, S.V.Manakov, S.P.Novikov and L.P.Pitaevskii, *Theory of Solitons. The Inverse Scattering Method* (Nauka, Moscow, 1980).

Annihilation of Topological Chiral Solitons

M.M. Musakhanov[1] *and I.V. Musatov*[2]

[1]Physics Department, Tashkent State University, Vuzgorodok, 700 095 Tashkent, Uzbekistan

[2]Research Institute of Applied Physics, Tashkent State University, Vuzgorodok, 700 095 Tashkent, Uzbekistan

Abstract. We investigate annihilation process of nucleons in chiral soliton model. Soliton – antisoliton pair is shown to decay into mesons at range of about 1fm. Contribution of annihilation channel into elastic scattering is discussed.

1. Introduction

The chiral models like the Skyrme model [1] seem to be quite appropriate for description of low - energy hadron physics. They represent a theory for interacting pion field, which has topological soliton solutions. The chiral theory may be considered as being a low - energy limit of QCD, which is believed to be a theory of strong interaction. The chiral Lagrangian contains only physical degrees of freedom (meson fields) and realizes fundamental symmetries of initial QCD Lagrangian. In particular, it reproduces the phenomena of spontaneous breaking of chiral symmetry. The connection of chiral theory with QCD was proved by calculation in the $N_c \to \infty$ limit [2].

In paper [3] analogous model, derived from low - energy effective action of instanton model of QCD vacuum, was proposed.

In chiral theory nucleons are identified with low - lying rotating states of topological soliton (Skyrmeon).

When the nucleon and antinucleon came close to each other, they interact by means of their external pion clouds. Of course, because of pion exchange, the nucleon states can transform into other higher rotational states (for example, Δ - isobar). If distance between centres of nucleon and antinucleon becomes less then some critical radius, this state looks like excited state of pion field over perturbative vacuum rather then nucleon – antinucleon pair. Then this states decays into pions.

So, important parameter of theory is the critical radius R_0, which is to be identified with the radius of annihilation. At large distances ($R > R_0$) solitons can be described by collective coordinates – coordinates of centres and parameters of orientation. At small distance ($R < R_0$) the collective coordinates lose their sense and total pion field of the resulting configuration must be considered as being a fluctuation over

Springer Proceedings in Physics, Vol. 67 Nonlinearity with Disorder
Editors: F. Abdullaev · A.R. Bishop · S. Pnevmatikos

trivial vacuum. The critical radius R_0 will be shown to be determined from the form of classical soliton – antisoliton potential and appears to be $R_0 \sim 1$fm.

On the other hand, the phenomenology of the low - energy nucleon – antinucleon interactions gives the value of the radius of annihilation of about $0.8 \div 1.2$fm[4,5,6].

Natural way of describing the process of soliton – antisoliton annihilation is the path integral method. Using of it we construct in this paper the low - energy amplitude of $N\bar{N}$ - annihilation and obtain expression for contribution of the annihilation channel to the amplitude of low - energy elastic scattering. The crucial point of our discussion is the necessity of redefining of semiclassical quantization procedure at small distances.

The paper is organized as follows. In section 2 we reproduce main formulae of chiral soliton model. In section 3 we give some arguments for redefinition of collective integrating measure at small distances. In section 4 it is shown how the annihilation amplitude can be evaluated.

2. General formulae

We will consider $SU(2)$ – chiral theory.

The Lagrangian of chiral model has the form of nonlinear σ – model that describe isovector pion fields $\vec{\pi}$:

$$L = \frac{1}{4} f_\pi^2 \, Tr\, L_\mu L^\mu + \frac{1}{32e^2} Tr[L_\mu, L_\nu]^2, \tag{1}$$

where $L_\mu = U^+ \partial_\mu U$ and $SU(2)$ -valued matrix U may be represented as $U = e^{i\vec{\tau}\vec{\Phi}} = \frac{1}{f_\pi}(\sigma + i\vec{\tau}\vec{\pi})$. The soliton solution with unit barion number is obtained with the hedgehog ansatz [1]:

$$U_{N,0}(\vec{x}) = \exp(i\vec{\tau} \cdot \frac{\vec{x}}{|\vec{x}|}\, \theta(|\vec{x}|)), \tag{2}$$

where θ is a scalar function. We will use of following determination of antiskyrmeon solution: $U_{\bar{N}} = U_N^+$.

The equations of motion stay invariant under translation of centre of the soliton and under rotation, giving with multiplying U by any constant $SU(2)$ – matrixes A: $U = A^+ \, U_0 \, A$.

3. Quasi-zero mode integration

Let us consider quantization of static soliton- antisoliton configuration. We start from configuration corresponding to well separated soliton-antisoliton pair, using of "production ansatz":

$$U(\vec{x};\, \vec{R}_N, \vec{R}_{\bar{N}}, \vec{\omega}_N, \vec{\omega}_{\bar{N}}) = \\ U_N(\vec{x} - \vec{R}_N, \vec{\omega}_N,) \cdot U_{\bar{N}}(\vec{x} - \vec{R}_{\bar{N}}, \vec{\omega}_{\bar{N}}) = e^{i\vec{\tau}\vec{\Phi}_{N\bar{N}}(\vec{x})}, \tag{3}$$

$$U_N(\vec{x}, \vec{\omega}) = A^{+}(\vec{\omega}_N) \cdot U_{N,0}(\vec{x}) \cdot A(\vec{\omega}_N),$$

where $A(\vec{\omega}) = e^{i\vec{\tau}\vec{\omega}}$ is the matrix of the orientation of soliton. In the $|\vec{R}_N - \vec{R}_{\bar{N}}| \to \infty$ limit the configuration of field $\vec{\Phi}_{N\bar{N}}$ is the solution of equation of motion with energy $E = 2M_N$ and yields the local minimum of action. Note that absolute minimum in the zero topological sector is achieved on trivial configuration $\vec{\Phi} = 0$.

When one quantizes fluctuations of pion field around classical field (3) $\vec{\phi}(\vec{x}, t) = \vec{\Phi}(\vec{x}, t) - \vec{\Phi}_{N\bar{N}}(\vec{x})$, $\vec{\pi} = f_\pi \vec{\phi}$ in semiclassical approximation he faces with the well known problem of the existing of flat directions (valleys) in functional space. It means that there are zero frequencies in the spectrum of operator $\delta^2 S \,/\, \delta\vec{\Phi}^2$. Because of this fact, the integration along such direction is not Gaussian and must be carried out exactly. Usual way to treat zero modes is to rewrite these integrals (using the Faddeev – Popov technique) as path integral over collective coordinates $\vec{R}_i(t)$, $\vec{\omega}_i(t)$[7]

$$\int D\vec{\pi}\; e^{iS(\vec{\Phi}_{N\bar{N}}, \vec{\pi})} = \int \prod_{i=N,\bar{N}} D\vec{R}_i(t) D\vec{\omega}_i(t) \int D\vec{\pi}'\; e^{iS(\vec{\Phi}_{N\bar{N}}, \vec{\pi})}. \tag{4}$$

Boundary conditions corresponding to soliton – antisoliton pair are implied. Prime in the right side denotes that integration is carried out over functional subspace of nonzero modes. We will be interested in zero modes only, regarding nonzero ones as giving rise just to renormalization of the soliton mass.

Factorization of the integration measure over collective coordinates of well-separated soliton- antisoliton pair onto product of measures of individual solitons $d\mu = d\mu_N \cdot d\mu_{\bar{N}}$ in Eq.(4) is caused by the fact that $S(\vec{\Phi}_{N\bar{N}})$ is invariant under independent translations of centres and under rotations of soliton or antisoliton. When static field configuration describes the soliton and antisoliton placed at large but finite distance it does not satisfy the equation of motion because of its nonlinearity.

Corresponding action now is the function of the relative distance $\vec{R} = \vec{R}_N - \vec{R}_{\bar{N}}$ and of the matrix of relative orientation $A = A_N^+(\vec{\omega}_N) \cdot A_{\bar{N}}(\vec{\omega}_{\bar{N}}) = e^{i\vec{\tau}\vec{\omega}} = a_0 + i\vec{\tau}\vec{a}$:

$$S_{N\bar{N}} = S_N + S_{\bar{N}} - V_{N\bar{N}}(\vec{R}, \vec{\omega}). \tag{5}$$

Eq.(5) defines the soliton – antisoliton potential $V_{N\bar{N}}$. It was evaluated numerically in Ref.[8] as a function of the relative distance and of two orientation parameters $\vec{a}^2$ and $T = \frac{\vec{a}\cdot\vec{R}}{|\vec{a}|\cdot|\vec{R}|}$ (See Fig 1).

We see that now some valleys in functional space is not flat. However, at sufficiently large R the variational derivative $\delta^2 S \,/\, \delta\Phi^2$ taken at variation of $\vec{\Phi}$ along valley is much less then that along transverse direction. Therefore, quasi-zero modes can be treated by the same way as zero ones [9], with account of the potential $V_{N\bar{N}}$. We conclude that measure of integration over all collective coordinates corresponding to both zero and nonzero modes at large R reads:

$$\int \prod_{i=N,\bar{N}} D\vec{R}_i(t) D\vec{\omega}_i(t) \times$$

$$\exp\left\{ i \int dt \left(-2M_N + \frac{M_N}{2}(\dot{\vec{R}}_N^2 + \dot{\vec{R}}_{\bar{N}}^2) + \frac{I_N}{2}(\dot{\vec{\omega}}_N^2 + \dot{\vec{\omega}}_{\bar{N}}^2) - V_{N\bar{N}}(\vec{R}, \vec{\omega}) \right) \right\}. \tag{6}$$

Evidently, it has a form of path integral describing scattering of two quantum particles – nucleon and antinucleon, interacting with each other by means of potential $V_{N\bar{N}}$.

Expression (6) needs some redefinition at small distances between centres of soliton and antisoliton. The main observation is that the field configuration (3) at $R \to 0$ is not distinguished from meson field fluctuation around trivial vacuum $\vec{\Phi} = 0$. In fact, since there is strong attraction in the system when R becomes less then radius of soliton R_0 [8] (See Fig.1), the quantization procedure breaks due to loss of stability along quasi-zero mode direction. This configuration is far from being a solution of equation of motion and cannot be used as stationary phase point in perturbation theory expansion. Collective coordinates method (and, generally speaking, soliton – antisoliton picture) in this situation is not adequate. Moreover, when soliton and antisoliton come close, regions with opposite topological charges are nearly overlapping. Therefore, a region in functional space corresponding to the soliton – antisoliton configurations at $R < R_0$ is to be regarded as perturbative

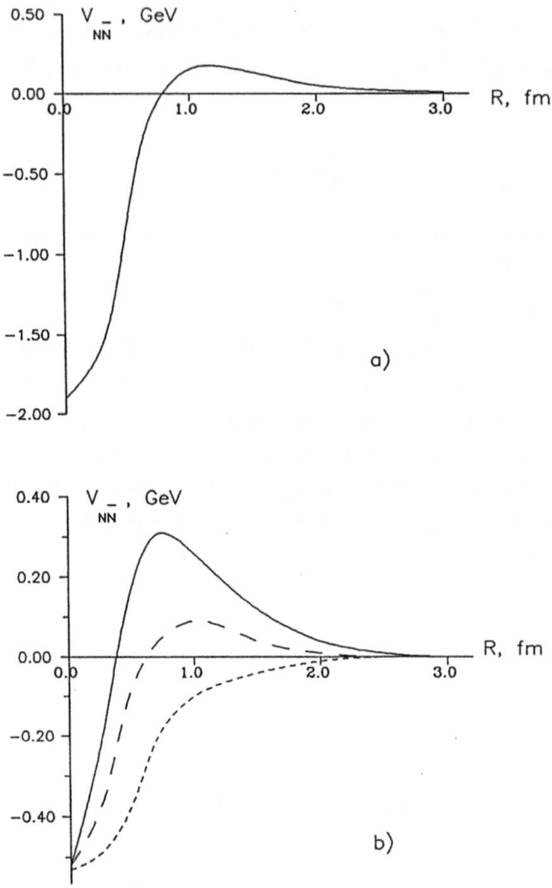

Figure 1: The classical skyrmeon – antiskyrmeon potential $V_{N\bar{N}}$ [8] as a function of relative distance R at various values of the orientation parameters:
a) $\vec{a}^2 = 0,\ T = 0$;
b) $\vec{a}^2 = 0.5$; solid line: $T^2 = 0$, dashed line: $T^2 = 0.5$, dotted line: $T^2 = 1.0$.

quantum fluctuations over trivial vacuum $\vec{\Phi} = 0$. However, these fluctuations are already taken into account in path integral analogous to left side of Eq. (3) but with zero boundary conditions. To avoid double count, this region must be excluded from Eq. (6) (ansatz proposed by Jevicki [10]):

$$d\mu = d\mu_N d\mu_{\bar{N}} \cdot e^{-iV_{N\bar{N}}(\vec{R},\vec{\omega})} \cdot \Theta(R - R_0). \tag{7}$$

Here $\Theta(x)$ is the standard Θ – function and R_0 is naturally to consider as the radius of nucleon. Hence we have get the short-range pure imaginary $N\bar{N}$ potential, which satisfactorily agrees with phenomenological data, provided that we put $R \sim 0.8 \div 1.2$fm.

Assembling Eqs.(6) and (7) we see that contribution of annihilation channel into scattering amplitude is provided by trajectories coming from $R(t = -\infty) = \infty$ to sphere $R(t_0) = R_0$ at some moment t_0 and outgoing from sphere $R(t_0') = R_0(t_0 > t_0')$ to $R(+\infty) = \infty$.

4. Annihilation

Let us come to evaluating of the annihilation amplitude. We shell consider the amplitude of soliton – antisoliton annihilation to vacuum in presence of external source:

$$Z(J, \Psi_{N\bar{N}}) = \int D\vec{\Phi} e^{i\left(S(\Phi)+\int d^3\vec{x}dt\vec{\Phi}(\vec{x},t)\vec{J}(\vec{x},t)\right)} \cdot \Psi^{in}_{N\bar{N}}. \qquad (8)$$

$\Psi^{in}_{N\bar{N}} = e^{i(P_N X + P_{\bar{N}} Y)} . \mathcal{D}(\vec{\omega_N})\mathcal{D}(\vec{\omega_{\bar{N}}})$ is the in-state wave function of nucleon – antinucleon pair, where $\mathcal{D}$is the wave function of soliton rotating state. Note that $\mathcal{D}$ is fixed by the spin – isospin quantum numbers of barion.

It is implied that integration in Eq. (8) is carried out over fields that satisfy boundary conditions corresponding to presence of soliton – antisoliton pair at $t = -\infty$ and pion radiation condition at $t = +\infty$. Process of annihilation proceeds through two stages. At the first one soliton and antisoliton move from in-state ($R = \infty$) and at a moment t_0 turn up at a critical distance $R(t_0) = R_0$. Then at the second stage field configuration $\Phi(\vec{x}, t) = \Phi(\vec{x}; \vec{R}_i(t_0), \vec{\omega}_i(t_0))$(See Eq. (3)) decays into pions. Summing up amplitudes with different t_0 and extracting variables related to overall orientation we obtain the amplitude of $N\bar{N}$ annihilation into n pions (in CM frame):

$$\int d\Omega\, \delta(|\vec{R}| - R_0)\langle n\,\vec{\pi} \mid \vec{\Phi}(\vec{x}, t_0;\, \vec{R}, \vec{\omega})\,\rangle\, \langle\, \vec{R}, \vec{\omega}, t_0 \mid \Psi^{in}_{N\bar{N}}\,\rangle, \qquad (9)$$

where $d\Omega$ denotes the integration measure over all variables: $d\Omega = \int d^4X d^4Y \int dt_0 d^3\vec{x} \int d^3\vec{R}\, d^3\vec{\omega}$.

The matrix element

$$\langle\, \vec{R}, \vec{\omega}, t_0 \mid \Psi^{in}_{N\bar{N}}\,\rangle = \langle\, \vec{R}, \vec{\omega} \mid e^{-H_{int}(t_f - t_0)} \mid \Psi^{in}_{N\bar{N}}\,\rangle \qquad (10)$$

has a sense of value of the $N\bar{N}$ quantum-mechanical wave function at the point $\vec{R}, \vec{\omega}$ and at time t_0 . Of course, it contains admixture of $N\bar{\Delta}$, $\Delta\bar{N}$, $\Delta\bar{\Delta}$ - states. It can be evaluated, for example, by means of solution of the Lippman – Shwinger equation with long-range $N\bar{N}$ potential defined in Eq. (5) and will be done elsewhere.

Now, we wish to compute the amplitude of decay of the state $\vec{\Phi}$. We can do this by using of the method of the oscillator quantization [11]. So, if interaction of pion in final state be neglected, field $\vec{\Phi}$ may be presented as a set of quantum oscillators $\vec{\Phi}_{\vec{k}}$ numbered by $\vec{k}$ with frequency $\omega_k = |\vec{k}|$, where $\vec{\Phi}_{\vec{k}}$ is the Fourier component of field

$$\vec{\Phi}(\vec{x},t) = \int \frac{d^3\vec{k}}{(2\pi)^{3/2}} \left(\vec{\Phi}(\vec{k})e^{-i(\vec{k}\vec{x}-\omega_k x_0)} + \vec{\Phi}^{+}(\vec{k})e^{+i(\vec{k}\vec{x}-\omega_k x_0)}\right). \quad (11)$$

Our task reduces to evaluation amplitude of transition from the state with definite coordinate $\vec{\Phi}_{\vec{k}}$ into the state with definite number of quanta, which is a value of the oscillator wave function $\Psi_n(\vec{\Phi}_{\vec{k}})$. Finally, we obtain

$$\begin{aligned} &\langle n\,\vec{\pi} \mid \vec{\Phi}(\vec{x},t_0;\ \vec{R},\vec{\omega})\,\rangle = \\ &N \cdot \tfrac{1}{\sqrt{n!}} \cdot \textstyle\prod_i^n \sqrt{2\omega_i}\, f_\pi\, \vec{\Phi}(\vec{k}_i) \cdot \exp\left(i \sum_i^n(\omega_i t_0 - \vec{k}_i\vec{x}_i)\right). \end{aligned} \quad (12)$$

N is the normalization constant:

$$N = \exp\left\{-f_\pi{}^2 \int \frac{d^3\vec{k}}{(2\pi)^{3/2}} \omega_k\, \vec{\Phi}^{+}(\vec{k})\vec{\Phi}(\vec{k})\right\}. \quad (13)$$

Integration over $d^3\vec{x}$ and dt_0 yields δ^4 -function of momentum – energy conservation .

Turning back to formula Eq.(7), we now able to write the short-range $N\bar{N}$ – potential in some details. Repeating discussion of the end of Section 2 and of Section 3, we immediately obtain contribution of the annihilation channel into amplitude of elastic scattering:

$$\begin{aligned} &\int d\Omega d\Omega'\ \delta(|\vec{R}|-R_0)\,\delta(|\vec{R}'|-R_0) \cdot \langle\, \Psi^{out}_{N\bar{N}} \mid \vec{R}', \vec{a}', t_0' \,\rangle \times \\ &\langle\, \vec{\Phi}(\vec{x}', t_0';\ \vec{R}', \vec{\omega}') \mid \vec{\Phi}(\vec{x}, t_0;\ \vec{R}, \vec{\omega})\,\rangle \cdot \langle\, \vec{R}, \vec{\omega}, t_0 \mid \Psi^{in}_{N\bar{N}} \,\rangle. \end{aligned} \quad (14)$$

To evaluate remaining matrix element $\langle\, \vec{\Phi}(t_0') \mid\ \vec{\Phi}(t_0)\,\rangle$ one can also apply the method of Ref.[11].

The formulae (7),(12) and (14) are in accordance with the results obtained by using of coupled channels method in nonrelativistic quantum mechanics (See, for example, Ref.[12]). Namely, Ψ -function $N\bar{N}$ -system can be decomposed into barion and meson channels by means of projecting operators $\hat{P}_N$ and $\hat{P}_\pi$: $|\Psi\rangle = \hat{P}_N|\Psi\rangle + \hat{P}_\pi|\Psi\rangle$. Then product of matrix elements in Eq.(12) in those notation is $\langle n\vec{\pi}|\ V\ |\hat{P}_N|\Psi\rangle$. Procedure of excluding of meson channels gives rise to potential in $N\bar{N}$ -channel $\hat{P}_N\, U\, \hat{P}_\pi(H - E_\pi)^{-1}\hat{P}_\pi V \hat{P}_N$ that is to compare with our short – range potential in Eq.(7) and (14) . Here V denotes potential in Shródinger equation for the total wave function $|\Psi\rangle$.

5. Conclusion

We proposed the method of describing of the $N\bar{N}$ -annihilation, based on chiral soliton model. The crucial point was shown to be redefinition of the quantization procedure at small distances between centres of soliton and antisoliton. This leads to the occurrence of the pure imaginary contribution in nucleon – antinucleon potential. Radius of new interaction in our model is determined as range of region, where strong attraction violates a stability of the semiclassical quantization procedure. Of course, this definition is rather rough, but yet provides the usual in chiral models accuracy.

Large number of acts of the $N\bar{N}$ – decay into pions are known to go through formation of resonance states. Our model can be easily extended to incorporate the description of vector mesons [13]. To include strangeness into the model one must to consider $SU(3)$ flavor group instead $SU(2)$ [14].

References

[1] T.H.R.Skyrme, Nucl. Phys. **31** (1962) 556.

[2] E.Witten, Nucl. Phys., **B223** (1983) 422, ibid. 433.

[3] D.I.Dyakonov, V.Yu.Petrov, Nucl. Phys. **B272** (1986) 457.

[4] C.Amsler, Preprint CERN/EP-86-178, Geneva, 1986.

[5] B.O.Kerbikov, L.A.Kondrat'uk, M.G.Sapozhnikov, in: Proceedings of the XXIII "Particle Physics" LNPI Winter School, Leningrad, 1988, p.53.

[6] J.Haidenbauer, T.Hippchen, K.Holinde, J.Speth, Z. Phys. **A334** (1989) 467.

[7] J.-L.Gervais, A.Jevicki, Nucl. Phys. **B110** (1976) 93, ibid. 113.

[8] M.M.Musakhanov, A.Rakhimov. (to be published)

[9] I.I.Balitsky, A.V.Yung, Phys. Lett. **B168** (1986) 113; Nucl. Phys. **B274** (1986) 475.

[10] A.Jevicki, Phys. Rev. **D21** (1980) 992.

[11] R.P.Feynman, A.R.Hibbs. Quantum Mechanics and Path Integrals. NY: McGraw-Hill; 1965.

[12] A.M.Badalyan, M.I.Polikarpov, Yu.A.Simonov, Phys. Rep. **82** (1982) 31.

[13] U.G.Meißner, I.Zahed, Phys. Rev. Lett. **56** (1986) 1035.

[14] I.Zahed, G.E.Brown, Phys. Rep. **142** (1986) 1.

On Some Probabilistic Problems in the Theory of Quadratic Operators

T.A. Sarymsakov and N.N. Ganikhodzhaev

Institute of Mathematics, Uzbek Academy of Sciences, 700 143 Tashkent, Uzbekistan

A Central Limit Theorem and conditions which imply the ergodic principle are studied for quadratic process which are related to quadratic operators in the same way as Markov processes are related to linear operators.

1. Quadratic operators. A mapping P of the simplex $T^{n-1}=\{(x_1,x_2,\dots,x_n) : x_i\geq 0, \sum_{i=1}^{n} x_i=1\}$ into itself of the form

$$P: \quad x_k'=\sum_{i,j=1}^{n} P_{ij,k}x_i x_j, \text{ where } P_{ij,k}\geq 0, \quad \sum_{k=1}^{n} P_{ij,k}=1$$

is called a quadratic operator. Quadratic operators are very important in the study of different models in biology, chemistry (see, for example [1]).

Let us show that Boltzmann model in statistical mechanics considered in [2] can be also described by quadratic operators. Consider a uniform gas composed of spherical molecules with the same radius and mass. Let the velocity space of the molecules which we assume to be $\mathbb{R}^3$ be partioned into n regions $E_1, E_2,\dots,E_n$, $E_i\cap E_j=\emptyset$ for $i\neq j$ and $\bigcup_{i=1}^{n} E_i=\mathbb{R}^3$. Put for $i=1,2,\dots,n$

$$x_i(t)=\left\{\begin{array}{l}\text{fraction of molecules which have velocites}\\ \text{lying in region } E_i \text{ at time } t.\end{array}\right\}$$

Under appropriate assumptions one can assert that

$$\left\{\begin{array}{l}\text{number of l-m}\\ \text{collisions in the time unit}\end{array}\right\}=\mu_{lm}x_l x_m$$

for some $\mu_{lm}>0$. Denote by P^i_{lm} that fraction of E_l molecules which are scattered into E_i under collisions with E_m molecules. Evidenttly ($P^1_{lm}, P^2_{lm}, \ldots, P^n_{lm}$) $\in T^{n-1}$ for any pair l,m. Moreover

$$\left\{\begin{array}{l}\text{number of } E_l \text{ molecules scattered into} \\ \text{region } E_i \text{ due to l-m collisions per time unit}\end{array}\right\} = P^i_{lm}\,\mu_{lm} x_l x_m,$$

from which evidently

$$x_i(t+1) - x_i(t) = \sum_{l,m} a^i_{lm} x_l x_m \quad , \ i=1,2,\ldots,n$$

for $a^i_{lm} = 0.5((P^i_{lm}+P^i_{ml}) - (\delta_{il}+\delta_{im}))\mu_{lm}$, where δ_{ij} is Kronecker delta (see [2]). From this equality it's easy to obtain the following equality

$$x_i(t+1) = \sum_{l,m=1}^{n} P_{ml,i}\, x_l(t) x_m(t)$$

for $P_{ml,i} = a^i_{lm} + \delta_{ml,i}$, where

$$\delta_{ml,i} = \begin{cases} 1 & , \text{ if } m=l=i \\ 0.5 & , \text{ if } m \neq l \text{ and either } m=i \text{ or } l=i \\ 0 & , \text{ else} \end{cases}$$

The coefficients $P_{ml,i}$ satisfy the conditions:

$$1)\ P_{ml,i} \geq 0 \ ; \qquad 2)\ P_{ml,i} = P_{lm,i} \ ; \qquad 3)\ \sum_{i=1}^{n} P_{ml,i} = 1.$$

In the case of biological models it is natural to assume the time to be discrete and in case of Boltzmann's model to be continuous.

2. Quadratic processes. In [3,4] we introduced quadratic processes, which are related with quadratic operators as well as Markov processes - with linear operators.

Let ($E, \mathcal{F}$) be a measured spase, $\mathcal{M}$.- the family of all probability measures on ($E, \mathcal{F}$) and let $\{P(s,x,y,t,A)\}$ be a family of functions determined for $t-s \geq 1$, $x, y \in E$ and arbitrary measurable set $A \in \mathcal{F}$. If this family of functions satisfy the following conditions:

I) $P(s,x,y,t,A) = P(s,y,x,t,A)$ for all $x,y \in E$, $A \in \mathcal{F}$;

II) $P(s,x,y,t,.) \in \mathcal{M}$ for all $x,y \in E$;

III) $P(s,x,y,t,A)$ as a function of two arguments x and y is measurable with respect to $(E \times E, \mathcal{F} \otimes \mathcal{F})$ for any fixed $A \in \mathcal{F}$; for any initial measure $m_0 \in \mathcal{M}$ and arbitrary $s < \tau < t$ such that $t-\tau \geq 1$, $\tau - s \geq 1$ either

$$\mathbf{IV_A})\ P(s,x,y,t,A)=\int_E\int_E P(s,x,y,\tau,du)P(\tau,u,v,t,A)m_\tau(dv),$$

where

$$m_\tau(B)=\int_E\int_E P(0,x,y,\tau,B)m_0(dx)m_0(dy)\quad \text{for any } B\in\mathcal{F},$$

or

$$\mathbf{IV_B})\ P(s,x,y,t,A)=\int_E\int_E\int_E\int_E P(s,x,z,\tau,du)P(s,y,v,\tau,dw)\times P(\tau,u,w,t,A)m_s(dz)m_s(dw)$$

then the process defined by functions $P(s,x,y,t,A)$ is called respectively type A or type B quadratic process in accordance with the form of the fundamental equation $\mathbf{IV_A}$ or $\mathbf{IV_B}$. The functions $P(s,x,y,t,A)$ are called transition functions. In this definition the transition function $P(s,x,y,t,A)$ means the probability of the appearance of some element from the set $A\in\mathcal{F}$ in the moment t by the interaction of elements x and y at the moment s.

Equations $\mathbf{IV_A}$ and $\mathbf{IV_B}$ can be interpreted as different rules for the appearance of the "grandson" . For these equations one can give a chemical treatment also. Namely, the appearance of a particle in reactions passing in ordinary chemical kinetics are described by equation of type $\mathbf{IV_A}$, and the equations of type $\mathbf{IV_B}$ reflect the appearance of particles in processes of catalysis.

The condition

$\mathbf{V})\ P(t,x,y,t+1,A) = P(0,x,y,1,A)$ for all $t\geqslant 1$

is naturally consistent with the biologic consideration; it can be interpreted as homogeneity of processes for the time unit interval.

The condition $\mathbf{V}$) does not imply the homogeneity of the process in general.

The family of quadratic processes can be decomposed into three classes:

1) homogeneous processes, for which $P(s,x,y,t,A)$ depends only on t-s for any s and t such that $t-s\geqslant 1$;

2) homogeneous for unit interval of time which satisfy condition $\mathbf{V}$), but not homogeneouses;

3) procesess not belonging to the 1 and 2 classes, will be called homogeneous.

In what follows by quadratic processes we shall mean the processes of classes 2; the transition functions of quadratic processes of type A we shall denote by P, and for processes of type B - by $\tilde{P}$.

3. Differential equations. In papers [3,4] differential and integro - differential equations with respect to transition functions were deduced similarly to [5].

Let $E = \mathbb{R}$ and $A_z = (-\infty, z]$ for $z \in \mathbb{R}$. Put

$F(s,x,y,t,z) = P(s,x,y,t,A_z)$, $\tilde{F}(s,x,y,t,z) = \tilde{P}(s,x,y,t,A_z)$.

It is clear that F and $\tilde{F}$ are distribution functions of quadratic processes of type A and type B respectively. Let f (respectively $\tilde{f}$) be then density of distribution function F (respectively $\tilde{F}$). The following differential equations with respect to f and $\tilde{f}$ were deduced in [4]:

$$\frac{\partial f(s,x,y,t,z)}{\partial t} = N(t,z)\, f(s,x,y,t,z) + A(t,z)\, \frac{\partial f(s,x,y,t,z)}{\partial z} +$$

$$+ B^2(t,z)\, \frac{\partial^2 f(s,x,y,t,z)}{\partial z^2}\ ;$$

$$\frac{\partial \tilde{f}(s,x,y,t,z)}{\partial s} = -\tilde{A}(s,x,z)\frac{\partial \tilde{f}(s,x,y,t,z)}{\partial x} - \tilde{A}(s,y,z)\frac{\partial \tilde{f}(s,x,y,t,z)}{\partial y}$$

$$-\tilde{B}^2(s,x,z)\frac{\partial^2 \tilde{f}(s,x,y,t,z)}{\partial x^2} - \tilde{B}^2(s,y,z)\frac{\partial^2 \tilde{f}(s,x,y,t,z)}{\partial y^2}\ .$$

4. Averaging of quadratic processes . In [6] the following theorem was proved.

THEOREM 1. *Let* $P(s,x,y,t,A)$ *(respectively* $\tilde{P}(s,x,y,t,A)$*) be transition functions determining quadratic process of type* A *(respectively* B*) and* $m_0 \in \mathcal{M}$ *be a initial distribution. Then the averages*

$$Q(s,x,t,A) = \int_E P(s,x,y,t,A)\, m_s(dy), \qquad \left[\ \textit{respectively}\right.$$

$$\left.\tilde{Q}(s,x,t,A) = \int_E \tilde{P}(s,x,y,t,A)\, m_s(dy)\ \right]$$

are the collection of transition functions of some Markov process's with same initial distribution m_0.

Using this theorem it is natural to define quadratic Wiener, quadratic Poisson and other processes as quadratic processes, whose averages are the corresponding Markov processes.

THEOREM 2. *Quadratic process defined by densities of transition functions*

$$f(s,x,y,t,z)=2^{t-s-1}\exp\left[-\frac{4^{t-s-1}}{2^{2(t-s)-1}-1}\left(z-\frac{x+y}{2^{t-s}}\right)^2\right]\times$$

$$\times\left[(2^{2(t-s)-1}-1)\pi\right]^{-0.5}$$

is a diffusion process.

The diffusion models play an important role in the population genetics (see, for example [7]).

5.Quadratic chains. Let $E=\{1,2,\ldots,n\}$ and let $x^{(0)}=(x_1^{(0)},x_2^{(0)},\ldots,x_n^{(0)})\in T^{n-1}$ be an initial distribution on E. For arbitrary quadratic operator S put $x^{(k+1)}=Sx^{(k)}$, $k=0,1,2,\ldots$. We say that a sequence of random variables $\xi_0,\xi_1,\xi_2,\ldots$, defined on E forms a quadratic chain with discrete time, if $P(\xi_k=i)=x_i^{(k)}$ for all $i=1,2,\ldots,n$ and $k=0,1,2,\ldots$.

In the space of realization of quadratic chain $\{\xi_k\}$, $k=0,1,2,\ldots$; $\Omega=\prod_{i=0}^{\infty}E_i$, where $E_i=E$ for all i, define finite-dimensional distributions in the following way:
for any cylindrical set put

$$P_{x^{(0)}}(\xi_k=i_0,\ \xi_{k+1}=i_1,\ldots,\xi_{k+l}=i_l)= \tag{1}$$

$$=\sum_{m_1,m_2,\ldots,m_l} P^{[k,k+1]}_{i_0m_1,i_1}\,P^{[k+1,k+2]}_{i_1m_2,i_2}\cdots P^{[k+l-1,k+l]}_{i_{l-1}m_l,i_l}\times$$

$$\times\, x_{i_0}^{(k)}x_{m_1}^{(k)}x_{m_2}^{(k+1)}\cdots x_{m_l}^{(k+l-1)}$$

where $P^{[s,t]}_{im,j}=P(s,i,m,t,j)$ and $x^{(0)}$ is a initial distribution on E .

THEOREM 3. *The collection of finite-dimensional distribution (1) is compatible.*

Thus by Kolmogorov theorem's unique measure $P_{x^{(0)}}$ on space of realization Ω exists which is compatible with finite-dimensional distributions (1). This measure will be called quadratic measure. Theorem 1 implies

THEOREM 4. *Quadratic measure* $P_{x^{(0)}}$ *is a Markov measure.*

Let us consider the question of validity of the Central Limit Theorem to quadratic chains. By theorem 4 quadratic chains are usually nonstationary Markov chains, and for these chains the mentioned problem was studied in many works (see e.g.[8]). In the case of the quadratic chain the ergodicity coefficient introduced in [8] is connected in natural way with corresponding quadratic operator and the trajectory $\{x^{(k)}\}$ $k=0,1,2,\ldots$ of this quadratic operator. If the set E consists of two elements, i.e. n=2, then the behavior of the trajectory of the quadratic operator can be easily observed, and this enables us to investigate completely the problem of validity of Central Limit Theorem.

Let $E=\{1,2\}$ and $x^{(0)}=(x,1-x)$ be an initial distribution on E. Put $P_{11,1}=a$, $P_{12,1}=P_{21,1}=b$, $P_{22,1}=c$ and $\Delta=4(1-a)c+(1-2b)^2$. Then $0<\Delta<5$ and any value from this interval is attained: for $0<\Delta<4$ all trajectories converge to the fixed point and for $4<\Delta<5$ there exists a loop of second order to which all trajectories converge, except the trajectory which stationary coincides with the fixed point. For the corresponding quadratic chain the ergodicity coefficients $\alpha^{(k)}$, $k=0,1,2,\ldots$ can be calculated explicitly

$$\alpha^{(k)}=\min(1+b-c+(a-2b+c)x_1^{(k)};\quad 1-b+c-(a-2b+c)x_1^{(k)}).$$

From this and theorem 1 in [8] we have

THEOREM 5. *Let* $E=\{1,2\}$ and $P_{11,1}=a$, $P_{12,1}=P_{21,1}=b$, $P_{22,1}=c$ *define a quadratic operator on the space of distributions on* E. *Further, let* $\{\xi_k\}$, $k=0,1,2,\ldots$, *form a quadratic chain. If* $\Delta\neq 1$ *then for* $n\to\infty$

$$P\left[\frac{S_n-MS_n}{\sqrt{DS_n}}<t\right]\to\frac{1}{\sqrt{2\pi}}\int_{-\infty}^{t}e^{-y^2/2}dy$$

uniformly with respect to t, *where* $S_n=\xi_0+\xi_1+\ldots+\xi_n$.

6. Ergodic principle. We shall say that transition functions $P^{[m,n]}(x,y,A)$ satisfies the ergodic principle, if

$$\lim_{n\to\infty}\left[P^{[m,n]}(x,y,A) - P^{[m,n]}(u,v,A)\right]=0$$

for any $x,y,u,v \in E$ and $A \in \mathcal{F}$ (compare with [5]).

THEOREM 6. *Let* $P(s,x,y,t,A)$ *be a family of transition functions of type* A *or type* B *quadratic process . If there exist a measurable set* $A_0 \in \mathcal{F}$ *with positive measure and numbers* λ_n $(0<\lambda_n<1)$ *such that:*

(a) $P^{[n,n+1]}(x,y,A_0) > \lambda_n$ *for any* $x,y \in E$;

(b) $\sum_n \lambda_n = \infty$;

(c) $n \sum_{k=1}^{n} (1-\lambda_k) \to 0$

then for the family $\{P(s,x,y,t,A)\}$ *the ergodic principle is valid.*

REFERENCES

1. Kesten H. Quadratic transformation: A model for population grouth I. Adv. Appl. Prob. 1970, v. 2, No 1, P. 1-82.
2. Jenks R.D. Quadratic differential systems for interactive population models. J. Diff. Equation, 1969, v. 5, No 3, P. 497-514.
3. Sarymsakov T.A., Ganikhodzhaev N.N. Analytic methods in the theory of quadratic stochastic operators. Soviet Math. Dokl., 1989, v. 39, No 2, P. 369-373.
4. Sarymsakov T.A., Ganikhodzhaev N.N. Analytic methods in the theory of quadratic stochastic operators. J. of Theoretical Prob. 1990, v. 3, No 1, P. 51-70.
5. Kolmogorov A.N. On analytic methods in the probability theory. Uspekhi Mat. Nauk 1938, No 5, P 5-51.
6. Ganikhodzhaev N.N. On averaging of quadratic stochastic processes. Dokl. Acad. Nauk of Uzbek SSR, 1989, No 10, P. 7-9, (in Russian).
7. Kimura M. Diffusion models in population genetics. J. Appl. Prob., 1969, v. 1, P. 177-232.
8. Dobrushin R.L. Central Limit Theorem for non-stationary Markov chains. Teor. Veroyat. i Primenen.1956, v. 1, No 1, P. 72-89, (in Russian).

Superexponential Damping of Mean Field Propagating in Randomly Inhomogeneous Medium with Anomalously Large-Scale Nonuniformities

A.Yu. Val'kov, V.P. Romanov, and T.I. Tipyasova

St. Petersburg State University, Physics Department,
198 904 St. Petersburg, Russia

Abstract. Waves propagation in a medium with anomalously large-scale random nonuniformities is considered. An instantaneous-homogeneous model is analyzed in detail. It is shown that in this case a mean field decay is superexponential. The result is generalized for media with singular spectrum of correlation function type of $q^{-\nu}$, $2\leqslant\nu<3$, as $q\to 0$.

It is known that the mean field $\langle u(r)\rangle$, propagating in a random medium, decayes as the exponential function of r -Bugger's damping law [1]. However, recently great interest is attached of systems with anomalously large-scale fluctuations (such as liquid crystals, phase transition points, etc.) in which damping decrement calculated on a formal way is infinite (infrared divergence) [2]. The authors [3], by means of diagrams' infrared singularities summation for the correlation function $\psi(r) \sim 1/r$, $r\to\infty$, show that mean field asymptotic behavior in this case has the superexponential character $\exp(-\tau r \ln r)$. Our work is devoted to common analysis of the superexponential damping problem for stochastical inhomogeneous systems with anomalously large-scale of nonuniformities.

Let us consider a scalar harmonic wave field $u(\vec{r},t)=$ $=u(\vec{r})\exp(-\iota\omega t)$. The amplitude $u(\vec{r})$ obeyes the Helmholts' equation

$$\left[\Delta + k_0^2 \left(1 + \varphi(\vec{r}) \right) \right] u(\vec{r}) = 0, \qquad (1)$$

where $k_0=\omega\sqrt{\varepsilon_0}/c$-wave number, c-wave velocity, ω-frequancy, $\varphi(\vec{r})$-random field. Below it will be supposed that $\varphi(\vec{r})$ is Gauss' random field with $\langle\varphi\rangle=0$, $\langle\varphi(\vec{r}_1)\varphi(\vec{r}_2)\rangle=\psi(\vec{r}_1-\vec{r}_2)$, here the angular brakets denote the statistical average.

The mean Green's function of the equation (1) is given by the formula [1]

Springer Proceedings in Physics, Vol. 67 Nonlinearity with Disorder
Editors: F. Abdullaev · A.R. Bishop · S. Pnevmatikos

$$\langle G(r) \rangle = \frac{1}{4\pi^2 \iota r} \int_{-\infty}^{\infty} \frac{q\, e^{\iota q r}\, dq}{q^2 - k_0^2 - \Sigma(q)} \quad . \tag{2}$$

Here Σ is the self energy part. Diagrammatically Σ is represented by the sum of strongbound graphs:

$\sum =$ 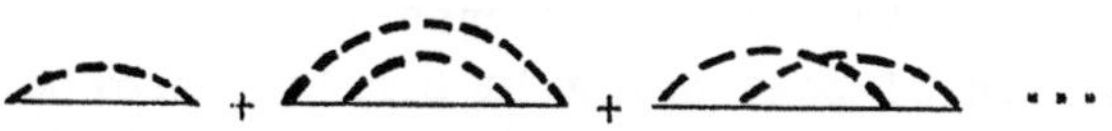 ... , (3)

where the line ——— corresponds to the Green's function $G_0(r)=(4\pi r)^{-1}\exp(\iota k_0 r)$ of the equation (1) when $\langle\varphi\rangle=0$. The line - - - corresponds to the correlation function $\psi(r)$.

Behaviour of $\langle G(r)\rangle$ is determined by singular points in the complex plane q of the spectrum $\langle G(q)\rangle=(q^2-k_0^2-\Sigma(q))^{-1}$.

1) If $\langle G(q)\rangle$ has a n-order pole in the point $q = k_{eff} = k' + \iota k''$ then its contribution in the asymptotic behaviour of $\langle G(r)\rangle$ is given by the expression

$$\langle G(r)\rangle \sim Z_p^{(n)} \frac{e^{\iota k' r}}{4\pi r} r^{n-1} e^{-k'' r} \quad , \tag{4}$$

where

$$Z_p^{(n)} = 2k_{eff} n \iota^{n-1} \left[2k_{eff}^{2-n}(\delta_{1n}+\delta_{2n}) - \frac{\partial^n \Sigma}{\partial k_{eff}^n} \right]^{-1}, \text{here } \delta_{mn} = \begin{cases} 0\ , & m \neq n \\ 1\ , & m = n \end{cases}$$

2) If $\langle G(q)\rangle$ has a ramification point in $q=k_{eff}$ type of $(q-k_{eff})^{-1+\nu}\ln^{\mu}(q/k_{eff}-1)f(q)$, then the corresponding contribution is

$$\langle G(r)\rangle \sim Z_R \frac{e^{\iota k' r}}{4\pi r} \ln^{\mu}(k_{eff} r) e^{-k'' r} r^{-\nu} \quad , \tag{5}$$

where $Z_R=2k_{eff}f(k_{eff})e^{-\iota\pi(\mu+\nu/2)}/\Gamma(1-\nu)$, $\Gamma(x)$- gamma function.

A situation is frequently encountered in which asymptotic behaviour of $\langle G(r)\rangle$ is determined by the first order pole (n=1 in (3)). In order to find its position in complex plane we have to resolve the despersion equation

$$q^2 - k_0^2 - \Sigma(q) = 0 \quad . \tag{6}$$

In the case $\Sigma \ll k_0^2$ we have $k_{eff} \approx k_0 + (2k_0)^{-1}\Sigma(k_0)$. If Bourret's approximation [4] is used for Σ (instead of the sum (3) only its first term is used) and the formula $\mathrm{Im}(p^2-k_0^2-\iota 0)^{-1} = \pi\delta(p^2-k_0^2)$ is taken into account then in lowest order we have

$$k'' = \frac{1}{2} k_0^3 \pi \int \frac{d\vec{q}}{(2\pi)^3}\, \psi(\vec{q})\ \ \delta\left[(\vec{k}_0-\vec{q})^2-k_0^2\right] \quad ,\ k' \approx k_0 \ . \tag{7}$$

It is necessary to note that in this case the value $2k''$ is equal to the total section of scattering in Born's approximation. There are some systems in which spectrum of the correlation function $\psi(q)$ has singularity type of $1/q^{\nu}$ as $q \to 0$.The case $\nu=2$ takes place for transferring fluctuations of an order parameter in nematic liquid crystals and magnetics.The case $\nu=1$ takes place for longitudinal fluctuation in these systems.The case $\nu=2-\eta$ takes place in second—order phase transition points (usual situation in physical systems is $\eta>0$, but sometimes the case $\eta<0$ is considered).

There is divergence of the integral in equation (7) as $q \to 0$, when formula (7) is used for the calculation damping decrement in the case $\nu \geqslant 2$. This divergence is caused by an anomalous large contribution to scattering comes from the large-scale nonuniformities. In order to analyze such a situation let us consider a model of a medium with fluctuations, which have infinite correlation radius. In this medium every point correlates with others equally independent of their position in space. The correlation function which corresponds to this model is

$$\psi(\vec{q}) = (2\pi)^3 b^2 \delta(\vec{q}) \qquad , \qquad \psi(\vec{r}) = b^2 . \tag{8}$$

In this case for the Bourret's approximation

$$\Sigma(q) = b^2 k_0^4 G_0(q) \tag{9}$$

and the singular points are first order poles $k_{(1,2)p} = k_0\sqrt{1 \pm b}$. In this approximation

$$\langle G(r) \rangle = \frac{1}{8\pi r} \sum_{j=1,2} \exp(\iota k_{jp} r) \ . \tag{10}$$

For this model we can easily find in closed form the solution in nonlinearity Kraichnan's approximation [6] (Σ = 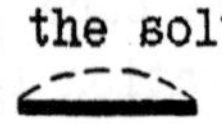 , where the line ▬▬ corresponds to the mean Green's function $\langle G \rangle$). In this approximation

$$\Sigma(q)=b_0^2 k_0^4 \langle G(q)\rangle \ , \tag{11}$$

and Dyson's equation becomes the square equation for $\langle G(q)\rangle$

$$\langle G(q)\rangle^{-1}= q^2 - k_0^2 - b^2 k_0^4 \langle G(q)\rangle \ . \tag{12}$$

The points $k_{(1,2)R}=k_0\sqrt{1\pm 2b}$ are the root type ramification points ($\mu=0, \nu=3/2$ in (5)) for the spectrum $\langle G(q)\rangle$. If $|k_{1R}-k_{2R}|r \ll 1$, then $\langle G(r)\rangle \sim G_0(r)$; but if $|k_{1R}-k_{2R}|r \gg 1$, then [7]

$$\langle G(r)\rangle \sim -\frac{1}{4\pi r}\ \frac{1}{\sqrt{2\pi b}\ b k_0^{9/4} r^{3/2}} \sum_{j=1,2} k_0^{3/4}\ \exp[\iota(k_{jR}r+(-1)^j\pi/4). \tag{13}$$

Thus, there is no usual exponential damping in Bourret's approximation and in Kraichnan's approximation. Both of these approximations display beatings for $b \ll 1$ and there is the weak algebraic decay $r^{-3/2}$ (on the background of the usual "geometric" decay r^{-1}) in Kraichnan's approximation .

The remarkable feature of the model which was considered is that one can find the exact expression for $\langle G(r)\rangle$ without any approximation since all integrations in the diagram expansion for $\langle G(q)\rangle$ in the cáse $\psi(\vec{q}) \sim \delta(\vec{q})$ can be done in explicit form.Result for $\langle G(q)\rangle$ is

$$\langle G(q)\rangle = G_0(q) \sum_{n=0}^{\infty} (2n-1)!!\,(G_0^2(q) b^2 k_0^4)^n \ . \tag{14}$$

In the derivation of formula (14),it was taken into account that in the considering situation each n-order diagram made equal contribution $G_0(q)(G_0^2(q)b^2k_0^4)^n$ in the spectrum $\langle G(q)\rangle$. Expansion on the powers of the value $(G_0^2 b^2 k_0^4)$ has radius of convergence equal to zero. But if we will take it as an asymptotic expansion then by using the identity

$$\sqrt{2\pi}(2n-1)!! = \int_{-\infty}^{\infty} dx\ \exp(-x^2/2)\ x^{2n}$$

we can carry out the sum. Then we obtain[8,9]

$$\langle G(q)\rangle = \frac{1}{\sqrt{2\pi}\, b} \int_{-\infty}^{\infty} \frac{\exp(-x^2/2b^2)}{q^2-k_0^2-xk_0^2-\iota o}\, dx \quad . \tag{15}$$

By making the Fourier transformation, we get for mean Green's function [5,8]

$$\langle G(r)\rangle = \frac{1}{\sqrt{2\pi}\, b} \int_{-\infty}^{\infty} dx\, \frac{1}{4\pi r}\, e^{\iota k_0 r \sqrt{1+x}}\, e^{-x^2/2b^2} \quad . \tag{16}$$

Note that last formula can be deduced on the base of almost obvious reasoning. The point is that the model with the correlation function $\psi(\vec{q})=(2\pi)^3 b^2 \delta(\vec{q})$ corresponds to the situation when the probability of any nonuniform fluctuation is equal to zero and only the uniform fluctuation $\varphi=\varphi_{|\vec{q}=0}$ is possible. In this case the distribution function for the state $x=\varphi_{|\vec{q}=0}$ is

$$\rho(x) = \frac{1}{\sqrt{2\pi}\, b}\, e^{-x^2/2b^2} . \tag{17}$$

Now note that the formula

$$G(r) = \frac{1}{4\pi r}\, e^{\iota k_0 r \sqrt{1+x}} \tag{18}$$

in fact defines Green's function of an arbitrary nonaveraged state with $\varepsilon=\varepsilon_0(1+\varphi)$. If we will do statistical averaging of the last formula with the distribution function (17), then we get the formula (16).

Let us use the steepest descent method in order to find the asymptotic behavior of the integral (16) when $k_0 r \gg 1$. By the substituting $\sqrt{1+x}=t$ the cubic equation $t_*^3+t_*+k_0 r b^2/r=0$ for stationary points is obtained. As it follows from the integration counter shape in the t-plane, the asymptotic expression of (16) will be defined by the root t_* such that $3\pi/2 < \arg t_* < 5\pi/3$. The uniform representation of b^2 is:

$$\langle G(r)\rangle \sim Z(r)\, \frac{1}{4\pi r}\, \exp(\iota\Phi(r)-h(r)) \;, \tag{19}$$

where

$$Z(r)=[(1+x_0^2)(1+3x_0^2)^{-1/2}(1+3x_0^2/4)^{-1/2}]^{1/2}\ , \tag{20-a}$$

$$\Phi(r)=k_0 r(1+3x_0^2/4)^{3/2}/(1+x_0^2)-\Phi_0\ , \tag{20-b}$$

$$h(r)=k_0 r x(1+3x_0^2/2)/[4(1+x_0^2)]\ , \tag{20-c}$$

$$\Phi_0=\tan^{-1}\left[\frac{x_0(5+3x_0^2)/2}{(1+x_0^2)(1+3x_0^2)^{1/2}+(1-x_0^2)(1+3x_0^2/4)^{1/2}}\right]- \tag{20-e}$$

$$-\tan^{-1}\left[\frac{x_0(1+3x_0^2/4)^{1/2}}{1+x_0^2/2}\right].$$

x_0-real root of the equation $x_0^3+x_0-k_0rb^2/2=0$. In particular, if $k_0rb^2 \ll 1$ then ,by taking in $z(r),\Phi(r),h(r)$ of the first terms their expansions on this small parameter, we obtain the approximate formula

$$\langle G(r)\rangle \approx \frac{1}{4\pi r}\exp(\iota k_0 r-k_0^2 r^2 b^2/8)\ . \tag{21}$$

In an analogous way for the mean field $\langle u(z)\rangle$,propagating along z,we have

$$\langle u(z)\rangle \sim \exp(\iota k_0 z-k_0^2 z^2 b^2/8)\ . \tag{22}$$

In the contrary limit $k_0rb^2 \gg 1$ exponent power in (19) is $\iota\Phi(r)-h(r)\sim 3b^{2/3}(k_0r)^{4/3}(\iota\sqrt{3}-1)/8\sqrt[3]{2}$, $h(r)\gg 1$ and the field becomes negligible.

Thus,the asymptotic behavior of the exact solution is principally different from the asymptotic behavior in Bourret's and Kraichnan's approximations.There are no beatings in exact solution and at the same time there is the unusual super-exponential damping. As it follows from (4), (5) such damping as (21), (22) cannot be obtained from the consideration of any simple singularities of $\langle G(q)\rangle$ laying in an area of limited value q,e.g. $\langle G(q)\rangle$ in the considered problem has its singularity only in $q=\infty$. However, owing to the integral (2) convergence $\langle G(q)\rangle$ has to go to zero as q go to infinity along the real axis. Conseqently, $q=\infty$ is an essential singularity point for $\langle G(q)\rangle$.

Let us discuss a physical mechanism of the unusual super-exponential damping in (21),(22). The normal exponential damping of the mean field $\langle u(z)\rangle \sim \exp(-\tau z)$ is caused by energy losses of incident field propagating along z due to scattering in all directions. In our model there is multiple forward scattering and there is no scattering in other directions. The resulting field is the superposition of the forward scattered waves which have stochastically shifted phases. The damping of the mean field is connected with the transformation of coherent waves into diffusive ones with the same direction of the propagation.

In the case of a medium with correlation function different from $\delta(\vec{q})$, there is both departure from incident direction of propagation (the scatteing in all directions) and wave phase shifts of the forward scattering waves.If a part of large-scale nonuniformities in the spectrum $\psi(q)$ is considerable so that the total section of scattering diverges in Born's approximation, then the resulting damping law will be intermediate between $\exp(-\tau r)$ and $\exp[-(\tau r)^2]$. It is natural to suppose that the spectrum $\langle G(q)\rangle$ in this situation will have only singularity in $q=\infty$,as it was in the case $\psi(\vec{q}) \sim \delta(\vec{q})$. The use of Dyson's resummation for the pertubance theory expansion is noneffective in such a situation. It is clear from the fact that all $(2n-1)!!$ n-order diagrams were equal to each other (both strongbound and weakbound).

If waves propagating along z through a stochastically random layer $0\leqslant z\leqslant z_0$ are considered,then it is effective to find the mean field in the exponential form

$$\langle u\rangle=u_0(z)\,\exp[\Phi_u(z)] \tag{23}$$

and to expand $\Phi_u(z)$ on the powers b^2

$$\Phi_u(z) = b^2 v_1(z) + b^4 v_2(z) + \dots . \tag{24}$$

In the lowest order in b^2 we obtain

$$\Phi_u(z)=\int\frac{d\vec{q}}{(2\pi)^3}K_z(\vec{q},z_0)\psi(\vec{q})\,, \tag{25}$$

where

$$K_z(q,z_0)=-(2\alpha)^{-1}k_0^3\left\{\frac{\sin^2[(\alpha-\beta)z_0/2]}{(\alpha-\beta)^2}+\frac{\sin^2[(\alpha+\beta)z_0/2]}{(\alpha+\beta)^2}+\right. \quad (26)$$

$$\left.+\frac{i}{2}\left[\frac{z_0}{\alpha-\beta}-\frac{\sin(\alpha-\beta)z_0}{(\alpha-\beta)^2}+\frac{z_0}{\alpha+\beta}-\frac{\sin(\alpha+\beta)z_0}{(\alpha+\beta)^2}\right]\right\},$$

where $\alpha=\sqrt{k_0^2-q_\perp^2}$, $\beta=k_0-q_\parallel$, $\mathrm{Im}\,\alpha\geqslant 0$; $z\geqslant z_0$. By using (23)-(26) for the correlation function

$$\psi(q)=b^2r_0^3(qr_0)^{-\gamma}f(qr_m), \quad (27)$$

where r_0 is a dimensional parameter, $f(0)\neq 0$, $0\leqslant\gamma<3$, we get for the limit $z\gg r_m$

$$\langle u(r)\rangle\approx U_0\exp[ik_0z-H(z)], \quad (28)$$

where

$$H(z)=\begin{cases}\sigma_B z/2 & 0\leqslant\gamma<1;\\ \sigma_B z/2-s_1\ln(z/r_m) & \gamma=1;\\ \sigma_B z/2+s_2(z/r\ )^{\gamma-1} & 1<\gamma<2;\\ s_3(z/r_0)\ln(z/r_m)+\sigma_1 z & \gamma=2;\\ s_2(z/r_0)^{\gamma-1}+\sigma_2 z & 2<\gamma<3;\end{cases} \quad (29)$$

Here

$$\sigma_1=\sigma_{01}+\sigma_0\ ; \qquad \sigma_2=\sigma_{02}+\sigma_0; \qquad ;\sigma_B=\frac{1}{4}k_0^2\int\frac{d\vec{q}_\perp}{(2\pi)^2}\psi(\vec{q}_\perp,0),$$

$$\sigma_{01}=s_3r_0^{-1}[\ln(2k_0r_m)+C-1],\quad \sigma_{02}=s_3r_0^{-1}(2k_0r_0)^{2-\gamma}/(2-\gamma),$$

$$\sigma_0=\frac{\pi}{2}k_0^3\int\frac{d\vec{q}}{(2\pi)^3}[\psi(q)-b^2r_0^3(qr_m)^{-\gamma}f(0)]\delta[(\vec{k}_0-\vec{q})^2-k_0^2],$$

$$s_1=2A_0f(0),\quad s_2=2A_0f(0)\Gamma(1-\gamma)\sin[\pi\gamma/2(2-\gamma)],\quad s_3=\pi A_0f(0),$$

$A_0=(k_0rb/4\pi)^2$, $C=0.5772...$- Euler's constant.

References

1. S.M.Rytov, Yu.A.Kravtsov, and V.I.Tatarskii, Introduction to Statistical Radiophysics, Part 2, Random Fields [in Russian], Nauka, Moskow (1978).
2. L.D.Landau and E.M.Lifshitz, Electrodynamics of Continuos Media [in Russian], Pergaman,1984.
3. L.Ts.Ajemian, A.N.Vasiljev,Yu.M.Pis'mak, Teor. Math. Fiz. 68, 323 (1986) [in Russian].
4. R.C.Bourret, Nuovo Cimento 26, 1 (1962).
5. N.A.Nemov, Opt. Spectr. 45, 617 (1978) [in Russian].
6. R.H.Kraishnan, J. Math. Phys. 2, 124 (1961).
7. N.A. Armand, V.N.Sekistov, Radiotech. Electr. 24, 615 (1979) [in Russian].
8. V.I.Tatarskii, Zh.Eksp.Teor.Fiz. 4, 1399 (1964) [in Russian].
9. A.L.Efros, Zh. Eksp. Teor. Fiz. 59, 880 (1970) [in Russian].
10. L.Ts.Ajemian, A.N.Vasiljev, M.M.Perekalin, H.Y.Reittu, Zh. Eksp.Teor.Fiz. 95, 1278 (1989) [in Russian].

Part II

Physical Applications

Nonlinearity, Disorder, the Spread of Neolithic Farming, and the Origin of the Indo-European Languages

M.H. Cohen

Exxon Research and Engineering Company, Clinton Township, Route 22 East, Annandale, NJ 08801, USA

The replacement of mesolithic hunters and gatherers by neolithic farmers in Europe is a very clean example of technology-based growth and change in human society. Recently, a remarkable quantitative model, the wave-of-advance model, has been put forward by Ammerman and Cavalli-Sforza (AC)[1,2,3] to describe this transition and its genetic consequences. Renfrew has elaborated the model to provide first an explanation of the origin of the Indo-European languages [4] and subsequently of the other two members of the Nostratic superfamily of languages.[5] The Fisher equation employed by AC does not, however, incorporate the spatial inhomogeneity of the food production potential of the neolithic farming technology within Europe. We generalize the Fisher equation to include the effects of such heterogeneity on demic diffusion and population growth. Our most important result is the prediction of internal boundaries, subsistence boundaries, around, e.g., mountain ranges and in the North within which the farming population density vanishes. These boundaries offer the possibility of explaining within the framework of the wave-of-advance model, such anomalies as the Basque language.

1. Introduction

Hunters and gatherers (HG) were widespread over the world by 12,000 B.P.,[6] their distribution limited primarily by glaciation or its consequences. However, they disappeared from Europe starting around 8000 B.P.,[7] lingering within the alpine regions for about 500 years longer than in the surrounding regions. Their disappearance coincided with the spread of neolithic farming into Europe.[8] The boundary of the region occupied by farmers advanced with roughly constant speed $v \simeq 1$ km/yr.[3,9]

There are two mechanisms by which farming might have spread and hunting and gathering disappeared. The first is simply by the diffusion of technology. The indigenous hunters and gatherers could have learned of farming from the invading farmers, adopted it, and competed successfully with the farmers for lebensraum. This, as we shall discuss in more detail below, would have led to substantial genetic and linguistic heterogeneity within Europe.

The second is by population displacement. The indigenes might not have adopted farming. There would have ensued a 50-fold increase in the population density of the invading farmers over that of indigenous HG. The indigenous stock would have been overwhelmed, leading to relative genetic and linguistic uniformity.

The facts, as they are summarized by AC [2] and Renfrew,[4,5] accord far better with uniformity and the population displacement mechanism. The problem of the neolithic transition in Europe thus becomes one of developing a quantitative theory of the population displacement mechanism. In §II, we review a model, the wave-of-advance model, introduced by AC in 1979 [1] to provide that quantitative basis. They employ the Fisher equation, which we discuss in §III. That equation is incomplete as regards the spatial heterogeneity of population growth. In §IV, therefore, we refine the model, introducing simple population-biology and cultural considerations which provide a basis for introducing spatial heterogeneity. Perhaps the most important result is the establishment of internal subsistence boundaries across which there is no net population flow and on and within which the density of farmers vanishes. These boundaries surround regions ill-suited to neolithic agriculture such as mountains or are northern limits to population spread. In the final section, §V, we list those elements of the theory still to be constructed and suggest that the subsistence boundaries offer a natural explanation for the existence of such anomalies as the Basque language.

2. The Wave of Advance Model

There are two ways in which a population of neolithic farmers could displace a pre-existing mesolithic population of hunters and gatherers. The first is by migration involving rapid, long-distance movements of substantial numbers of persons and attendant abrupt changes in the archeological record. The second is by slow, gradual, and continued movement into new areas with subsequent growth of population density N to the saturation population density N_s, the carrying capacity of the terrain reachable with the neolithic food-production technology. AC argue that there is no evidence for migration in the archeological record [1,2] and that the record instead supports the second way [7,8], which they term demic diffusion. They propose a wave of advance of the neolithic farmers at constant speed v to have occurred via demic diffusion and subsequent population growth. Relative genetic uniformity follows immediately, as later gene distributions evolved from the gene distribution of the small seed population at the origin of the wave of advance.

Renfrew then pointed out that the wave-of-advance model implies that the neolithic farmers would speak languages forming a family which evolved from the single language originally spoken by the seed population at the origin of the wave of advance. He then made the natural identification with the Indo-European language family; that is, the neolithic farmers carried the Indo-European languages into Europe and into the steppes of Asia.[4] In this way, he accounted for the linguistic uniformity of Europe.

It is important to emphasize that by the time they were displaced, the indigenous hunter-gatherer (HG) population must have become linguistically and genetically heterogeneous. Presumably, the world's population of HG had, by mesolithic times, spread out from a single focus, presumably in Africa [10], via an analogous wave of advance. They must have reached their carrying capacity everywhere except at the frontier of their advance well before the start of the wave of advance of the neolithic farmers (NF) into Europe. There is ethnographic evidence that HG's tend to move primarily within relatively stable bounded territories once their carrying capacity is reached [11]. Conditions are therefore present for substantial genetic and linguistic drift within territories or regions over the interval of 5000 to 7000 years or longer between the two waves of advance, HG and NF. Thus, in the NF wave of advance, a relatively homogeneous exogenous population overwhelmed a heterogeneous indigenous population.

Archeolinguistic analysis [12] provides evidence that the HG spoke nostratic in Western Asia by 14,000 BP and that nostratic began to break up into the nostratic superfamily of languages around 13,000 BP. The individual members of the superfamily Afro-Asiatic, Indo-European, and Dravidian or Elamite-Dravidian were established as individual languages by 11,000 BP for the first and earlier than 9000 BP for the latter two. Afro-Asiatic began to break up into a family of languages around 9000 BP.

Renfrew [5] and AC [2] emphasize that the neolithic agricultural economy is well established by 9000 BP within three separate nuclear zones of the fertile crescent, the Levant, Anatolia, and Mesopotamia. Renfrew takes the wave-of-advance model one step further [5] by proposing that there were three separate waves of advance of neolithic farmers emerging from the fertile crescent: one from the Levant which carried the Afro-Asiatic language family into the Arabian peninsula and northern Africa; one from Mesopotamia which carried the Elamite-Dravidian family into Persia and the Indian subcontinent, as well as the original one proposed by AC which carried the Indo-European family into

Europe and thence through the Balkans and north of the Black Sea eastward into Asia.

3. The Fisher Equation

AC utilized the Fisher equation (FE) to provide a quantitative basis for their wave-of-advance model. This equation was introduced by Fisher in 1937 to describe the advance of favorable genes into a population [13], and it has received substantial attention since [14]. The FE is

$$\frac{\partial N}{\partial t} = D\nabla^2 N + \frac{1}{\tau} N(1 - (N/N_S)) , \tag{3.1}$$

where the first term on the rhs describes the population flux arising from population or demic diffusion, i.e., from random movements of individuals or small groups, with D the diffusion constant, and the second term describes population growth. The equation clearly is nonlinear, fulfilling one of the requirements for discussion at this workshop. It has two fixed points, an unstable one at $N = 0$ and a stable one at $N = N_S$. Under the action of the two terms on the rhs, population will spread and grow from any region of population density $N \leq N_S$.

The spread is in fact wave-like, and solutions of the FE provide beautifully simple examples of the wave-of-advance proposed by CA and by Renfrew. If one considers the one-dimensional case and supposes that N depends only on the combination $x - vt$, i.e., is a wave propagating in the positive x direction with speed v, one finds that locally stable solutions of (3.1) exist for all $v \geq 2\sqrt{(D/\tau)}$. This family of solutions in fact describes solitary waves with N increasing monotonically from 0 to N_S as x decreases over a region of width proportional to v. The solution has the limiting form

$$N/N_s = [1 + e^{(z - z_0)/v}]^{-1} \tag{3.2}$$

as $v \to \infty$, where $z = x - vt$, and becomes

$$N/N_s = [1 + e^{(z - z_0)/\sqrt{6}}]^{-2} \tag{3.3}$$

in the special case $v = 5/\sqrt{6} \simeq 2.04$.

Which of these solitary waves is reached asymptotically is intimately related to how quickly the initial population distribution approaches zero or one at $x = \pm\infty$ [14]. When an initial population density has compact support, the asymptotic solution is the wave of minimum velocity, $v = 2\sqrt{(D/\tau)}$. This provides justification for the implicit assumption by AC that the velocity of the wave of advance was given approximately by $2\sqrt{(D/\tau)}$. In two dimensions, D may be written as $\frac{1}{4}L^2\nu$, where L^2 is the mean square random displacement and ν the displacement frequency. AC assumed, again implicitly, that $\nu = 1/\tau$ so that $v = L/\tau$. By taking 25 km for L and 25 years for τ, they arrive at their observed rate of advance of NF into Europe, 1 km/yr.

In addition to noise, e.g., temporal fluctuations in N_S and τ, there is spatial heterogeneity in the real world. Such inhomogeneity is systematic, e.g., variations of N_S and τ with latitude, and random, e.g., local variations in altitude and soil composition. While temporal fluctuations are certainly important, we shall see that spatial fluctuations can have a profound influence on the wave-of-advance. Accordingly, we focus on these in the following sections, introducing disorder and fulfilling the second and final requirement for presentation at this workshop. To understand how to introduce spatial heterogeneity, one must derive anew a suitable generalization of the Fisher equation from consideration of the relevant population biology and ethnology.

4. Refining the Model

A fully developed wave-of-advance model would include three populations, the neolithic farmers, the indigenous hunter-gatherers, and those of HG stock who have adopted farming as well as the interactions between them. The task of introducing spatial heterogeneity into a model containing only a single population is, however, sufficient for the present paper. In the following, we develop in turn explicit, simple models of the various elements of the whole model and combine them ultimately into a generalization of the Fisher Equation containing position-dependent parameters which allow the introduction of heterogeneity into the wave-of-advance model.

Food Production. A population density N will produce food at a rate F per unit area. F is proportional to N at low values of N and saturates at F_0, the ultimate food-production capacity per unit area reachable in a given location with the food-production technology employed. A very simple interpolation between these limits is

$$F = F_0 N/(N_0 + N) , \tag{4.1}$$

where $N_0^{-1} = A$ is the area exploited per person in the limit of low population density, $N \to 0$. A is technology dependent. From (4.1) it follows that the food production per person is

$$f = F_0/(N_0 + N) . \tag{4.2}$$

Births and Deaths. In the absence of diffusion, population change is simply the difference between birth and death rates, $\dot{N}_B$ and $\dot{N}_D$,

$$\frac{dN}{dt} = \dot{N}_B - \dot{N}_D , \tag{4.3}$$

where

$$\dot{N}_B = bN \quad \text{and} \quad \dot{N}_D = dN \tag{4.4}$$

with b and d the rates per person. For b, we assume a form

$$b = b_0 f/(f_0 + f) \tag{4.5}$$

which interpolates between a linear dependence on food production at low food production and saturates at the limiting value b_0. We regard f_0 to be of biological origin and to be culture insensitive, whereas b_0 is subject to societal control and is culture dependent. Eqs.(5.2) and (5.5) imply a density-dependent birth rate,

$$b = \frac{b_1 N_1}{N_1 + N} ; \quad b_1 = \frac{b_0 s}{1 + s} ; \quad N_1 = N_0(1 + s) ; \tag{4.6}$$

where

$$s = f_1/f_0 ; \quad f_1 = F_0 A \tag{4.7}$$

with f_1 the food production per person as $N \to 0$. s, the ratio of this food production capacity f_1 per person in a given location via a given technology to the biological requirement f_0 for successful reproduction, is thus a measure or index of the fertility of a location.

For d, on the other hand, we assume the form

$$d = d_0(f_0 + f)/f \ , \tag{4.8}$$

which interpolates between a death rate which goes to infinity as the food supply vanishes and one which saturates at d_0 for large f. d_0 is largely culture and technology insensitive. Eqs.(4.2) and (4.8) imply a density-dependent death rate,

$$d = d_1(N_1 + N)/N \ ; \ d_1 = d_0(1 + s) \tag{4.9}$$

It will be of importance that

$$b_1 d_1 = b_0 d_0 \tag{4.10}$$

independent of s and having only relatively weak culture or technology sensitivity.

Thus far, we have introduced the parameters b_0, d_0, f_0, F_0, and A, or various combinations of these. However, F_0 and A are not independent. What is important in a society is their product $f_1 = F_0 A$ or, equivalently, $s = f_1/f_0$. As F_0 increases, s will saturate through reduction of A at some maximum value s_m beyond which further food production is of no value to the society,

$$s = f_1/f_0 = F_0 A/f_0 \xrightarrow[F_0 \to \infty]{} s_m; \ A \xrightarrow[F_0 \to \infty]{} s_m f_0/F_0 \tag{4.11}$$

s_m is culture dependent.

On the other hand, as F_0 decreases towards zero, A can be increased to compensate for it only to some limiting value A_M determined by the technology and by human biology. A_M permits us to introduce a more appropriate measure of the fertility of a locale,

$$\sigma = F_0 A_M/f_0 \ ; \ s \xrightarrow[F_0 \to 0]{} \sigma \tag{4.12}$$

Interpolating between the limiting values (4.11) and (4.12) of s, we obtain

$$s = s_m\sigma/(s_m + \sigma) \ ; \ A = N_0^{-1} = A_M(s_m + \sigma)/s_m \tag{4.13}$$

We now have the locale-independent parameters b_0, d_0, s_m, and A_M and, in addition, one locale-dependent parameter, s or σ, through which spatial heterogeneity will enter the theory. However, it is important to note that s and σ can be weakly culture/technology-dependent through compensation of F_0 and A_M in the product $F_0 A_M$.

Saturation. If one inserts (4.6) and (4.9) into (4.4) and then into (4.3), one obtains a simple generalization of the logistic equation for N, to be discussed below, which implies a long time saturation value N_s of N given by

$$N_s = N_1(r - 1); \ r = \sqrt{(b_1/d_1)} = r_0 s/(s + 1) \tag{4.14}$$

$$r_0 = \sqrt{(b_0/d_0)} \tag{4.15}$$

Clearly, for a society to be able to establish itself anywhere, r_0, which is weakly technology or culture sensitive, must exceed unity, an elementary result of population biology. For survival in a particular location, r must exceed unity in that location. More explicitly,

$$s_m > s > (r_0 - 1)^{-1} \tag{4.16}$$

for subsistence in a particular location to be feasible.

The death rate at saturation is, correspondingly,

$$d_s = \sqrt{(b_0 d_0)}\ , \tag{4.17}$$

weakly culture or technology sensitive.

Comparison of Farmers With Hunters and Gatherers. AC [2] and Renfrew [4,5] take a value of 50 for the ratio of N_s for neolithic farmers to that for mesolithic hunters and gatherers, citing archeological and enthnological evidence. From the previous subsection, one can obtain the expression

$$N_s = A_M^{-1}[\sigma(r_0 - 1 - \frac{1}{s_m}) - 1] \tag{4.18}$$

for N_S. Moreover, from the arguments of that subsection, s_m, σ, and r_0 differ little between the two cultures. Accordingly, the ratio of A_M in the two cultures must be about 50, which is consistent with ethnological findings.

The death rate ratio on saturation d_s^{NF}/d_s^{HG} should reduce to $\sqrt{(b_0^{NF}/b_0^{HG})}$ from (4.17) and the cultural insensitivity of d_0. Ethnological findings on lifetime fertility figures [2] suggest that d_s^{NF}/d_s^{HG} should therefore differ by no more than about 30%. CA cite [15] evidence for comparable variation in life expectancy past the age of 15 among a group of cultures including farmers and HG over a period of 70,000 yrs. from the middle paleolithic to 400 BP. This provides confirmation of our argument that A_M is the sensitive parameter in the model.

Population Kinetics. The arguments of subsection B above can be collected into the growth equation

$$\dot{n} = \frac{1}{\tau}\ \frac{n_1 + n}{n_0 + n}\ n(1 - n) = \frac{1}{\tau(n)}\ n(1 - n)\ , \tag{4.19}$$

$$\frac{1}{\tau} = \frac{d_1}{n_0}\ ,\ n_0 = \frac{1}{r - 1}\ ,\ n_1 = \frac{r + 1}{r - 1}\ . \tag{4.20}$$

Eq.(4.19) is similar to the logistic equation, the growth equation underlying the Fisher equation, except that the growth rate $\tau(n)^{-1}$ is now n dependent. The parameters n_0 and n_1 depend on $r \epsilon (1,\infty)$ in such a way that

$$n_0 \leq n_1\ ,\ o \leq n_0 \leq \infty\ ,\ 1 \leq n_1 \leq \infty\ . \tag{4.21}$$

Given the initial condition n = o, t = -∞, the solution of (4.21) is

$$\frac{[n/(n_1 + n)]^{(r + 1)^{-1}}}{[(1 - n)/(n_1 + n)]^{\frac{1}{2}}} = e^{(t - t_0)/\tau}\ , \tag{4.22}$$

which has the properties

$$n \xrightarrow[t \to -\infty]{} n_1^{-\frac{1}{2}\, n_0}\ e^{(t - t_0)(r + 1)/\tau} \tag{4.23a}$$

$$n \xrightarrow[t \to \infty]{} 1 - (1 + n_1)^{n_1^{-1}} e^{-(t - t_0)\, 2/\tau} \tag{4.23b}$$

$$n \xrightarrow[r \to 1]{} [1 + e^{(t - t_0)\, 2/\tau}]^{-1} \tag{4.24a}$$

$$n \xrightarrow[r \to \infty]{} \begin{cases} \tanh (t - t_0)/\tau , & t > t_0 \\ 0 , & t < t_0 \end{cases} \tag{4.24b}$$

The solution passes continuously from the logistic curve (4.24a) at $r = 1$ to the truncated hyperbolic tangent at (4.24b) as r increases from 1 to ∞, with the point of inflection moving down continuously from $n = \frac{1}{2}$ at $r = 1$ to $n = o$ at $r = \infty$. Of course, r remains finite and of order unity in realistic models.

Population (Demic) Diffusion. The considerations of birth rates, death rates, food production, etc., discussed thus far do not depart in any profound way from the standard considerations of population biology and ethnology, except perhaps for the way in which they are integrated. However, in previous considerations of population diffusion, only random movement has been considered, leaving out an essential part of the response to a heterogeneous environment. Let $\vec{J}$ be the population or demic current density. It contains two parts, a diffusion current $\vec{J}_D$ and a "forced" current $\vec{J}_F$.

$$\vec{J} = \vec{J}_D + \vec{J}_F \tag{4.25}$$

The diffusion current is given by Fick's law

$$\vec{J}_D = - D \vec{\nabla} N , \tag{4.26}$$

where D, the diffusion coefficient, is culture and technology dependent. It is larger for HG than for NF; it is larger for nomadic pastoralists than for sedentary farmers; and it is larger for cultures possessing the horse than for those not. An expression for D has already been given in §3, and its magnitude discussed for NF. We do not expect D to vary with local conditions because the movement of population in a wave of advance is very much slower than the rate at which individuals can transport themselves through almost any terrain. Recall that the observed velocity of advance is 1 km/yr.

$\vec{J}_F$ is the term omitted in previous work. It is the response of a population to variations in the carrying capacity of its environment. One can obtain an explicit expression for it through a precise analogy with Einstein's derivation of the Einstein relation between the diffusion coefficient and mobility of a classical particle [16]. The result is

$$\vec{J}_F = ND\vec{\nabla}(\ln N_s) = Dn\vec{\nabla}N_s \tag{4.27}$$

Inserting (4.26) and (4.27) into (4.25) gives desired result for the demic flow,

$$\vec{J} = - D\, N_s \vec{\nabla} n \tag{4.28}$$

From the continuity equation, the divergence of $\vec{J}$ contributes the following term to $\frac{\partial n}{\partial t}$,

$$\left(\frac{\partial n}{\partial t}\right)_{FLOW} = \frac{D}{N_s} \vec{\nabla} \cdot N_s \vec{\nabla} n = D\, [\nabla^2 n + \vec{\nabla}(\ell n N_s) \cdot \vec{\nabla} n] \tag{4.29}$$

The second term on the right in (4.29) is the new term.

Population Dynamics, Homogeneous Case. The second term in the rhs of (4.29) drops out in the spatially uniform case. Adding (4.29) to (4.19) leads to the corresponding generalization of the FE.

$$\frac{\partial n}{\partial t} = D \nabla^2 n + \frac{1}{\tau} \frac{n_1 + n}{n_0 + n} n(1 - n) \tag{4.30}$$

Scaling x and t as in §3 simplifies this to

$$\frac{\partial n}{\partial t} = \nabla^2 n + \frac{n_0 + n}{n_1 + n} n(1 - n) \tag{4.30'}$$

where $t/\tau \to t$ and $x/(D\tau)^{\frac{1}{2}} \to x$. The appropriate boundary condition in a realistic case is the vanishing of the normal derivative of n at all barriers impenetrable to demic diffusion.

As before, there are solitary wave solutions of (4.30′) in 1D with $n(x,t) = g(x - vt) = g(z)$, $g(\infty) = 0$ und $g(-\infty) = 1$. These satisfy the equation

$$g'' + vg'' + \frac{n_1 + g}{n_0 + g} g(1 - g) = 0 . \tag{4.31}$$

There exist solutions to (4.31) satisfying the boundary conditions for all $v \geq 2\sqrt{r + 1}$. These solutions are stable. In the limit $v \to \infty$, the form of g is given by (4.22) for n with $(t - t_0)/\tau$ on the rhs replaced by $(z - z_0)/v$. Nothing else is yet known, but presumably an initial state with compact support would evolve into a wave of minimal speed, $v = 2\sqrt{r + 1}$.[14]

Spatial Heterogeneity, Disorder. Europe, as do all other habitable regions of the globe, presents a spatially heterogeneous environment to any food production technology. It has a random coast line, river system, and lake distribution, affecting diffusive movement and imposing geographic boundary conditions. It has random topography, soil constitution and microclimates, affecting local food production capacity. In addition, there are systematic climate changes from south to north and coast to interior affecting food production capacity and nutritional needs.

The first of two major effects of this spatial inhomogeneity is to introduce position-dependence into the parameters of the demic equation N_S, τ, n_0, and n_1 through their implicit dependence on F_0 or σ. D is assumed to be unaffected, i.e., constant, as discussed above. Accordingly, the full demic equation is

$$\frac{\partial n}{\partial t} = \frac{D}{N_s} \vec{\nabla} \cdot N_s \vec{\nabla} n + \frac{1}{\tau} \frac{n_1 + n}{n_0 + n} n(1 - n) . \tag{4.32}$$

The second major effect of spatial nonuniformity is to introduce a new kind of boundary with quite profound implications for understanding the past. We have discussed the usual boundary condition imposed by impenetrable geographical barriers to movement such as shorelines, etc., on which the normal component of current flow $\vec{J}$ vanishes through the vanishing of $\vec{\nabla} n$. The new boundary is a subsistence boundary, the locus of all points at which σ approaches from above the critical value

$$\sigma_c = s_m^{-1}[s_m(r_0 - 1) - 1] \tag{4.33}$$

at which $r \to 1$ and, most importantly, $N_S \to 0$. N vanishes on such a boundary, but n can remain finite as can $\vec{\nabla}n$. The important fact is that $\vec{J} \to 0$ because of the factor N_S in (4.28). There is no flow across a subsistence boundary and no population beyond it. We shall return to the significance of subsistence boundaries in the next and final section.

There are three relevant length scales in our model. The first two refer to the two kinds of spatial inhomogeneity. L_C is the correlation length of random local fluctuations in σ or F_0, and L_R is the scale of regional variation of climate. Rough estimates are $\lesssim 10^2$ km for L_C and $\lesssim 10^3$ km for L_R in Europe. A reasonable simplification is afforded by taking $L_C << L_R$. The third scale is the demic scale, the width of the front of the wave of advance, $L_D = 2(D\tau)^{\frac{1}{2}}$. The previous assumption that $\nu = \tau^{-1}$ implies $L_D = L = 25$ km. Again, a reasonable simplification is to take $L_D << L_C << R$. Thus, σ can be taken as a stationary random variable within L_R, and the variation of N_S can be ignored so that it drops out of the equation. Thus, Eq.(4.30) can be used in what amounts to a non-linear, parabolic, ray-optics limit provided the subsistence boundaries are retained.

5. Discussion; What Next

Having developed the demic equation in the last section, together with its boundary conditions, and the ordering of the length scales on which inhomogeneity occurs, the next task is the development of an explicit two-dimensional theory of solitary-wave propagation including the attendant stability considerations. Next, the theory should be extended to include all three components of the population, as mentioned in the introduction to the last section. Next, allowance should be made for learning, i.e., the advance of food production technology, which leads to the slow advance of subsistence boundaries. Finally, simulations should be carried out for a realistic food production model of Europe.

However, before all this is carried out, it is already possible to glimpse some of the implications. For example, the Alps presented a subsistence boundary reached about 6000 BP and overcome by 5500 BP.[7] The mesolithic HG did not disappear within this boundary at 6000 BP, but remained until 5500 BP. In the absence of the subsistence boundary, the wave-of-advance could have moved 500 km over those 500 years.

Now the overwhelming of the mesolithic HG by the NF implies that the ~ 10^2 year period over which the overwhelming occurs is too short for the indigenous population to preserve itself by adopting farming. On the other hand, when a persistence boundary exists for the NF across which they do not move but within which the HG can remain at saturation, the possibility arises that it is the HG which learn, adopt, and adapt neolithic agriculture successfully to their local environment, preserving thereby their genetic constitution and their language.

I do not claim this to be the case for the Alps, a complex case, but it offers an intriguing, speculative explanation of the persistence of Basque as an isolated language in Europe. This would require the existence of a subsistence boundary around the Pyrenees across which no NF advance occurred for long enough for the mesolithic indigenes within it to advance their food-production technology and increase their population density enough to avoid being overwhelmed. In this picture, the Basque language is a descendant of the mesolithic language spoken around the Pyrenees, just as Nostratic was spoken in the fertile crescent.

More generally, one would expect regions of high elevation to present barriers, not to the movement of NF of, e.g., Indo-European stock, but to their food production technology. The grains brought out of Anatolia and gradually adapted as more northerly climes are reached would still be ill-adapted to high, mountainous country at the latitude of the wave of advance. Indeed there is an intriguing correlation between isolated languages and mountainous regions, e.g., Basque and the Pyrenees, the languages of the

Caucasus, and the languages of the Urals, as well as the presence of Finno-Ugric at the North Eastern limit of the wave of advance.

It should be noted that Renfrew has understood qualitatively the concept of the subsistence boundary and has discussed the slowing of the wave-of-advance in northern Europe in terms of the added time required to adapt the neolithic crops.[4] However, the concept of internal boundaries around mountainous regions, their linguistic implications, and the corresponding linguistic implications of the slowly moving northern boundary all are new.

Another interesting point relates to conjectures concerning the route by which the Indo-European languages reached Eastern Europe and the Asian steppes. Renfrew, an archeologist, argues for a route across the Aegean, northward through the Balkans, and eastward around the Black Sea. The linguists, Gamkrelidze and Ivanov [17], argue for a route northward through the Caucasus. A subsistence boundary around the Caucasus would rule out the latter or, at the least, diminish its relative importance. The subsistence boundary hypothesis is the more economical, providing as well an explanation of the Caucasus languages.

REFERENCES

(1) A. J. Ammerman and L. L. Cavalli-Sforza in Transformations: Mathematical Approaches to Culture Change, eds., C. Renfrew and K. L. Cooke (Academic Press, New York, 1979), p. 275.

(2) A. J. Ammerman and L. L. Cavalli-Sforza, The Neolithic Transition and the Genetics of Populations in Europe (Princeton University Press, Princeton, 1984).

(3) A. J. Ammerman and L. L. Cavalli-Sforza, Man **6**, 674 (1971).

(4) C. Renfrew, Archeology and Language (Cambridge University Press, New York, 1988).

(5) C. Renfrew, Scientific American, October 1989, p. 106.

(6) Ref. 2, Fig. 1.1.

(7) ibid., Fig. 4.6.

(8) ibid., Fig. 4.5.

(9) ibid., Fig. 4.2.

(10) C. B. Stringer, Scientific American, December 1990, p. 98.

(11) Australian aboriginal society at the time of first contact provides a clear example.

(12) V. Shevoroshkin, The Sciences, May/June 1990, p. 20; A. Militarëv and V. Shnirelman, The Problem of Proto-Afrasian Home and Culture (Nanka Publishers, 1988).

(13) R. A. Fisher, Annals of Eugenics, London **7**, 355 (1937).

(14) D. A. Larson, SIAM J. Appl. Math. **34**, 93 (1978).

(15) Ref. 2, Fig. 5.1.

(16) A. Einstein, Ann.d. Phys. **17**, 549 (1905).

(17) T. V. Gamkrelidze and V. V. Ivanov, Scientific American, March 1990, p. 110.

Role of Disorder on the Dynamics of a Nonlinear Model for DNA Thermal Denaturation

M. Peyrard

Université de Bourgogne, Physique Non Linéaire,
Ondes et Structures Cohérentes, Faculté des Sciences,
6 blvd. Gabriel, F-21000 Dijon, France

The dynamics of thermal denaturation of DNA is a good example in which nonlinearity coexits with disorder. The amplitude of the motions is so high that bonds break and the base sequence is inhomogeneous since it contains the genetic code. Using a simple nonlinear model, we study the role of local inhomogeneities or of extended disorder on the dynamics of the localized excitations and on the denaturation rate by numerical simulations at constrained temperature. Approximate analytical results are obtained for the trapping of the breatherlike excitations by isolated defects and the statistical mechanics of the disordered molecule.

1. Introduction

During the transcription of DNA, *i.e.* during the reading of the genetic code, the two strands of the DNA double helix separate locally from each other, forming the so called "transcrition bubble". Therefore the formation of this bubble corresponds to a local denaturation of the molecule. It is induced by an enzyme, the RNA-polymerase, but shows some similarities with the thermal denaturation of the molecule. When a DNA molecule is slowly heated, first large fluctuational openings are observed, then local denaturation bubbles appear. If the heating goes on, the total denaturation is achieved through the growing of these bubbles. Thus, the investigation of the thermal denaturation of DNA is a first step toward the understanding of the transcription.

Since the denaturation involves the breaking of some chemical bonds, it is essentially a *nonlinear* process. The idea that nonlinear excitations could play a role in the dynamics of DNA has been developped for many years [1]–[6] however, in these studies, the *creation* of nonlinear excitations that could open locally the double helix is not investigated. Recently, on the basis of a simple model for the DNA molecule, we suggested that an energy localization due to nonlinear effects could occur according to a mechanism analogous to self focussing in plasmas, and cause

Editors: F. Abdullaev · A.R. Bishop · S. Pnevmatikos

ATTATCAAAAAGGATCTTCACCTAGATCCTTTTAAAT-
TAAAAATGAAGTTTTAAATCAATCTAAAGTATATAT-
GAGTAAACTTGGTCTGACAGTTACCAATGCTTAATC-
AGTGAGGCACCTATCTCACGGATCTGTCTATTTCGT-
TCATCCATAGTTGCCTGACTCCCCGTCGTGTAGATA-
ACTACGATACGGGAGGGCT

Figure 1 : Sequence of fragment of *esterichia coli* DNA. Only the bases of one strand are shown. A, T, G, C, design respectively $A-T$, $T-A$, $G-C$ and $C-G$ base pairs

the denaturation [7]. In this model, the inhomogeneities due to the base sequence are neglected and the localization is a pure nonlinear effect.

However, real DNA is not homogeneous because two types of base pairs A–T (Adenine-Thymine) and G–C (Guanine-Cytosine) are present to form the genetic code. Your existence as an highly evolved organism reading this paper attests that the genetic code is certainly not *disorder*. However, looking at the DNA sequence displayed in fig. 1 suggests that treating mathematically the inhomogeneities in DNA as random disorder is a legitimate approximation. We shall adopt this approximation in this work. A well known property of disorder in linear systems is to induce a localization of the excitations according to the mechanism proposed by Anderson [8]. Since we are interested in an energy localization mechanism which induces a local denaturation, we must determine to what extend the disorder in the base pair sequence contributes to this localization and how it inteferes with the nonlinear localization found in the homogeneous model. This is the aim of this paper. In section II we present the model and its properties in the homogeneous case. Section III describes numerical studies with a model including defects localized in a few places along the molecule or random disorder. Section IV contains an analytical analysis on the numerical results and a discussion.

2. The DNA model in the homogeneous case

In order to describe the denaturation, the model includes for each base pair only one degree of freedom, the stretching y of the hydrogen bonds connecting the two bases. Following previous investigations [7] [9] [10], the potential for the hydrogen bonds is approximated by a Morse potential so that the hamiltonian of the model is

$$H = \sum_n \frac{1}{2}\left(\frac{dy_n}{dt}\right)^2 + \frac{1}{2}k_n(y_n - y_{n-1})^2 + D_n\left[\exp(-\alpha y_n) - 1\right]^2 . \quad (1)$$

The index n denotes a base pair. An harmonic coupling is assumed between neighboring base pairs. Since mass fluctuations between different pairs are only of the order of 3 %, a common mass m has been assumed for all pairs. The inhomogeneities are introduced in the coupling constants k_n and the dissociation energy D_n of the Morse potentials while the parameter α of these potentials has been chosen uniform along the molecule to reduce the number of free model parameters.

For an homogeneous molecule, k_n and D_n do not depend on n and the statistical mechanics of the model (1) can be investigated by the transfer operator technique [7]. This analysis shows that the mean stretching of the hydrogen bonds $\langle y \rangle$ exhibits a sharp increase leading to a divergence when the temperature approaches the "melting" temperature

$$T_m = 2\sqrt{kD}/\alpha k_B \quad (2)$$

where k_B is the Boltzmann constant. This rise is consistent with the thermal denaturation of DNA, but the statistical mechanics does not tell us *how* it occurs, for instance does it start locally, or is it an homogeneous process?

In order to achieve a better understanding of the denaturation process, particularly in the presence of inhomogeneities, we have performed numerical simulations of the system described by hamiltonian (1) at constrained temperature. We use the Nose scheme [11] which introduces an extended hamiltonian adding a non local variable and its conjugate to the physical variables y_n in order to represent the thermal bath. This scheme gives *exact* equilibium properties for a canonical ensemble [11], but, since we are interested in the thermal denaturation process, the simulations are not performed exactly at equilibrium. Instead, we impose a slow temperature ramp (typically from 200 K to 500 K). We have checked that the slope of the ramp does not affect the results. In some of the simulations, the ramp has been interrupted by long temperature plateaux in order to verify that the state reached at given temperature on the ramp is identical to the equilibrium state at the same temperature on a plateau.

The appropriate choice of the parameters for such a simplified model of DNA is a controversial point [12]. If we use the Morse potential parameters determined from small amplitude lattice dynamics calculations [10], the melting temperature T_m is far too high unless we chose a very small coupling constant k. This discrepancy has two different origins. First, the spectroscopic experiments probe only small amplitude motions wheras the thermal denaturation is concerned by the whole potential curve. Second, in our simple model, the interaction between two bases in a pair is an *effective* interaction beween the groups that form one DNA unit. It includes for instance the repulsive interaction between the phosphate groups that softens the hydrogen bond interaction between the bases. An investigation to determine the parameters of this effective potential is under way. In this work, we have tested two parameter sets which are expected to embrace the range of realistic parameters. Both yield $T_m = 400$ K, but, in the first one this is achieved with a rather large dissociation energy $D = 0.0875$ eV and a small coupling constant $k = 0.031$ eV/Å^2, while in the second case, D is only $D = 0.02188$ eV and k has been raised to $k = 0.124$ eV/Å^2. The other model parameters are $\alpha = 4.45$ Å^{-1} and $m = 300$ atomic mass units in both cases. The simulations are performed with perodic boundary conditions, and, except when indicated otherwise, they include 256 base pairs.

The melting process in these two cases is shown in Fig. 2 (a) and (b). A large amount of information on the denaturation process in the DNA model is encoded in these figures. Let us now decode this information. At low or intermediate temperature (upper or middle parts of the figures) patterns of darker regions (i.e. larger stretching) appear on both figures, but their shapes are very different from each other in the two cases. Moving along an horizontal line in the figures, *i.e.* moving along the DNA chain, we notice that the hydrogen bond stretching exhibits large fluctuations. Dark regions of large stretching are separated by lighter domains where the stretching is weak. This shows that the mean stretching $\langle y \rangle$ calculated by the statistical mechanics is due to localized large amplitude excitations and not to an homogeneous increase along the molecule. This is consistent with the fluctuational openings that are observed experimentally and could be responsible for drug intercalation [13]. The lines formed by the dark regions attest that the localized excitations are long lived because they can be followed as time evolves. If one follows one of these lines, we can notice that it is interrupted regularly and looks like a dotted line (this is more visible in Fig 2 (a)). This is due to *an internal breathing* of the localized excitation that oscillates between a large amplitude (black dots in the figure) and a small amplitude state (light dots) in a regular manner.

These results are consistent with the small amplitude expansion of the equation of motion presented in ref. [7]. The equation of motion deriving from hamiltonian (1)

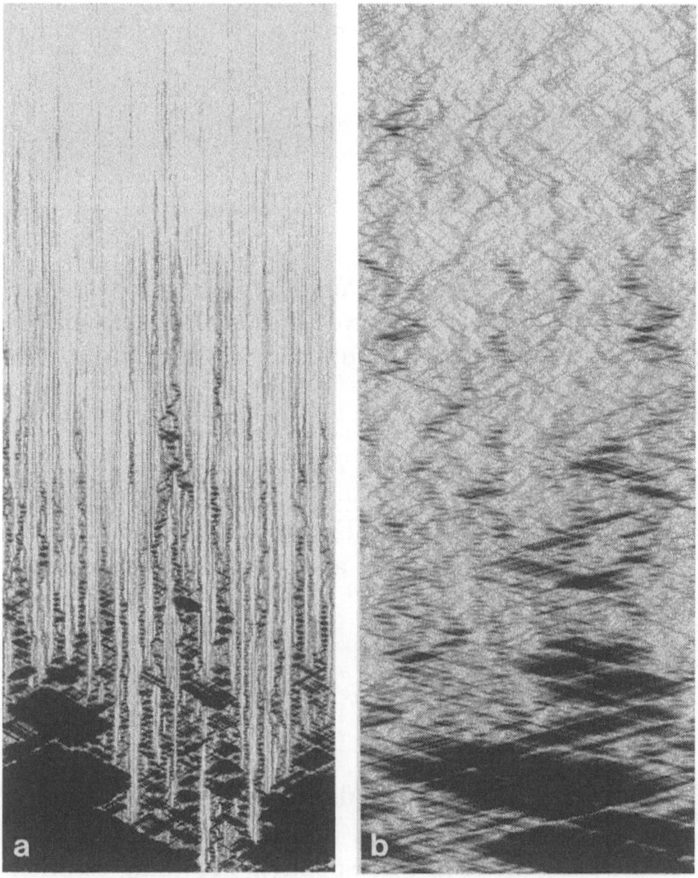

Figure 2 : Evolution of the base stretching y along the DNA model during heating for the first (a) and second (b) model parameter sets. The horizontal axis indicates the position along the 256 cells of the molecule and the vertical axis corresponds to temperature (or time since time t and temperature T are linearly related by the equation of the temperature ramp $T = T_0 + st.$). Temperature increases from top to bottom in the figures. In part (a), temperature ranges from 200 K (top) to 800 K (bottom) and in part (b), it ranges from 200 K (top) to 500 K (bottom). The base pair stretching y_n is shown by a grey scale extending from 0 (white) to 5 Å(black). Large black regions correspond to bubbles where the hydrogen bonds connecting the bases in a pair are completely broken.

$$m\frac{\partial^2 y_n}{\partial t^2} - k(y_{n+1} + y_{n-1} - 2y_n) - 2\alpha D e^{-\alpha y_n}\left(e^{-\alpha y_n} - 1\right) = 0\,, \qquad (3)$$

can be expanded in the continuum limit for small y as

$$m\frac{\partial^2 y}{\partial t^2} - ka^2\frac{\partial^2 y}{\partial x^2} + 2D\alpha^2 y - 3D\alpha^3 y^2 + \frac{7}{3}D\alpha^4 y^3 = 0\,, \qquad (4)$$

where a is the distance between two base pairs. In these equations, the indices n have been dropped for k and D as we consider the homogeneous molecule. A solution of Eq. (4) can be obtained via a multiple scale expansion as

$$y = \varepsilon F_1(X,\tau)e^{i\omega t} + c.c. + \varepsilon^2\left[F_0(X,\tau) + F_2(X,\tau)e^{2i\omega t} + c.c.\right]\,, \qquad (5)$$

with $X = \varepsilon x, \tau = \varepsilon t$. The functions F_0 and F_2 can be expressed in terms of F_1 as $F_0 = 3\alpha|F_1|^2$ and $F_2 = -\alpha F_1^2/2$ and F_1 is a solution of the Nonlinear Schrödinger (NLS) equation

$$i\frac{\partial F_1}{\partial s} + \frac{c_0^2}{2\omega}\frac{\partial^2 F_1}{\partial X^2} + 2\omega\alpha^2|F_1|^2 F_1 = 0\,, \qquad (6)$$

where $s = \varepsilon\tau$, $c_0^2 = ka^2/m$, and ω corresponds to the lowest frequency of the linear motions, $\omega = 2D\alpha^2/m$.

The solitons of the NLS equation correspond to breathing solutions in y which provide an approximate description of the localized oscillations observed in the simulations. In the case of Fig. 2 (b), the nonlinear excitations are mobile along the DNA molecule as shown by their oblique trajectories in the space-time coordinates of the figure. For the weak coupling case of fig. 2 (a), the localized excitations have a larger amplitude, they are narrower and they are completely trapped by the discreteness effects, generating the vertical lines in the figure.

As temperature increases toward the denaturation temperature, the NLS description ceases to be valid because the small amplitude expansion breaks down. The breathing motions stop and some of the localized excitations stay in an open state, generating the denaturation bubbles that are oserved experimentally. From this point, the denaturation spreads out in a rather small temperature range which corresponds to the sharp increase of $\langle y\rangle$ given by the statistical mechanics analysis. The denaturation is generated by an increase in the size of a few bubbles that tend to cover the whole molecule. The dynamics of their growth is interesting.

It is easier to follow it in the weak coupling case (fig. 2 (a)) because the limits of the bubbles are very sharp in this case. It should be noticed that, around a given temperature, i.e. in a given region of the figure, the bubble boundaries move with a *well defined speed.* Therefore a localized denaturation extends as a triangular region in the diagram. However a bubble boundary can be reflected by a region of low stretching (light grey region on the figure). After the scattering, the bubble boundary preserves the modulus of its speed. These results strongly suggest the existence of a cross-over temperature for which the system switches from a NLS description with breatherlike excitations to another description where the excitations are the bubble boundaries. We don't have yet an analytical description of this new regime, but its dynamics is certainly important to determine the denaturation temperature because the scattering of the bubble boundaries cause the growth or shrinking of the denaturated regions.

Before investigating the role of disorder, it is important to notice that discreteness effects have a strong influence on the denaturation temperature. In fig. 2 (a), the temperature ranges from 200 K to 800 K while in fig. 2 (b), similar denaturation rates are obtained in a temperature domain extending only from 200 K to 500 K. In both cases, the statistical mechanics result obtained by the transfer operator method predicts $T_m = 400$ K. Although the results obtained with a rather small sample (256 base pairs) exhibit large fluctuations, this is rather well verified in the stong coupling case (fig. 2 (b)) because at 430 K, 80% of the sample is denaturated. In the discrete, weak coupling case, the same denaturation rate is only achieved at 730 K, showing a large disagreement with the theoretical results. Although the discreteness effects can formally be treated in the transfer operator approach [14], we have not been able to carry the analytical calculation for the particular case of the Morse potential considered here. A numerical treatment of the transfer operator would be necessary in this case.

Since the strong coupling case leads to more realistic denaturation temperatures, and because we do not want discreteness effects to interfere with the effect of disorder we are interested in, all the calculations presented in the following have been performed in the strong coupling case (second parameter set).

3. Numerical studies thermal denaturation in the presence of disorder

3.1 Localized inhomogeneities

The two types of base pairs $A-T$ and $G-C$ are connected respectively by 2 and 3 hydrogen bonds. As mentionned earlier, the on-site potential in our simple model is not only due to these hydrogen bonds, however, we can expect rather large variations of this potential along the molecule when the base sequence changes. Simple qualitative arguments suggest that the $A-T$ regions where the on-site potential is weaker could play the role of nucleation sites to initiate the denaturation. Numerical simulations show that this naive approach is not correct. Figure 3 shows an example with the same gray level representation as in fig. 2. The model and simulation parameters are exactly the same as in fig. 2, but the dissociation energy of the Morse potential has been reduced from 0.02188 eV to 0.010 eV in two regions of the molecule extenting over 10 and 20 base pairs. In the low temperature range (upper part of fig. 3), the stretching is larger in the defect regions as expected. But the localized excitations generated in these "weak" regions stay trapped on the defects. The trapping is almost perfect for the larger defect while for the smaller one, the large amplitude excitations escape temporarily from the defect. Therefore the defects *do not* play the role of nucleation centers for the denaturation because the bubbles created in these regions *cannot grow*. When the temperature is raised further (lower part of fig. 3), instead of assisting the denaturation, the two regions with a reduced dissociation energy D inhibit it. At 500 K, the highest temperature reached in the simulation, the denaturation rate is smaller in the presence of the defects. than for the homogeneous molecule.

3.2 Disordered molecule.

The results obtained for local inhomogeneities are confirmed when we consider a disordered molecular segment. Two types of $A-T$, $G-C$ sequences have been investigated. The first type is simply obtained by picking at random the values of D_n among the two values $D_1 = 0.01188$ eV and $D_2 = 0.03188$ eV chosen so that the mean value of D is the same as in the homogeneous case. In this sequence, there is no correlation from one base pair to the next one and the characteristic length of the disorder is $\ell_c = 1$ unit. To generate the second sequence, once a value of D_1 or D_2 has been selected it is used for Δn base pairs before a new value of

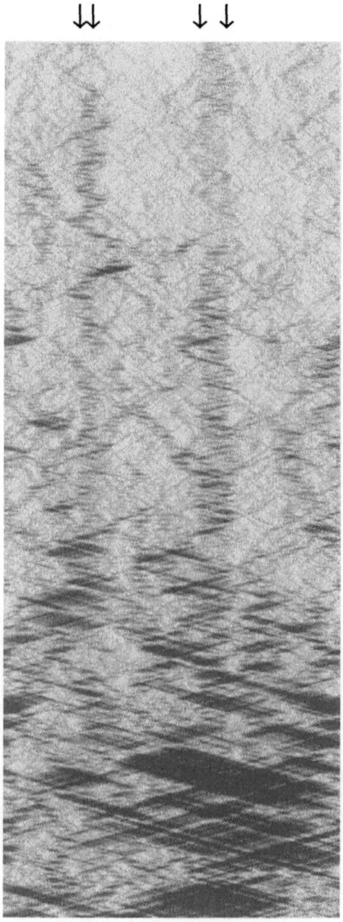

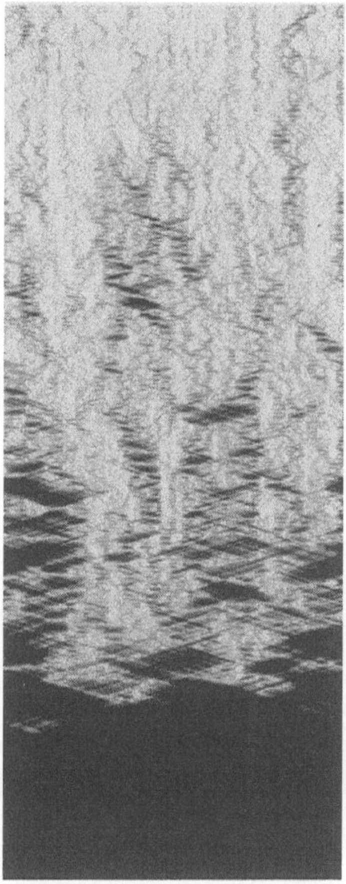

Figure 3 (left) : Evolution during heating of the base stretching y along a DNA model with two localized defects. The dissociation energy of the Morse potential is reduced from $0.021 eV$ to $0.010 eV$ in the two regions delimited by the couples of arrows on top of the figure. The temperature ranges from $200K$ to $500K$ as in fig. 2(b) and the grey scale is the same as before.

Figure 5 (right) : Evolution during heating of the base stretching y along a DNA model with disorder in the coupling constants. The values of k_n have been randomly chosen among the two values $k_1 = 0.062$ eV/Å^2 and $k_2 = 0.186$ eV/Å^2. All other parameters are the same as in fig. 2 (b).

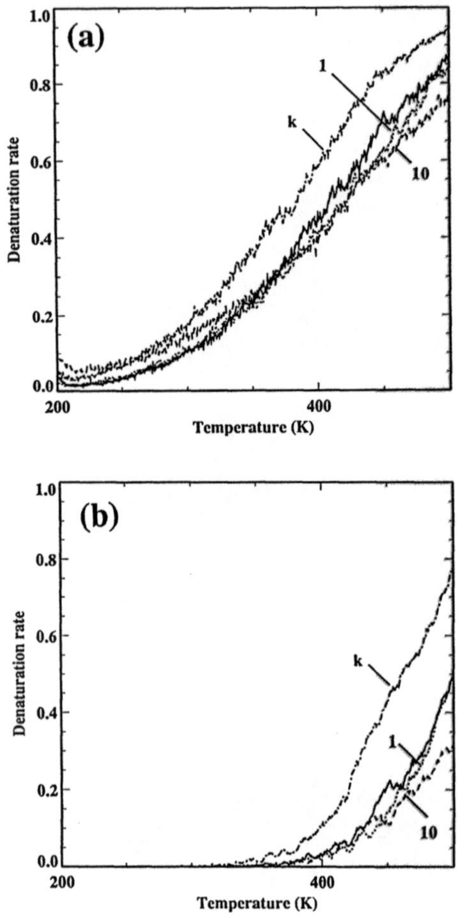

Figure 4 : Denaturation curves for DNA models with disorder. The denaturation rate is calculated by counting the number of base pairs which have a mean stretching exceeding a given threshold. In fig. (a), the threshold has been chosen as 1 Åand in fig. (b) it is equal to 5 Å. Consequently fig. (a) is more sensitive to small amplitude motions whereas fig. (b) probes the base pairs which are completely broken. The figures show the denaturation curves for the homogeneous molecule (full line), the two types of sequences with disorder in D and a characteristic length of 1 (dotted line labelled 1) or 10 (dotted line labelled 10), and a third disordered case where D is homogeneous and equal to 0.02188 eV but k_n is randomly chosen among the two values $k_1 = 0.062$ eV $/\text{Å}^2$ and $k_2 = 0.186$ eV $/\text{Å}^2$ (dotted line labelled k).

D is drawn. Δn itself is random in the range 0, 2ℓ and has an average equal to $\ell_c = 10$.

In order to get better statistics and generate meaningful denaturation curves, the simulations performed on a Sun computer with 256 base pairs have been repeated with 16,384 base pairs (2^{14}) on a Connection Machine using one processor for each base pair. The results are presented in fig. 4 (a) and (b). The importance of such a large number of base pairs is clear from the picture because the noise is still important on some of the curves. The fluctuations at high temperature are caused by the opening or closing of large denaturation bubbles. When comparing there curves with the smooth denaturation curves obtained experimentally, one must keep in mind that that the smallest experimental samples contain millions of base pairs. In spite of the simplicity of our model and the exceptional computing power of the connection machine, this is beyond the present numerical possibilities so that we cannot expect accurate measurements of the denaturation temperature from the simulations.

Figure 4 shows that disorder in D has only a weak effect on the denaturation curve. However the 2 types of sequence lead to slightly different results. For a characteristic length of the disorder $\ell_c = 1$, the disorder in D is always almost negligible, while for the larger characteristic length $\ell_c = 10$, at low temperature, the denaturation rate is larger than in the homogeneous case while it becomes smaller at high temperature. This is consistent with the results obtained for isolated defects because at low temperature the regions with a smaller on site potential are the source of large amplitude motions, hence the higher denaturation rate, whereas at high temperature the dominant effect of disorder is to prevent the denaturation bubbles from expanding.

We have also investigated the effect of disorder in the coupling constant k by considering a third disordered case where D is homogeneous and equal to 0.02188 eV but k_n is randomly chosen among the two values $k_1 = 0.062$ eV/Å^2 and $k_2 = 0.186$ eV/Å^2. In this case the effect of disorder is much more important. In the low temperature range, the denaturation rate of the sequence disordered in k is slightly smaller than for a sequence disordered in D with a characteristic length $\ell_c = 10$, but when the temperature is raised, the denaturation extends much faster in the sequence which has disordered coupling constants. Some insight on the mechanism can be gained by derived from the gray scale diagram of a 256 base pairs simulation (fig. 5). At low temperature, the oblique lines seen in fig 2 which correspond to the motion of the nonlinear excitations, are absent from fig. 5. This shows that the disorder in k tends to trap the excitations as does the disorder in D. But when large denaturation bubble start to form at high temperature, they grow up very fast, and in a short chain with 256 base pairs the denaturation is then very sharp.

4. Analysis and discussion

In spite of the simplicity of our DNA model, we have not been able to obtain a complete analytical interpretation of the results in the presence of disorder. However, the partial results that can be derived for the dynamics of the nonlinear excitations in the presence of localized defects or on the statistical mechanics of the disordered model shed some light on the numerical results.

4.1 Influence of a localized inhomogeneity

The variations of k_n or D_n from one base pair to another can be introduced in the evolution equation for the stretching by treating them as a space variation $\varepsilon f(x)$ around the average value of k or D. The continuum limit generalisation of Eq. (3) with disorder becomes then

$$m\frac{\partial^2 y}{\partial t^2} - a^2\frac{\partial}{\partial x}\left\{k[1+\varepsilon_k f_k(x)]\frac{\partial y}{\partial x}\right\} - 2\alpha D[1+\varepsilon_D f_D(x)]e^{-\alpha y}\left(e^{-\alpha y}-1\right) = 0\,, \qquad (7)$$

As in the homogeneous case, we can look for a solution as a small amplitude expansion given by Eq. (5). The NLS equation is generated by the $\exp(i\omega t)$ term in the expansion. In the homogeneous case all the contributions to this term are of order ε^2 because the zeroth order contributions cancel when ω is chosen to be at the lower edge of the dispersion curve of the lattice, $\omega = 2D\alpha^2/m$, and no first order term appears in the expansion. When D depends on space, the cancellation of the zeroth order terms is no longer possible everywhere. Therefore a zeroth order term appears in the $\exp(i\omega t)$ factor. Moreover, the inhomogeneities in k generate a first order term. These two extra terms are the leading terms of the perturbation due to disorder. In the inhomogeneous case, the NLS equation derived from the small amplitude expansion is then

$$i\frac{\partial F_1}{\partial s} + \frac{c_0^2}{2\omega}\frac{\partial^2 F_1}{\partial X^2} + 2\omega\alpha^2|F_1|^2F_1 = -\frac{c_0^2\varepsilon_k}{2\omega\varepsilon}\frac{\partial f_k}{\partial x}\frac{\partial F_1}{\partial X} + \frac{\omega\varepsilon_d}{2\varepsilon^2}f_D(x)F_1(X)\,. \qquad (8)$$

The r.h.s. shows that, in the presence of disorder, a small amplitude expansion is only valid for a very weak modulation of the model parameters ($\varepsilon_k < \varepsilon$, $\varepsilon_D < \varepsilon^2$). In our numerical simulation, we have used $\varepsilon_k = \varepsilon_D \approx 0.3$. Although this amount of disorder is too high for the expansion to be quantitatively correct, the simulations show that it can still provide a qualitative understanding of the dynamics of the nonlinear excitations in the model. In the particular case of an abrupt change in the model parameter described by a function $f(x) \sim [\theta(x) - \theta(x - L)]$ where $\theta(x)$ is the step function and L designs the size of the defect, the r.h.s. of Eq. (8) can be treated as a perturbation in the adiabatic approximation [15]. The results show that the inhomogeneities provide and effective potential that can trap the breathers in agreement with the numerical simulations. Thus the small amplitude expansion can provide a description of the dynamics of the molecule in the low temperature range, even in the presence of disorder. As noticed before, it breaks down at higher temperatures when the bubble boundaries become the dominant excitations in the system. Since we have no proper analytical description of this regime in the homogeneous case, a basis for an analytical analysis of disorder in this regime is still lacking.

4.2 Statistical mechanics of the disordered nonlinear chain

The statistical mechanics of an inhomogeneous system can be formally treated with the transfer operator method along the lines suggested by Fogel [16]. The partition function of a chain of N base pairs with periodic boundary conditions

$$Z = \int_{-\infty}^{+\infty} dy_0 dp_0 \prod_{n=1}^{N} dy_n dp_n e^{-\beta H(\{p_n, y_n\})} \delta(y_N - y_0) \delta(p_N - p_0) \,, \qquad (9)$$

(with $\beta = 1/(k_B T)$) can be separated into the momentum part

$$Z_p = (2\pi m k_B T)^{N/2} \qquad (10)$$

and the displacement part Z_y. Since the coupling involves only nearest neighbor interactions, Z_y can be expressed in the form

$$Z_y = \int_{-\infty}^{+\infty} dy_0 \prod_{n=1}^{N} dy_n e^{-\beta f_n(y_n - y_{n-1})} \delta(y_N - y_0) \,. \qquad (11)$$

In an inhomogeneous system, the functions f_n depend on the particular site n so that the transfer operator depends on n. For DNA, we have to consider the two types of base pairs, $A-T$ and $G-C$. For the type of sequences that we have considered, the introduction of one $G-C$ base pair in an $A-T$ sequence results in a change of the corresponding value of D and in *one* coupling constant k. Therefore the outcome is that one transfer operator $\mathbf{K}'$ is introduced in a chain of the transfer operators $\mathbf{K}$ of the $A-T$ sequence.

To treat the disordered system, it is convenient to introduce a bra-ket notation. The transfer operator $\mathbf{K}$ acting on a ket $|f\rangle$, $\mathbf{K}|f\rangle = |g\rangle$ corresponds to the functionnal expression

$$\int K(y,y')f(y')dy' = g(y). \tag{12}$$

If we define its eigenkets by

$$\mathbf{K}|p\rangle = e^{-\beta\epsilon_p}|p\rangle \ , \tag{13}$$

the partition function Z_y given by Eq.(11) can be expressed as

$$Z_y = \sum_p \langle p|\mathbf{K}_n \dots \mathbf{K}_1|p\rangle \ . \tag{14}$$

Under this form, the result for the homogeneous chain where all the $\mathbf{K}$ are the same, and are indeed diagonal on the basis of their eigenkets, is immediately obtained,

$$Z_y = \sum_p \langle p|e^{-N\beta\epsilon_p}|p\rangle = \sum_p e^{-N\beta\epsilon_p} \ , \tag{15}$$

and an extension to an inhomogeneous chain can be derived easily. For this purpose, we select one type of base pair (for instance $A-T$) as the reference type *i.e.* we chose to expand on the eigenkets of its transfer operator $\mathbf{K}$. The partition function is now of the form

$$Z_y = \sum_p \langle p|\mathbf{K}_N\mathbf{K}_{N-1} \dots \mathbf{K}'\mathbf{K} \dots \mathbf{K}' \dots \mathbf{K}_1|p\rangle \ . \tag{16}$$

In this expression, unit 1 has been assumed to be of type $A-T$ but this is not a restriction because we can chose the unit labelled 1 due to the periodic boundary conditions. Let us denote by M the number of $G-C$ units in the chain of N base pairs, and by (l_i-1) the number of base pairs between the $(i-1)^{th}$ and the ith $C-G$ unit ($\sum l_i = N$, see fig. 6). When

Figure 6 : Schematic picture of a disordered DNA chain defining the notations used for the calculation of the partition function. The $A-T$ base pairs are indicated by the triangles and the $G-C$ base pairs are indicated by the circles.

the evaluation of Z_y is performed according to Eq. (16), starting from unit 1, since the $\mathbf{K}'$ operators are not diagonals on the basis contrary to the $\mathbf{K}$ operators, each time a $\mathbf{K}'$ is encountered, a summation on a new index is introduced so that we get

$$Z_y = \sum_{p_1}\sum_{p_2}\ldots\sum_{p_M} e^{-\beta\epsilon_{p_1}[l_{M+1}+l_1-1]}\, e^{-\beta\epsilon_{p_2}[l_2-1]}\, e^{-\beta\epsilon_{p_3}[l_3-1]}\ldots e^{-\beta\epsilon_{p_M}[l_M-1]}$$
$$\langle p_1|\mathbf{K}'|p_M\rangle\,\langle p_M|\mathbf{K}'|p_{M-1}\rangle\,\ldots\langle p_2|\mathbf{K}'|p_1\rangle\,. \tag{17}$$

Therefore the results contains summations on all the eigenstates of the transfer operator $\mathbf{K}$, including its continuum spectrum. For the homogeneous case, the summation that appears can be truncated because the lowest eigenvalue ϵ_0 dominates the result in the thermodynamics limit due to the factor N in Eq. (15). This is no longer possible in the disordered case because the summations are only carried between two $G-C$ units. Consequently, although Eq. (17) gives a formal solution for the partition function of the disordered molecule, in the particular case of our model with the Morse potentials, only an approximate expression has been obtained. The idea of the approximation is that in DNA the transfer operators $\mathbf{K}$ and $\mathbf{K}'$ differ only by the potential coefficients and coupling constant k, but have the same general expression. Thus, the non diagonal terms of $\mathbf{K}'$ on the eingenstates of $\mathbf{K}$ can be expected to be small. The lowest approximation is simply to neglect these terms and take into account only the difference in the diagonal terms between $\mathbf{K}$ and $\mathbf{K}'$. In this approximation, the partition function Z_y becomes

$$Z_y = \sum_p e^{-\beta\epsilon_p(M-N)}\, e^{-\beta\epsilon'_p(M)}\,, \tag{18}$$

where ϵ'_p is defined by $\exp(-\beta\epsilon'_p) = \langle p|\mathbf{K}'|p\rangle$. Within this approximation, the thermodynamics limit simply gives

$$Z_y = e^{-\beta\epsilon_0(M-N)}\, e^{-\beta\epsilon_0'(M)} . \tag{19}$$

The calculation of $\langle y\rangle$ can proceed along the same path. In the disordered system, we must perform a thermal average at a given site and then an average on the sites of the molecule. Chosing to label N the particular site that we investigate (as before the label is free since we assume periodic boundary conditions), the value of $\langle y_N\rangle$ is obtained as

$$\langle y_N\rangle = \frac{1}{Z_y}\sum_{p_{M+1}}\sum_{p_1}\cdots\sum_{p_M}\langle p|y|p\rangle e^{-\beta\epsilon_{p_{M+1}}[l_{M+1}+l_1-1]}\,\langle p_{M+1}|\mathbf{K}'|p_M\rangle$$
$$e^{-\beta\epsilon_{p_M}[l_M-1]}\,\langle p_M|\mathbf{K}'|p_{M-1}\rangle\,\ldots e^{-\beta\epsilon_{p_1}[l_1-1]} \tag{20}$$

This expression shows that $\langle y_N\rangle$ depends on the position of site N with respect to all other $G-C$ units in the molecule, as expected. Using the same "quasi-diagonal $\mathbf{K}'$ approximation", $\langle y_N\rangle$ is given by

$$\langle y_N\rangle = \frac{1}{Z_y}\sum_p\langle p|y|p\rangle e^{-\beta\epsilon_p(M-N)}\, e^{-\beta\epsilon_p'(M)} . \tag{21}$$

Within this approximation, the average over the molecule gives a very simple expression

$$\langle y\rangle = (N-M)\langle y\rangle_A + M\langle y\rangle_G , \tag{22}$$

where $\langle y\rangle_A$ is the mean value of y in a pure $A-T$ chain and $\langle y\rangle_G$ is the mean value of y in a pure $G-C$ chain.

In the continuum limit the eigenvalues of the transfer operator can be derived by solving a Schrödinger-like equation for a particle in a Morse potential[7]. With the model parameters that we have used, the transfer operator has only one discrete eigenvalue. At low temperature, this eigenvalue is far from the lowest eigenvalue in the continuum and thus the approximation which neglects its coupling with the continuum is valid. In this temperature range the numerical simulations show that, in a given part of the molecule, $\langle y\rangle$ depends stongly on the local sequence of base pairs and is larger for the $A-T$ base pairs. This is in agreement with the "quasi-diagonal $\mathbf{K}'$ approximation". It should be noticed that the existence of different denaturation temperatures for different regions of a DNA molecule is also observed experimentally. The approximation breaks down at high temperatures when the discrete eigenstate of the transfer operator approaches the dissociation energy of the Morse potential. The coupling with the continuum can no longer be neglected and our numerical result suggest that it is much larger for a disorder in k than for a disordred in D. It is not surprising to find that the role of the off-diagonal terms is larger when the disorder affects the coupling between sites, but our approximation does not allow us to estimate quantitatively this effect.

4.3 Discussion and conclusion

The results presented in this paper can be analyzed either in terms of the physics of DNA denaturation or in terms of the properties of nonlinear excitations in disordred systems.

Considering the physics of DNA, our results exhibit two opposite effects of the disorder on the thermal denaturation. In the low temperature range the regions where the dissociation energy of the Morse potentials is reduced promote a local denaturation. On the contrary, in the high temperature range where the mobility of the boundaries of the denaturation bubbles is essential for the denaturation, the disorder has a slight inhibitory effect. However the overall influence of the inhomogeneities is small in most of the investigated cases, except when the disorder is introduced in the coupling constant between adjacent base pairs. In our simulations, the amount of disorder is about 30%, a high value for real DNA. Even though the number of hydrogen bonds in a base pair varies from 2 to 3, the potential of the model is an effective potential which includes many other interactions that reduce the relative change from one type of base pair to another. Therefore, the results obtained for our simple DNA model suggest that the inhomogeneities do not have a strong effect on DNA denaturation.

Considering the dynamics of nonlinear excitations in disordred systems, it should be noticed that, in spite of our attempt to study nonlinearity *WITH* disorder, we have in fact observed nonlinearity *AND* disorder. Depending on the temperature range, one of the two dominates the other. In the low temperature domain, the disorder governs the dynamics. The nonlinearity is indeed present since the localized excitations that we observe in the disordered systems are similar to the one found in the homogeneous system, but they are strongly localized by the disorder and hardly move. On the contrary, in the high temperature range, the effect of the disorder becomes very small. As noticed above, the mobility of the bubble boundaries is slightly reduced in the disordered model, but this effect stays marginal. The dynamics of the model and its denaturation rate are very similar to the properties of the homogeneous sytem. The only significant difference is observed when the disorder is introduced in the coupling constant. Consequently the simulations of a system in thermal equilibrium confirm the results obtained for isolated solitons[17]: large amplitude nonlinear excitations are only weakly affected by disorder.

ACKNOWLEDGEMENTS

Part of this work has been performed during a stay in the Center for Nonlinear studies of the Los Alamos National Laboratory. I would like to thank D. K. Campbell for enlightening discussions during this visit and for the opportunity he gave me once more to spend some time in the stimulating atmosphere of the CNLS. I want to thank S. Pope and E. Loth for their assistance in the use of the Connection Machine, and A. Bishop and Yuri Kivshar for fruitful discussions. Some of the numerical calculations have been performed on a Telmat T-Node financed by the Parallel Computing Action of the CEC (grant PCA 4214).

REFERENCES

1. S. W. Englander, N. R. Kallenbach, A. J. Heeger, J. A. Krumhansl and S. Litwin, Proc. Nat. Acad. Sci. USA **777**, 7222 (1980)
2. S. Yomosa, Phys. Rev. A**27**, 2120 (1983) and A**30**, 474 (1984)
3. S. Takeno and S.Homma, Prog. Theor. Phys. **77**, 548 (1987)
4. Chun-Ting Zhang, Phys. Rev. **A35**, 886 (1987)
5. V. Muto, J. Halding, P.L. Christiansen and A.C. Scott, J. Biomol. Struct. Dyn. **5**, 873 (1988)
6. V. Muto, A.C. Scott and P.L. Christiansen, Physics Letters A **136**, 33 (1989)
7. M. Peyrard and A.R. Bishop, Phys. Rev. Lett **62**, 2755 (1989)
8. P. W. Anderson, Phys. Rev.**109**, 1492 (1958)
9. Y. Gao and E. W. Prohofsky, J. Chem. Phys. **80**, 2242 (1984)
10. Y. Gao, K. V. Devi-Prasad and E. W. Prohofsky, J. Chem. Phys. **80**, 6291 (1984)
11. S. Nose, J. Chem. Phys. **81**, 511 (1984)

12. L.L. Van Zandt, Phys. Rev. A **40**, 6134 (1990), M. Techera, L. Daemen and E.W. Prohofsky, Phys. Rev. A **42**, 5033 (1990), Van Zandt, Phys. Rev. A **42**, 5036 (1990)

13. E.W. Prohofsky,K.C. Lu,L.L. Van Zandt and B.P. Putnam, Physics Letters A **70**, 492 (1979), H. Teitelbaum and S.W. Englander, J. Mol. Biol. **92**, 55 (1975)

14. S. E. Trullinger and K. H. Sasaki, Physica D **28**, 181 (1987)

15. Y. Kivshar and B. Malomed, Rev. Mod. Phys. **61**, 763 (1989) and private communication

16. M. B. Fogel, *Nonlinear order parameter fields : I Soliton dynamics, II Thermodynamics of a model impure system.* phD Cornell University (1977)

17. Yu. Kivshar, S.A. Gredeskul, A. Sanchez and L. Vasquez, Phys. Rev. Lett. **64**, 1693 (1990)

Nonlinear Dynamics in a Double Chain Model of DNA

K. Forinash[1], *A.R. Bishop*[2], *and P. Lomdahl*[2]

[1]Indiana University Southeast, New Albany, IN 47150, USA
[2]Los Alamos National Laboratory, Los Alamos, NM 87545, USA

Abstract. We investigate numerically the mechanics of a simple lattice model of DNA. The model consists of two linear mass chains connected by a Morse potential representing the hydrogen bonding between the sugar- phosphate backbones. Effects on discrete breather type solutions caused by random noise and mass defects, representing interaction with the local environment, are studied.

I Introduction

Although an understanding of the chemical changes involved in transcription and replication of DNA and RNA rests on fairly secure footing [1], the dynamics of these processes are still being examined. This is due in part to the complexity of the problem and the probability that nonlinear vibrational modes are somehow involved [2]. Several rather simple models have been proposed [3,4,5,6] which seek to explain the open states of DNA and RNA as nonlinear excitations along the chain stretching or even breaking the hydrogen bonding between the sides of the chain. Although it is not established that open states of DNA initiate transcription or replication, since open states appear randomly along chains which are not replicating, it does seem plausible that open states do play some nucleative role in the replication process, especially in the melting regime. Improvements have been made in many of these models as additional complications such as more realistic potentials and discreteness effects are taken into account. One such model which has met with some success in predicting a correct melting temperature for DNA using reasonable bonding parameters as input is due to Peyrard and Bishop [7]. They examined the statistical mechanics of a model consisting of two chains of masses connected linearly along their length and connected via a Morse potential between the two chains. The work described here investigates the dynamics of their model by numerical simulation. In addition, the effects of various mass defects representing interaction of the chain with its surroundings have been studied. Additional studies along these lines, including finite temperature effects, may be found in refs. (8,9,10).

II The model

The model consists of two chains of masses connected by linear springs along their length with the addition of nonlinear coupling between masses of each chain. The Hamiltonian for the system is

Springer Proceedings in Physics, Vol. 67 Nonlinearity with Disorder
Editors: F. Abdullaev · A.R. Bishop · S. Pnevmatikos

$$H = \Sigma_n \ 1/2 \ m \ (u'^2_n + w'^2_n) + 1/2 \cdot k \ [(u_n - u_{n-1})^2 + (w_n - w_{n-1})^2] + V(u_n - v_n) \ , \qquad (1)$$

where u is the top chain displacement from equilibrium and w is the bottom chain displacement. Here k represents the linear coupling strength along the top and bottom chains. The corresponding velocities are u' and w'. The Morse potential was chosen to represent the inter-chain hydrogen bonding for the model DNA:

$$V(u_n - w_n) = D\{ \exp[-a(u_n - w_n)] - 1 \}^2 \ . \qquad (2)$$

The parameters a and D were chosen to correspond to realistic values for the inter-chain hydrogen bond in the DNA molecule.

The variable changes $x_n = (u_n + w_n)/\sqrt{2}$ and $y_n = (u_n - w_n)/\sqrt{2}$ separate the Hamiltonian into in-phase and out-of-phase components. Only the out-of-phase motion stretches the hydrogen bond. The Hamiltonian for out-of-phase motion leads to an equation of motion

$$m\, \partial^2 y_n/\partial t^2 - k\ (y_{n+1} + y_{n-1} - 2y_n)^2 - 2\sqrt{2}Da \exp(-\sqrt{2}a y_n)\ [\exp(-\sqrt{2}a y_n) - 1] = 0 \ . \qquad (3)$$

All figures shown were generated by time stepping through the integration of the equations of motion for the connected chains using a fifth order Runga-Kutta method [12]. Energies were calculated directly from the model and were conserved in all cases (except the random noise cases discussed below) to better than 0.001%. The masses are constrained to move only in the vertical direction and the ends of the chains were left free in order to more accurately reflect biological conditions. The effect of also including longitudinal motion has been examined recently by Muto et. al. [8].

III Solutions and Stability

Remoissenet [11] has shown that the Hamiltonian (1) above leads, in leading nonlinear order and using a multiple-scale expansion method, to a nonlinear Schrodinger equation with discrete breather solitary waves of the form

$$y_n(t) = \varepsilon A \operatorname{sech}\chi \cos(Kna - \Omega t) - 2\alpha e^2 A^2 \operatorname{sech}^2\chi + (\alpha A^2\varepsilon^2/3)\operatorname{sech}^2\chi \cos[2(Kna - \Omega t)]\} \ , \qquad (4)$$

where

$$A = [(u_e^2 - 2u_e u_c)/2PQ]^{1/2} \qquad (5a)$$

$$\chi = \varepsilon(na - V_e t)/L_e \ , \qquad (5b)$$

$$\Omega = (\varepsilon u_e/2P)(V_g + \varepsilon u_c)^{1/2} + \omega \ , \qquad (5c)$$

$$K = \kappa + \varepsilon(u_e/2P) \ , \qquad (5d)$$

$$P = (ka^2/2m\ \omega)\ [\cos(\kappa a) - (k/m\ \omega^2)\sin^2(\kappa a)]\ , \tag{5e}$$

$$Q = (\omega_o^2/\ 2\omega)\ \{4\ \alpha^2 - 2\ \alpha^2/[3+(16k/m\ \ \omega_o^2)\ \sin^4(\kappa\ a/2)]\ \ -3\beta\}\ , \tag{5f}$$

$$\omega^2 = \omega_o^2 + \ 4(k/m)\sin^2(\kappa\ a/2)\ , \tag{5g}$$

$$\omega_o^2 = 4Da^2/m\ , \tag{5h}$$

$$\alpha = -6\sqrt{2}\,a/\ \varepsilon\ , \tag{5i}$$

$$\beta = 7a^2/\ 3\varepsilon^2\ . \tag{5j}$$

Here $V_g = (ka/m\ \omega)\ \ \sin(\kappa\ a/2)$ is the group velocity, $L_e = 2P/(u_e^2 - 2u_eu_c)^{1/2}$, is the breather width, a is the lattice spacing, κ is the linear carrier wave vector, u_e is the envelope velocity, u_c is the carrier wave velocity and ε is an arbitrary scaling parameter which controls the (coupled) amplitude and width of the breather. The parameters a, D and k determine the strength of the coupling.

The numerical simulation used these solutions as initial conditions and found them to be reasonably stable for a wide range of coupling parameters including the biologically significant values $D=0.33$ eV, $a=1.8Å^{-1}$ in the Morse potential and linear spring constant $k=0.003$ eV/$Å^2$. Values examined ranged from 10 to 0.5 for a; 10 to 0.2 for D and 20 to 0.003 for k. It is important physically that for the biological parameters quoted above the breathers are very narrow and discrete lattice pinning effects are expected [13]. Pinning was indeed seen in our simulations (Fig. 1) and for this reason slightly larger values of the parameters were chosen for most of the simulations so that effects not dominated by pinning could be examined. It remains to be seen whether discreteness effects will ultimately dominate biological conditions in realistic potentials and chemical and geometric environments.

Equation (3) is not integrable in closed form with the result that the anharmonic phonons described by equation (4) are not true breather solitons and decay into other modes eventually. The amplitude of the breathers described in this paper decreases by approximately 25% over a time interval of 100 time units (10,000 steps at intervals of 0.01).

The adjustable parameter ε in eq (4) determines the maximum amplitude and also the width of the initial breather. The masses in the chain were launched with velocities corresponding to derivatives of their position and as a consequences, ε also has some effect on the velocity of the moving breather. Large amplitude breathers are unstable, breaking up after after 20 or so time units.

The carrier wave vector, κ, controls the shape of the breather. For the case $0< \kappa <1$, coefficients in the nonlinear Schrodinger equation are in the correct range for a breather solution [14]. Other signs and values of κ, giving rise to other types of nonlinear pulses such as "dark" solitons, asymmetric envelopes etc., were not examined. The parameter κ also has the largest effect on the velocity of the breather.

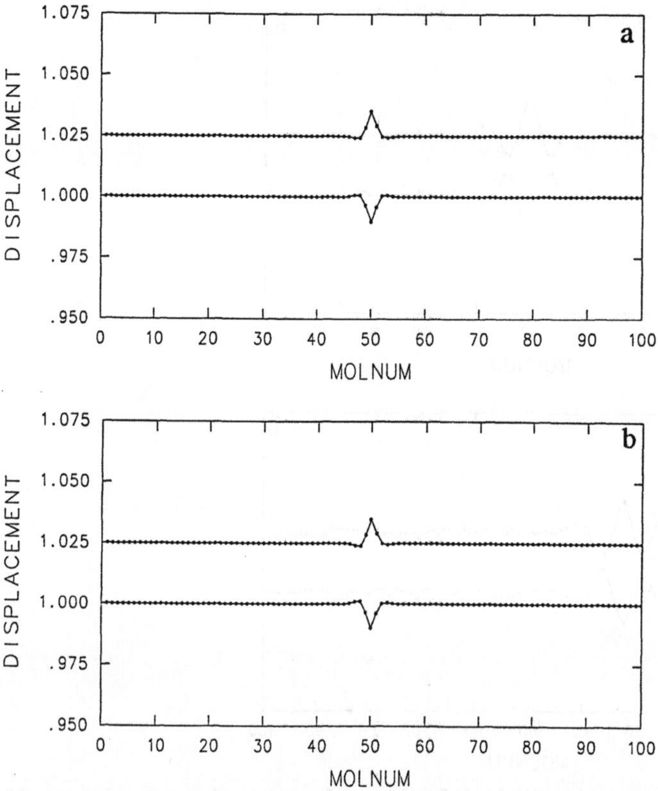

Fig. 1 A breather pinned by the discreatness of the latice. Figure 1a is the initial condition and figure 1b is at 97.6 time units. Parameter values are k = 0.01, κ = 0.1, D = 0.5, m = 1.0, a = 1.5 and ε = 0.4.

Larger values of κ (with the restriction $0< \kappa <1$) cause more initial velocity asymmetry and therefore greater breather speeds along the chain.

Stability of the breathers was tested upon the introduction of random noise both as a simple initial condition and also in the case of continuous application of random noise. In the case of initial random noise the breathers were as stable as non perturbed initial conditions. This was true up to the point where random noise began to mask the breather shape. Energy was conserved in all of these cases. Stability was also tested for the introduction of random noise every 20 time steps, approximately representing a chain in contact with a thermal bath. A quantitative measure of stability is difficult but in these cases the breathers appeared quite stable for reasonably long time runs, typically 100 time units (Fig. 2). Energy was not conserved exactly in these trials. The effects of thermal noise on single and double chains has been extensively examined by Muto et. al. [8].

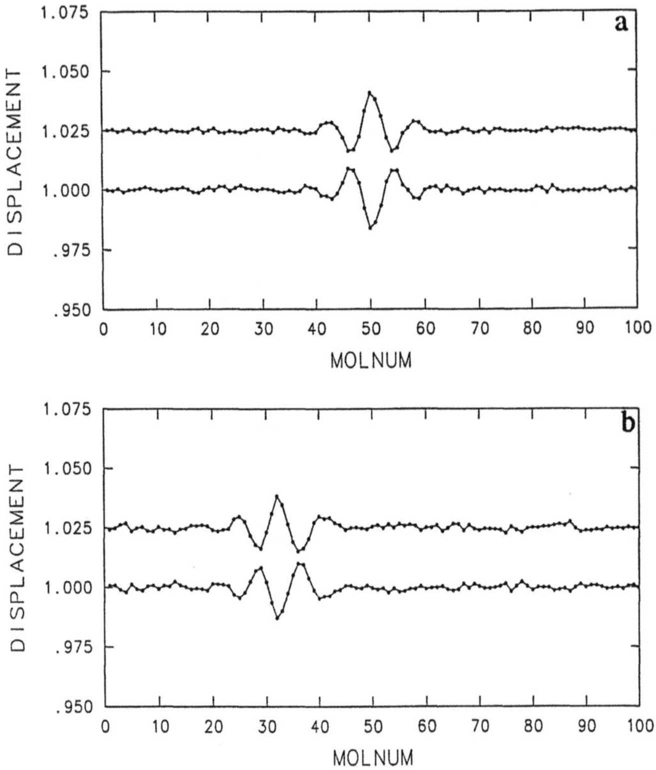

Fig. 2 A breather moving to the left with random noise introduced every 20 time steps. Figure 2a is the initial configuration and figure 2b is at 77 time units. Parameter values are $k = 1.0$, $\kappa = 0.95$, $D = 1.0$, $m = 1.0$, $a = 1.5$ and $\varepsilon = 0.01$.

IV Single mass defects

In order to examine the role of large mass defects along the chain, single mass defects varying between 0.25 and 4 times the original mass were placed at the center of a breather and the time evolution was followed. Any defect greater than approximately $\pm 10\%$ of the original mass was a sufficient disturbance to cause the breather to split up relatively quickly (Fig. 3). Breathers with defects less than approximately 5% appeared to be stable for the time scales examined.

A more quantitative approach for examining the relationship between linear defect modes and breather modes is to calculate single particle velocity power spectrums. This was done using the maximum entropy method [12] to find the single particle power spectrum of the center (defect) mass on the chain. Linear theory predicts for a single mass defect in a linear discrete chain with free boundary conditions that lighter defects will give rise to discrete (local mode) frequencies *above* the band of (extended) normal modes [15].

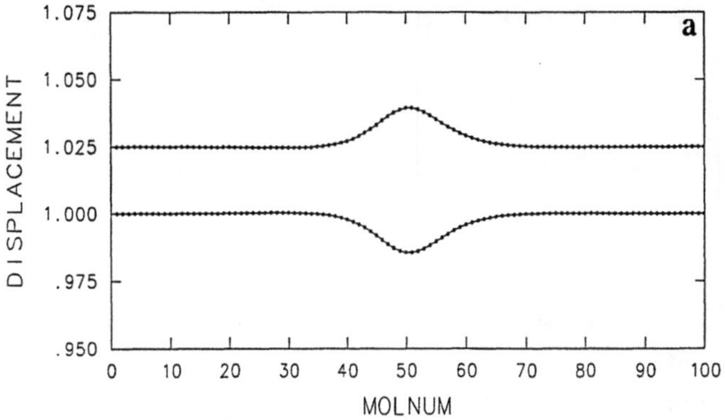

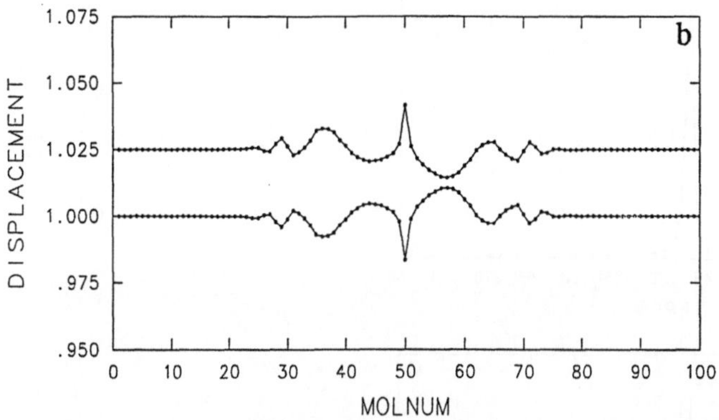

Fig. 3 A nearly stationary breather as an initial condition (3a) and at 74 time units (3b) showing breakup due to a mass defect of m = 4 located at the center of the chain. Parameter values are $k = 1.0$, $\kappa = 0.1$, $D = 1.0$, $m = 1.0$, $a = 1.5$ and $\varepsilon = 0.01$.

Assuming the (linear) coupling to be complicated enough (more than one linear coupling or second nearest neighbor interaction) so that the normal mode band does not extend to zero frequency, heavier defects will give rise to frequencies *below* the band. For only nearest neighbor linear interactions the heavier impurity modes occur in the band as resonances. The band of linear normal mode frequencies for a linearized Morse potential is give by:

$$\omega = \left[(k/m)(\lambda - 2\cos\phi)\right]^{1/2}$$

where $\lambda = 2 + 4Da^2/k$. (6)

Linear modes for single mass defects on the double chain system were examined by displacing a single mass defect by a small distance and finding the single particle frequency spectrum. Frequencies for breathers without defects were found to lie within in the linear band of normal modes. When single mass defects were added to the breather there was some interaction of modes as can be seen in Fig. 4.

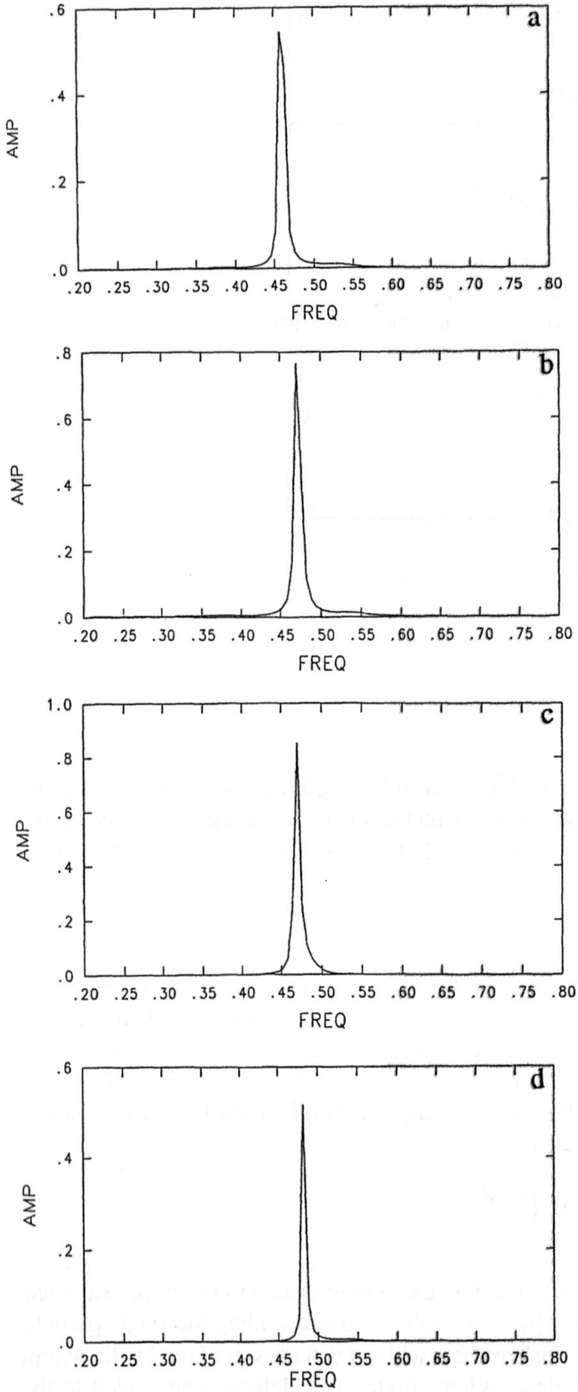
a
AMP
.0
.2
.4
.6
.20 .25 .30 .35 .40 .45 .50 .55 .60 .65 .70 .75 .80
FREQ
b
AMP
.0
.2
.4
.6
.8
.20 .25 .30 .35 .40 .45 .50 .55 .60 .65 .70 .75 .80
FREQ
c
AMP
.0
.2
.4
.6
.8
1.0
.20 .25 .30 .35 .40 .45 .50 .55 .60 .65 .70 .75 .80
FREQ
d
AMP
.0
.2
.4
.6
.20 .25 .30 .35 .40 .45 .50 .55 .60 .65 .70 .75 .80
FREQ

Fig. 4

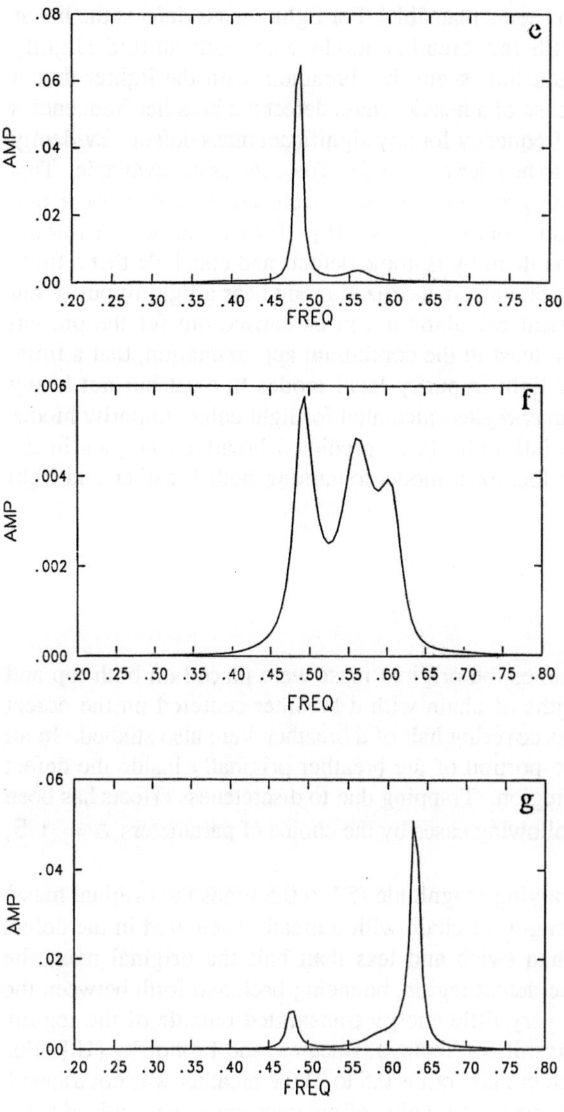

Fig. 4 Single particle frequency spectrum for the center (defect) mass on the chain. Defect ratios of 1.25 (Fig. 4a), 1.15 (Fig. 4b), 1.05 (Fig. 4c), 1.00 (no defect)(Fig. 4d), 0.95 (Fig. 4e), 0.85 (Fig. 4f), and 0.75 (Fig. 4g) are shown.

The following interpretation seems plausible. For lighter mass defects the linear defect modes appear along with the breather mode with both shifted slightly downwards in frequency. It would thus seem that breathers with the lighter defect frequency do not exist. For the case of a heavier mass defect the breather frequencies becomes locked with the defect frequency for any significant mass defect. Evidently breathers with frequencies equal to heavier mass defect frequencies are available. This is in agreement with arguments presented by Kosevich and Kovalev for a one dimensional model of the Frenkel-Kontorova type [16]. They examine anharmonic localized modes and the effect on them by isotopic defects and conclude that a finite minimum energy is needed to excite a self-localized mode near a light impurity but not near a heavy defect. Identical calculations can be carried out for the present double chain system showing, at least in the continuum approximation, that a finite minimum energy is needed for light impurity local modes to exist but not heavy impurity local modes. Minimum energies calculated for light defect impurity modes in a continuum approximation fall well above predicted breather energies in the present model, so that a single localized mode containing both breather and light impurity is not possible.

V Multiple mass defects

Multiple mass defects of two times the original mass were placed on both top and bottom chains for varying lengths of chain with a breather centered on the defect region. Uncentered mass defects covering half of a breather were also studied. In all of these cases the energy of the portion of the breather originally inside the defect region remained trapped in the region. Trapping due to discreteness effects has been excluded in this and all of the following cases by the choice of parameters $a = 1.5$, $D = 1.0$ and $k = 1.0$.

Multiple mass defects of varying magnitude (50 to 0.5 times the original mass) were placed along a 40 mass length of chain with a breather centered in the defect region. For defects greater than twice and less than half the original mass the breather remained trapped in the defect region, bouncing back and forth between the boundaries of the region with very little energy transmitted outside of the region. This is in agreement with the results of Techera, Daemen, and Prohofsky [10]. For regions with weaker defects (defect mass ratios 0.5 to 2) the breather was not trapped. Some reflection of energy and some modification of breather shape was noticed when the breather penetrated these weaker defect regions. The value of ε (and therefore the amplitude) did not appear to change the trapping of the breather in defect regions up to amplitudes large enough to cause break-up. Fig. 5 shows the reflection of a moving breather from a defect region.

A sinusoidally varying mass defect of $m(1+0.1\sin(\omega x))$ was also placed on the chain with a breather as initial conditions. When the wavelength of the defect region is much smaller than the width of the breather, the breather remains stable with small changes in its structure. A defect region of large wavelength relative to the breather width will also result in stable breathers with a coherent "particle"- like motion, well described by perturbation theory or energy arguments. In the

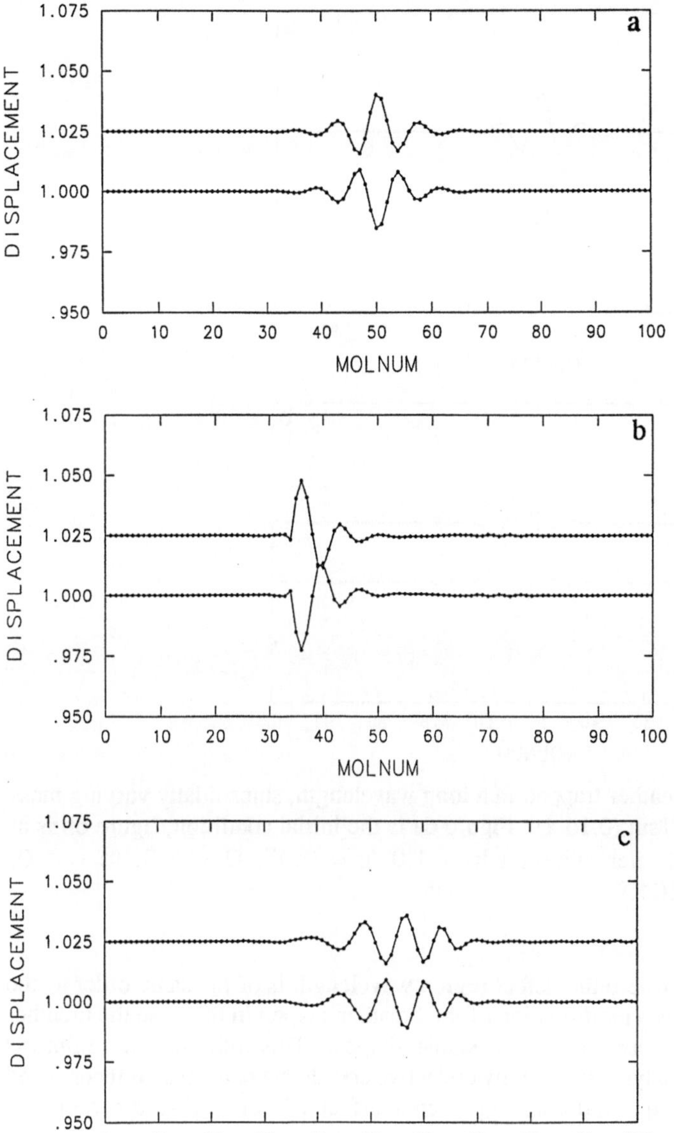

Fig. 5 Reflection of a moving breather from a defect region. Mass defects of m = 2 are loaded on both chains from mass 25 to 35. Figure 5a is initial condition, figure 5b is at 76.6 time units and figure 5c is at 159 time units. Parameter values are k = 1.0, κ = 0.95, D = 1.0, m = 1.0, a = 1.5 and ε = 0.01.

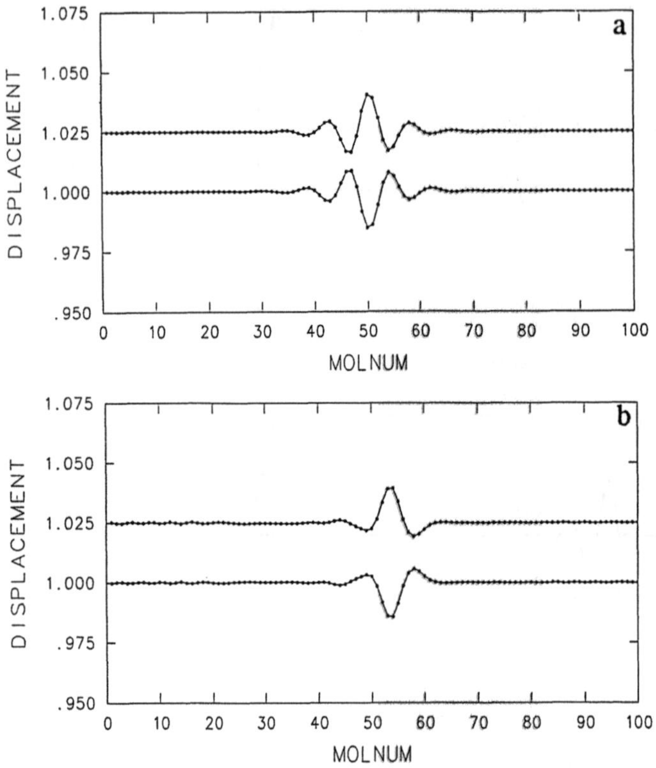

Fig. 6 Moving breather trapped in a long wavelength, sinusoidally varying mass defect of m = 1.0 +0.1sin (0.16 x). Figure 6a is the initial condition, figure 6b is at 190.0 time units. Parameter values are $k = 1.0$, $\kappa = 0.95$, $D = 1.0$, $m = 1.0$, $a = 1.5$ and $\varepsilon = 0.01$.

intermediate regime, where the defect region wavelength is of the same order as the breather width, there is a modification of the breather shape. In this case the breather oscillates irregularly between several stable shapes. This intermediate regime is difficult to describe analytically (e.g. by collective coordinate perturbation theory) and is perhaps the most qualitatively significant effect of combining disorder and nonlinearity.

Sinusoidally varying defect regions of large wavelength with smaller variances $m(1+0.01\sin(\omega x))$ do not trap moving breathers, whereas regions with larger defects $m(1+0.1\sin(\omega x))$ do trap breathers (Fig. 6). As the breather propagates some energy shifts periodically between potential and kinetic energy while maintaining constant total energy. For defects of ± 0.01 this variation in energy from kinetic to potential is larger than the total energy difference between regions with defects of +0.01m and regions of - 0.01m. For defects of ± 0.1m the variation in energy from kinetic to potential is smaller than the total energy difference between regions with defects of +0.1 and - 0.1m . Thus breathers have the required energy for propagating into weak mass defect regions but not into strong defect regions. It is also interesting to note that the breathers remain trapped in regions of higher mass (and therefore higher energy) but break up when initially placed in regions of lower mass. This is the opposite of what might be expected and is possibly due to the absence of appropriate energies for breathers in the lighter regions which start out with energies corresponding to a heavier region, as in the case of single mass defects discussed above.

VI Conclusion

Real DNA molecules and their interactions with their environment are vastly more complicated than a simple mass chain model can represent. Longitudinal motion along the chain and effects of the secondary structure of DNA, for example, has not been included here. However it is clear that nonlinear self focusing of energy provides a possible mechanism for the initiation of open states of DNA involved in melting and replication processes. It is also interesting to note from the present results that self localized coherent states are affected by small mass defects and in fact cannot be sustained in the presence of large defects. Thus a mechanism exists whereby small inhomogeneities representing genetic coding can affect the dynamics of opening and closing of the double chain system. The present model is thus intended as a step towards understanding some of the nonlinear dynamics which may be occurring along the chain during or preceding the processes of chain opening and closing.

One of us (K.F.) is grateful for support from the Theoretical Division and Center of Nonlinear Studies, Los Alamos National Laboratory, where most of this work was performed. Work at Los Alamos is performed under the auspices of the U.S. D.O.E.

References

[1] B. Alberts, D. Bray, J. Lewis, M. Raff, K. Roberts, and J.D. Watson, *Molecular Biology of the Cell* (Garland, New York,1983).

[2] E. W. Prohofsky, *Phys. Rev.* **A 38**, 1538 (1988).

[3] S. W. Englander, N. R. Kallenbach, A. J. Heeger, J. A. Krumhansl, and S. Kitwin, *Proc. Natl. Acad. Sci. U.S.A.* **777**, 7222 (1980).

[4] S. Yomosa, *Phys. Rev.* **A 27**, 2120 (1983); **A 30**, 474 (1984).

[5] S. Takeno and S. Homma, *Prog. Theor. Phys.* **77**, 548 (1987).

[6] Chung-Ting Zhang, *Phys. Rev.* **A 35**, 886 (1987).

[7] M. Peyrard and A. R. Bishop, *Phys. Rev. Lett.* **62**, 2755 (1989).

[8] V. Muto, P.S. Lomdahl, and P. L. Christiansen, Los Alamos pre-print LA-UR 89-4027.

[9] V. Muto, A. C. Scott, and P. L. Christiansen, Los Alamos pre-print LA-UR 89-4116.

[10] M. Techera, L.L. Daemen, and E.W. Prohofsky, Purdue University Preprint.

[11] M. Remoissenet, *Phys. Rev.* **B 33**, 2386 (1986).

[12] W. H. Press, B. P. Flannery, S. A. Teukolsky, and W. T. Vetterling, *Numerical Recipes* (Cambridge University Press, Cambridge, 1989).

[13] J. F. Currie, S. E. Trullinger, A. R. Bishop, and J. A. Krumhansl, *Phys. Rev.* **B 15**, 5567 (1977).

[14] S. Pnevmatikos, N. Flytzanis, and M. Remoissenet, *Phys. Rev.* **B 33**, 2308 (1986).

[15] R. F. Wallis and A. A. Maradudin, *Prog. Theo. Phys.* **24**, 1055 (1960).

[16] A. M. Kosevich and A. S. Kovalev, *Sov. J. Low Temp. Phys.* **1**, 742 (1975).

Energy Transfer in α-Helical Proteins by Fermi Resonance

M.A. Collins and D.L. Clarke

Research School of Chemistry, Australian National University,
GPO Box 4, Canberra, ACT 2601, Australia

Abstract. The transport of vibrational energy by Fermi resonance interactions is reported in three systems which model α-helical proteins. The degree of disorder in the classical and quantum dynamics is compared, and used to determine conditions under which classical mechanics is qualitatively accurate. Efficient energy transfer is demonstrated under physically reasonable conditions.

1. Introduction

This paper describes a mechanism whereby α-helical segments of proteins may provide the means of transporting energy (in this case vibrational energy) between sites of chemical reactivity. We do not attempt to prove that such a mechanism actually operates in nature: that is an experimental question. Here we demonstrate that such a mechanism is feasible in three model systems: a simple model of a hydrogen bonded chain of amide or peptide groups; a realistic simulation of a N-methylacetamide chain; and a large scale simulation of a realistic model of α-poly(L-alanine). Details of these studies have been or will be published[1,2,3]. Here, much of the detail (including references) is avoided in favour of an overview of the physical mechanism and critical features of the dynamics. Classical and quantum dynamics are considered for the simple model and used to determine the initial conditions for which classical dynamics reproduces essential features of the quantum dynamics.

The idea that vibrational energy, from the hydrolysis of adenosine triphosphate (ATP), can propagate through the hydrogen bonds of proteins as a localised pulse has been considered previously, albeit via a very different mechanism[4]. However, this localised excitation has been shown, in classical simulations, to be unstable at temperatures as high as room temperature and to be trapped at low temperature[5]. We propose a new mechanism, based on Fermi resonance processes. The paper is divided in five parts: a short description of the biological context; the physical origin of the Fermi resonance processes; a simple mathematical model; simulation of vibrations of hydrogen bonded amides (peptides) and the energy transport dynamics that result; and finally the simulation of energy transfer in an α-helical protein.

2. The biological context

Many important functions of cells are carried out within and at the surface of membrane-bound proteins. Pertinent examples include the ion pumps, such as the

Na^+/K^+ and Ca^{2+} pumps, which maintain ion concentrations within the cell against a concentration gradient. The direct supply of energy to these pumps is provided by the hydrolysis of ATP to adenosine diphosphate (ADP):

$$ATP^{4-} + H_2O \rightarrow ADP^{3-} + HPO_4{}^{2-} + H^+ \qquad (1)$$

(under normal physiological conditions). The detailed mechanism of the whole process is not clearly understood. However, in recent years the structure of membrane bound proteins has become more clearly understood. In a very few cases, X-ray diffraction provides a complete structure. In many other cases, the complete residue sequence provides a picture of the secondary structure. One conclusion from these studies is that interlocking α-helices comprise the predominant secondary structure in the membrane spanning regions of such proteins. These helices span the membrane, a distance of about 30 Å. The function of such bundles of α-helices is unclear, though there is evidence that ion channels are formed within them.

It is known that the hydrolysis of ATP takes place near the inner surface of such membrane-bound proteins. No definitive structure for these binding sites is available. However, X-ray diffraction has shown that phosphate groups bind strongly to enzymes with a βαβ secondary structure, via the formation of one (or two) hydrogen bond between a phosphate oxygen and an NH group at the terminus of the α-helical segment. This is not surprising since the dipole moment of an α-helix is equivalent to a charge of approximately +1/2 at the NH terminus. Given the predominance of α-helices in the secondary structure and the affinity of the NH terminus for binding poly-phosphate moieties, we adopt a working hypothesis that the binding of ATP to membrane-bound proteins involves, in part, the hydrogen bonding of a phosphate oxygen to an NH group at the terminus of an α-helix. This α-helix could form with others a membrane-spanning bundle enclosing an ion channel. At some point in this channel, particular residues form the active ion gates and binding sites. It seems evident that the site of ATP hydrolysis and these sites are not adjacent. Energy must be transferred between them.

Exactly how much energy is released in the kinetic process of cleaving the γ phosphate group from ATP is not known. However, it is probably in excess of 3000 cm^{-1}. If the phosphate group next to the bond cleavage is hydrogen bonded to an NH group then this might provide a pathway by which energy is transferred to the protein. The NH stretching modes of the protein, around 3300 cm^{-1}, and the NH in-plane bending modes around 1550-1650 cm^{-1}, are well placed in frequency to accept energy of this magnitude. Hence, we adopt a further speculative hypothesis: that the hydrolysis of ATP at a membrane-bound protein leads initially to the excitation of NH stretching and bending modes localised at the NH terminus of an α-helix.

3. The Fermi resonance mechanism

To understand the consequences of exciting an α-helix in this way, it is necessary to understand the characteristic vibrations of the amide or peptide group, in particular the three high frequency modes: the Amide-A, Amide-I and Amide-II modes. The

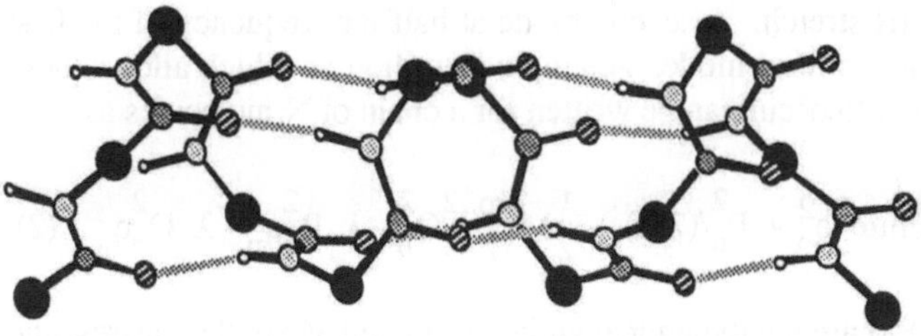

Figure 1. A section of α-helix. For clarity, substituents on the alpha carbon are not shown.

Amide-A mode is a localised NH stretching vibration. The Amide-I mode is characterised largely by C=O stretching and to a lesser extent NH in-plane bending. The Amide-II mode involves NH in-plane bending and CN stretching. The lower frequency Amide-III mode also contains NH bending. Hydrogen bonding occurs between NH and CO groups on adjacent amide groups in the condensed state, in a head to tail fashion. In the α-helix, three hydrogen bonding chains or spines are set up between peptide groups on adjacent turns, as illustrated in Fig 1.

This hydrogen bonding directly couples modes involving the NH and CO stretching. A strong feature of the vibrational spectroscopy of amides is the large frequency shift of some modes from their values in the gas phase to those in condensed phases where hydrogen bonding occurs. The NH stretching modes shift from about 3500 cm^{-1} (gas phase) to 3280-3310 cm^{-1} in peptides and condensed secondary amides. The Amide-I mode shifts from near 1730 cm^{-1} (gas phase) to 1650-1670 cm^{-1} (condensed phases). This is due partly to a weakening of the C=O diagonal force constant on hydrogen bonding. Either O....H contraction or NH stretching could lead to a further weakening of the C=O force constant. Thus we expect significant cubic coupling terms in the potential energy, which are linear in the NH (or O....H) stretching coordinate and quadratic in the Amide-I mode on the adjacent amide group.

The NH stretching vibration is kinematically coupled to the Amide-I mode on the same molecule, since the later mode involves NH in-plane bending. This coupling has been studied in recent years in systems with CH and OH groups, and is basically the coupling of a pendulum's period (CNH bending) to its length (NH stretching).

The following scenario becomes possible: upon excitation of the NH stretch, energy will be transferred into the Amide-I mode (and possibly the Amide-II mode) by an intramolecular (stretch-bend) Fermi resonance and therefore into the CO stretch. This group is hydrogen bonded to the NH on the next peptide group, and is in 2:1 resonance with its stretch. Energy will be transferred by an intermolecular Fermi resonance into the next molecule, and so on. A flow of energy, by nonlinear resonances, becomes possible along the hydrogen bonded spine of an α-helix.

4. A simple model

Consider a simple model. Assign each trans secondary amide group in a hydrogen bonded chain just two relevant internal vibrational modes. One is a high frequency mode representing the NH stretch. A second mode at half the frequency of the first represents (primarily) the Amide-I mode. A simple Hamiltonian which allows these Fermi resonance processes to occur can be written for a chain of N molecules as

$$H = \sum_{n=1}^{N} p_n^2/(2m) + \frac{1}{2} m\omega^2 q_n^2 + P_n^2/(2M) + \frac{1}{2} M(\frac{\omega}{2})^2 Q_n^2 - \lambda_1 P_n^2 q_n - \lambda_2 Q_n^2 q_{n-1} \quad (2)$$

,where (q_n,p_n) and (Q_n,P_n) are the dynamical variables, m and M are the masses, and ω and $\omega/2$ are the frequencies for the two modes on each molecule; λ_1 and λ_2 are coupling constants (positive for the mechanisms outlined above). Defining the usual annihilation operators a_n and A_n (for the high and low frequency modes) and conjugate creation operators a_n^* and A_n^*, the Hamiltonian of Eq.(2) is then

$$H = \sum_{n=1}^{N} \hbar\omega(a_n^* a_n + \frac{1}{2}) + \hbar\frac{\omega}{2} (A_n^* A_n + \frac{1}{2}) + \alpha_1 (A_n - A_n^*)^2 (a_n + a_n^*)$$
$$- \alpha_2 (A_n + A_n^*)^2 (a_{n-1} + a_{n-1}^*) \quad . \quad (3)$$

We can assume that α_1 and α_2 are small. Eigenenergies of H which are correct to first order in the small parameters, α_1 and α_2, can be evaluated by perturbation theory. However, not all the terms in H make a contribution to the eigenenergies in first order. If we retain only these terms, we have

$$H = \sum_{n=1}^{N} \hbar\omega(a_n^* a_n + \frac{1}{2}) + \hbar\frac{\omega}{2} (A_n^* A_n + \frac{1}{2}) + \alpha_1 (A_n^2 a_n^* + A_n^{*2} a_n)$$
$$- \alpha_2 (A_n^2 a_{n-1}^* + A_n^{*2} a_{n-1}). \quad (4)$$

The model Hamiltonian has been reduced to one which describes a chain of sites, each containing two oscillators in 2:1 frequency ratio, where one quantum of a high frequency mode can be created or destroyed only as two quanta of the low frequency mode are destroyed or created (at different rates within and between sites).

Classical Soliton. The same Hamiltonian can be derived classically. Hamilton's equations for the annihilation operators are then

$$-i\hbar \frac{d a_n}{dt} = \alpha_1 A_n^2 - \alpha_2 A_{n+1}^2 \; ;$$
$$-i\hbar \frac{d A_n}{dt} = 2 \left(\alpha_1 a_n - \alpha_2 a_{n-1} \right) A_n^* \quad . \quad (5)$$

While it may not appear obvious a first glance, these equations are a generalisation of the equations of motion for the Toda lattice[1]. From Eq. (5),

$$\frac{d}{dt}\left(\frac{d A_n}{dt} / A_n^*\right) = \left[2\alpha_1\alpha_2 \left(A_{n+1}^2 + A_{n-1}^2 - 2A_n^2 \right) - 2(\alpha_1 - \alpha_2)^2 A_n^2 \right] \hbar^{-2} \; .(6)$$

If all A_n are real at some instant, they will remain so. Then Eq. (6) contains the derivative of $\ln(A_n)$. Defining $y_n = 2 \ln (A_n)$, Eq. (6) becomes

$$\frac{d^2 y_n}{dt^2} = \left[4\alpha_1\alpha_2 \left(e^{y_{n+1}} + e^{y_{n-1}} - 2e^{y_n} \right) - 4\left(\alpha_1 - \alpha_2 \right)^2 e^{y_n} \right] \hbar^{-2} . \quad (7)$$

When $\alpha_1 = \alpha_2$, Eq. (7) is just Newton's equation for the bond lengths in a one dimensional lattice with nearest neighbour exponential forces. This Toda lattice is well known because it is an integrable system which supports nonlinear sound waves, including solitons which are localised compression pulses. Here, we have localised pulses of excitation in both high and low frequency modes propagating together along the chain. This is an unexpected result, because these vibrations are optical modes not acoustic modes like those in the Toda lattice. The group velocity for a simple packet of optical modes is zero; yet here we see pulse propagation. The quantum dynamics make this result a little clearer.

Quantum dynamics. The ground state $| O >$ of the Hamiltonian, Eq. (4), is given by a product of ground states in each oscillator. Assuming periodic boundary conditions, an obvious ansatz for the excited states corresponding to single excitation of the high, or double excitation of the low, frequency modes is given by $| \Psi(k) >$:

$$| \Psi (k) > = N^{-\frac{1}{2}} \sum_{n=1}^{N} \exp(ikn) \left(c_k a_n^* + C_k A_n^{*2} \right) | O > . \quad (8)$$

A little algebra shows that this form is an eigenstate with an energy relative to the ground state of

$$E(k) = \hbar\omega \pm \left\{ 2\left[\alpha_1^2 + \alpha_2^2 - 2\alpha_1\alpha_2 \cos(k) \right] \right\}^{\frac{1}{2}} . \quad (9)$$

For the special case $\alpha_1 = \alpha_2 = \alpha$,

$$E(k) = \hbar\omega \pm 2\sqrt{2}\, \alpha \sin \left| \frac{k}{2} \right| . \quad (10)$$

Eq. (10) is remarkably similar to the energy band for longitudinal acoustic modes. In a single molecule, Fermi resonance interaction of two modes leads to a splitting of the degeneracy and two distinct energy levels. From Eq. (9), we see that the anharmonic terms in our Hamiltonian result in a splitting of the degenerate energy bands, with a wavenumber dependent splitting. Assuming that α_1 and α_2 have the same sign, the energy splitting is least at $k = 0$, when the gap is given by $2 \sqrt{2} (\alpha_1 - \alpha_2)$. In the limit $\alpha_1 = \alpha_2$, the gap vanishes entirely. Thus, the competing Fermi resonance mechanisms produce two noteworthy effects: a diminution of the Fermi splitting expected from either mechanism and a marked increase in the group velocity of excitations near the zone centre.

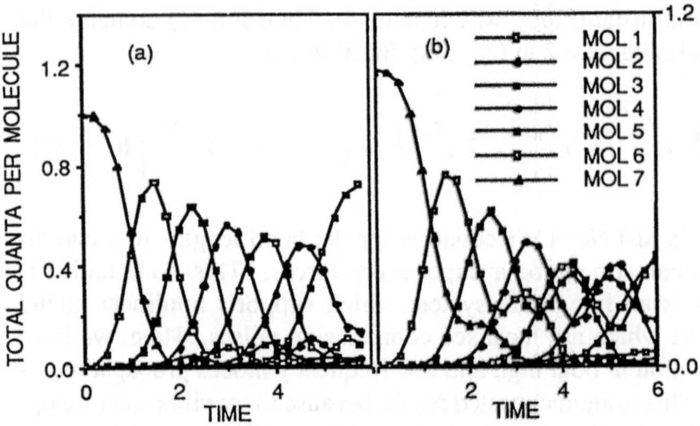

Figure 2. The quantum average total quanta per molecule ($a_n^*a_n + A_n^*A_n/2$) versus time in units of $\hbar/\alpha_1$, for an initial state with one high frequency quantum in molecule seven, for two values of $\beta = \alpha_2/\alpha_1$;(a) $\beta = 1$; (b) $\beta = 0.8$.

The eigenvectors of Eq. (8) form a complete basis for all states which evolve from the initial excitation of one high frequency quantum on a site. Figure 2 presents the time dependent redistribution of energy following such excitation of the terminal site of a seven member chain (seven peptides coupled by hydrogen bonds on an α-helix span the width of a cell membrane). Figure 2 shows that the vibrational energy is transferred along the chain as a pulse. When $\alpha_1 = \alpha_2$, the pulse has a simple form with one maximum and a width that increases with time. Despite this apparent spreading, the lattice behind the advancing pulse is left vibrationally cold. When α_1 and α_2 differ, the lattice behind the advancing peak retains some portion of the energy. The dynamics following two quantum excitation are very similar.

Classical energy redistribution: coherence versus chaos. Classical simulations provide a measure of the dynamics of a system which is too large or too complex to be studied quantitatively with quantum mechanics. In more detailed models of hydrogen bonded amides or proteins, quantum mechanics cannot be employed. The purpose of this section is to investigate how accurately classical mechanics can simulate the type of dynamics seen in Figure 2, so that it can be employed with some idea of its limitations and validity.

There are two important features of Figure 2 which we require a classical simulation to reproduce: the qualitative behaviour of the energy redistribution, and the time scale for this transfer process. Unfortunately, our studies show that classical mechanics cannot satisfy both requirements under the same initial conditions: due to the inability of classical mechanics to simulate the vacuum state of a chain.

When classical oscillators are coupled, energy transfer can occur between them even when they only have zero point energy. When many oscillators are coupled, very large fluctuations in the individual oscillators can occur, often with at least one quantum of energy appearing in a mode that should, quantum mechanically, be in the ground state. Figure 3 displays the average time dependence of the total quanta in each site of a seven member chain for which $\alpha_1 = \alpha_2$. Letting $\nu_n = a_n{}^*a_n$ and $\mu_n = A_n{}^*A_n$, initial conditions are given by $\nu_7 = \rho + 1$, $\mu_7 = \rho$, $\nu_n = \mu_n = \rho$, for n =

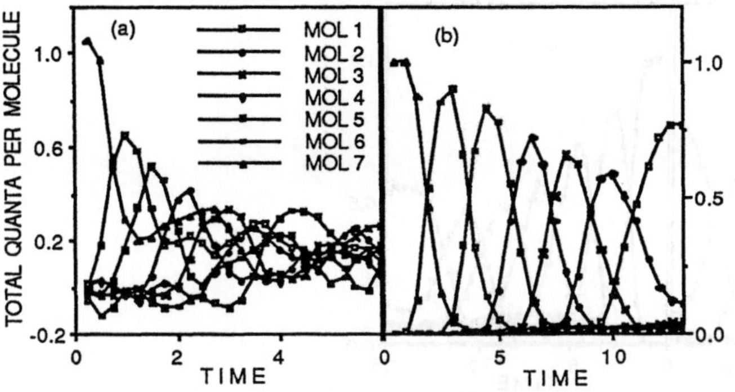

Fig. 3. The classical average, over 50 trajectories, of the total quanta per molecule versus time, in units of $\hbar/\alpha_1$, in a chain of seven molecules, for $\alpha_2 = \alpha_1$, with all v_n and μ_n set initially to ρ except $v_7 = \rho+1$, and (a) $\rho = 0$, (b) $\rho = -0.49$.

1,...,6; and random oscillator phases. While $\rho = 0$ represents the usual classical correspondence for a ground state oscillator, Figure 3 shows that for this initial condition, classical dynamics fails to simulate the quantum dynamics of Figure 2. Examination of individual trajectories shows the wild fluctuations referred to above. In order to force the set of coupled ground state oscillators to remain unexcited, their energy is reduced by setting ρ less than zero($\rho = -0.5$ represents no energy). Figure 3 shows that $\rho = -0.49$ in the initially unexcited oscillators results in a qualitative simulation of the quantum dynamics of Figure 2, although the time scale for energy transfer is slower by a factor of almost two.

A closely related observation of the unphysical effect of zero point energy in classical simulations has been reported by Lu and Hase[6]. They show that the classical simulation of CH overtone decay in benzene is more accurate when zero point energy is reduced in the same way as above. It seems likely that the classical dynamics of systems with nonlinear resonance becomes chaotic at very low energy. This leads to destruction of periodic orbits corresponding to regular motion of each oscillator at constant energy and to unphysical energy redistribution. Only at very low energy (below the zero point) does classical mechanics mimic the regular motion corresponding to simple eigenstates of the quantum dynamics.

Since the object of the classical simulations is to provide a convenient method of studying dynamics in large systems, we employ these "cold" initial conditions which give the best qualitative simulation of the quantum dynamics, while keeping in mind that the time scale seen for energy transfer may high by a factor near two. However, this leaves open the question of how one can best simulate classically the temperature dependence of such systems.

Figure 4 displays the classical averages corresponding to Figure 2 (with $\rho = -0.49$) for one quantum in the initially excited terminal high frequency mode, for different values of α_1/α_2. The coherent energy transfer evident for $\alpha_1 = \alpha_2$, gradually breaks down as the rates of intra and inter molecule energy transfer differ, in qualitative agreement with the quantum dynamics. When α_1 and α_2 are nearly equal, the initial

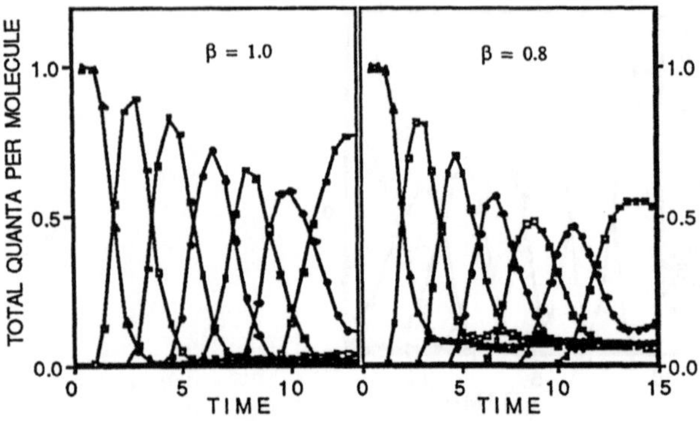

Fig. 4 Classical average number of quanta per molecule, corresponding to Fig. 2.

excitation rapidly forms into a pulse. This pulse slowly spreads in width as it propagates down the chain. From this and other measures, it appears as if the initial excitation naturally forms the type of soliton pulse derived above.

4. N-methylacetamide

N-methylacetamide (NMA) is the simplest amide whose structure and vibrational spectrum closely resemble those of the peptide group. We simulate the vibrational motion of a chain of seven hydrogen bonded NMA molecules to examine the validity of the above simple model in a realistic representation of a molecular system.

Three major approximations apply: classical mechanics with the "cold" initial conditions examined above; methyl groups are replaced by point masses; and the simulations are run in the two dimensions of the molecular plane.

The potential energy surface, V, is a valence force field: a Taylor series in internal displacement coordinates. To reduce the number of unique force constants, we make a division of the force constant matrix into two parts: an intramolecular part dealing with the coordinates within each NMA molecule in the chain and an intermolecular part binding the molecules into the chain. For a chain of N molecules

$$V = \sum_{n=1}^{N} V_{intra}^{(n)} + \sum_{n=1}^{N-1} V_{inter}^{(n,n+1)} + V_{anh}^{(n,n+1)} \quad . \tag{11}$$

V_{anh} refers to the inclusion is this model of anharmonic coupling terms across the hydrogen bond linking the amides. Following the discussion above, V_{anh} is taken to be

$$V_{anh}^{(n,n+1)} = c_{NHCO}\, \delta r_{NH}^{(n)} \left[\delta r_{CO}^{(n+1)} \right]^2 \quad . \tag{12}$$

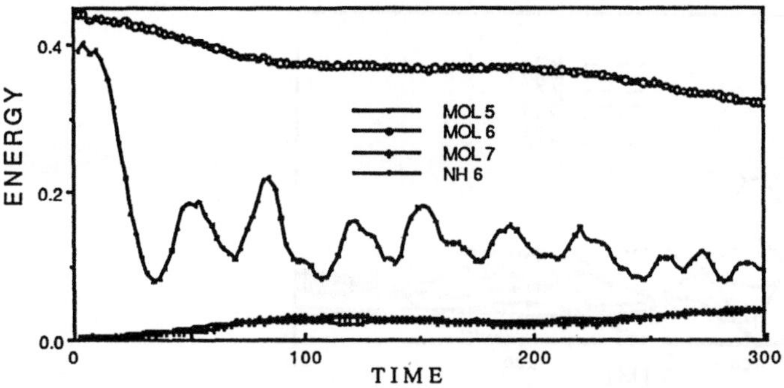

Figure 5. Average energy versus time for three neighbouring molecules and the initially excited NH stretch of molecule 6 when $c_{NHCO} = 0$ (300 time units ≈ 3 ps).

The complete potential surface was constructed in four steps: (i) V_{intra} was obtained from *ab initio* quantum chemical surfaces for NMA and scaled to reproduce gas phase vibrational frequencies of NMA and N–deuterated NMA (NMAd); (ii) V_{inter} was adapted from a semi-empirical potential; (iii) V_{intra} and V_{inter} were combined and the resultant potential scaled to reproduce vibrational frequencies of condensed NMA and NMAd; (iv) V_{anh} terms were added. With this potential energy function, classical energy transfer was examined as a function of the NH stretching frequency and the magnitude of the anharmonic coupling terms. Trajectories were performed with $V_{anh} = 0$, to characterise the intramolecular energy transfer out of the excited NH stretch. The fastest energy transfer out of the initially excited NH stretch was found at a harmonic frequency of 3340 cm^{-1} (with Morse oscillators to account for anharmonicity), and efficient energy transfer occurred over a frequency range of approximately 50 cm^{-1} about the optimum frequency. In Fig. 5, molecule six is initially excited (molecule seven has the free NH stretch). The energy in its NH stretch reaches a minimum of about 20% of the initial excitation at 35 time units (about 0.35 ps). The energy is transferred into both the Amide-I and Amide-II modes. This is consistent with a Fermi resonance between the NH stretch and NH bend. After approximately 2 ps, energy appears in the Amide-III modes, and finally energy randomisation becomes dominant. Fig. 5 indicates that, without any explicit anharmonic coupling across the hydrogen bond, energy flow out of the molecule is slow; and even this small flow of energy may be overestimated owing to the manner in which the intermolecular energy is apportioned[2].

When anharmonic coupling is present, Fig. 6 shows that energy flow is rapid and that a pulse of energy is evident, quite like that seen in the simple model.

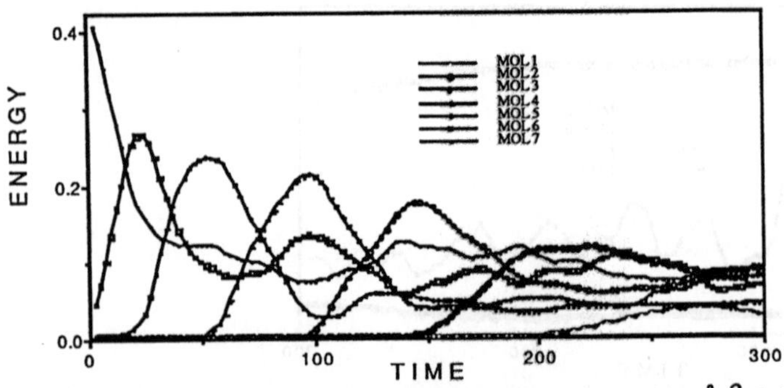

Figure 6 presents trajectory averages for $c_{NHCO} = -6$ aJ Å^{-3} following initial excitation of the Amide-I mode localised on molecule seven by two quanta.

5. α-Poly(L-alanine)

We model a section of helix 21 residues long[3]. Therefore, each of the hydrogen bonded chains running down the helix contains seven residues. This length is characteristic of α-helical sections in membrane-bound proteins. The form of the potential surface is given by a sum over peptide groups of the form:

$$V = \sum_{n=1}^{21} V_{intra}^{(n)} + V_{RDK}^{(n)} + \sum_{n=1}^{18} V_{inter}^{(n,n+3)} + V_{anh}^{(n,n+3)}, \tag{13}$$

where V_{intra} represents the scaled intra-peptide potential similar to that for NMA ; V_{RDK} describes the remaining backbone and torsion force constants; V_{inter} represents cross hydrogen bond interactions derived from Jorgensen and Swenson; and V_{anh} is a similar cross hydrogen bond cubic term involving c_{NHCO} as in the NMA surface. The NH stretching modes have been treated effectively as harmonic oscillators, following the NMA study. Two critical features of the surface, the NH stretch force constant and c_{NHCO} have been treated as parameters. With the NH stretch frequency and the cubic coupling parameter optimised at 3304 cm^{-1} and -3.5 aJ Å^{-3} respectively, 28 trajectories were performed over a long time (Fig. 7). The initial excitation was placed in one of the free NH stretch oscillators (residue 20), allowing for energy transfer along the complete hydrogen bonded chain. The energy propagates almost completely down the helix, with a very small amount of energy reaching the seventh residue after about 350 time units (3.5 ps). Once the energy pulse has passed a residue in Fig. 7, the residue does not interact strongly with its neighbouring groups even though it still has some energy in it. As a result, the energy pulse is seen to move relatively undisturbed to the end of the helix. The NH stretch energies also show the energy pulse moving to the end of the chain. These averages disguise a range of behaviour from energy trapping in a few trajectories to very coherent energy transfer. Such coherent transfer is shown in Fig.8.

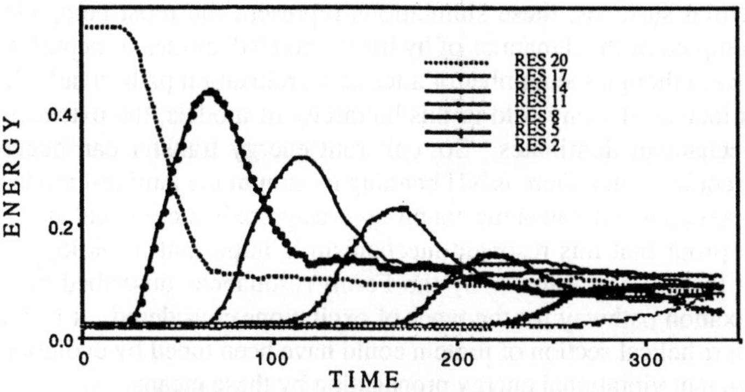

Figure 7. Average energy versus time following excitation of the NH stretch on residue 20, for ω_{NH} = 3304 cm^{-1} and c_{NHCO} = -3.5 aJ Å^{-3}, for peptide residues in the hydrogen bonded chain containing residue 20.

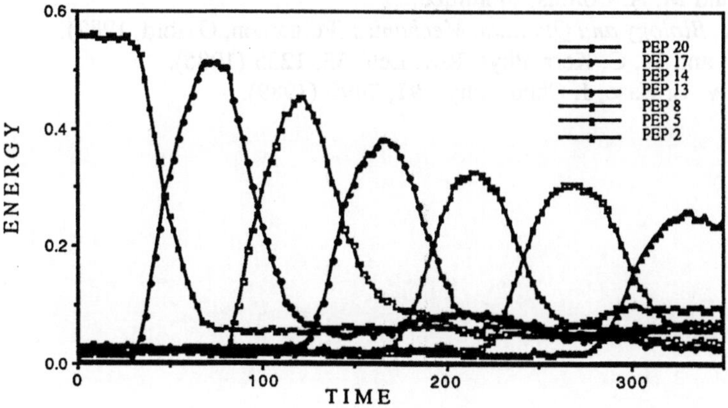

Figure 8. As for Figure 7, for a single trajectory showing efficient energy transfer.

6. Summary

We have presented a hierarchy of models for the simulation of energy transfer in hydrogen bonded amides and proteins from a simple heuristic model to a comprehensive simulation of a section of α-helix. The simple model clarifies the relationship between the Fermi resonance mechanism and atomic lattice systems. Both quantum and classical simulations show that stable 'soliton like' pulses of vibrational energy form from localised excitation and propagate rapidly. The solution of both the classical and quantum equations establish the conditions under which classical simulations provide a good qualitative description of the quantum dynamics. The simulations of NMA and α-Poly(L-alanine) provide a more serious platform for testing the biological hypothesis. By including the exact kinetic energy and comprehensive potential surfaces, these simulations represent the most complete characterisations attempted of the dynamics of hydrogen bonded amides or peptides. They illustrate that, even though the number of alternative relaxation paths available to the dynamics has increased many-fold in this hierarchy of models, the extended Fermi resonance mechanism dominates. So, coherent energy transfer can occur given two principal requirements: there is NH bending motion in the Amide-I modes and the anharmonic potential and kinematic couplings are approximately in balance.

While there is no proof that this resonant mechanism is important in biological systems, it is likely that energy transfer by the Fermi resonances described here forms a feasible relaxation pathway for the types of excitations considered. It is not inconceivable that an α-helical section of protein could have been tuned by evolution to provide highly efficient vibrational energy propagation by these means.

1 D. L. Clarke and M. A. Collins, J. Chem. Phys. **92**, 5602 (1990).
2 D. L. Clarke and M. A. Collins, J. Chem. Phys, in press.
3 D. L. Clarke and M. A. Collins, submitted.
4 A. S. Davydov, *Biology and Quantum Mechanics* (Pergamon, Oxford, 1982).
5 P. S. Lomdahl and W. C. Kerr, Phys. Rev. Lett. **55**, 1235 (1985).
6 D-H. Lu and W. L. Hase, J. Chem. Phys. **91**, 7490 (1989).

Chaotic Dynamics of Fluxons in Large-Area Josephson Junctions

F.Kh. Abdullaev and B.A. Umarov**

Ohio State University, Mathematical Department, OH, USA
*On leave from Heat Physics Department of the Uzbek Academy of Sciences, 700 135 Tashkent, Katartal Str. 28, Uzbekistan

The fluxon dynamics in inhomogeneous Josephson junctions under the action of ac driven bias current has been considered. The possibility of chaotic dynamics of a fluxon has been demonstrated, the criteria of chaos for various models of an inhomogeneity have been obtained.

1. Introduction

Last years the investigation of the fluxon dynamics in the Josephson junctions attracts much interest of scientists. It is connected with a possibility to control the fluxon parameters with the help of external forces or the changes of junction parameters [1-2].

In [3-5] the investigation has been carried out for one-dimensional junction. It has been proved that the fluxon can be trapped by the inhomogenities and oscillate near the inhomogenities, and emit linear waves. It has been also demonstrated that the dynamics of a soliton can becam chaotic under some conditions.

In [6-7] the study of large-area Josephson junctions has shown the existence of weakly damped oscillations of radial-symmetric fluxons in the field of the inhomogenity. In this work a model of the Josephson junction with varied frequency has been proposed where the frequency change has been obtained by switching separate pulses of the current that allowed to shift oscillations from one stable point of the phase trajectory to another one.

In this case it is necessary to take into account the fact that in such systems the transit motions are possible at sufficient velocities and so the study of the dynamics of fluxon near a separatrics separating the transit and trapped trajectories is of certain interest. As follows from the general theoretical consideration the appearance of complex nonlinear and chaotic regimes in the fluxon dynamics is possible in the given case. In this paper we consider the fluxon motion in large-area inhomogeneous Josephson junction at the presence of external variable current. Two concrete models of inhomogeneities are under consideration: a periodic grading and micro short-circuits.

By applying the Melnikov technique we are demonstrating the possibility of chaotic dynamics of fluxon and finding the criteria of its appearance.

2. The formulation of the problem and main equations.

Let us consider a motion of the radial symmetric vortex in inhomogeneous junction under the action of ac driven bias current. In this case the equation for the phase difference between two superconducting films in standard dimensionless variables [2] takes the form

$$U_{tt} - U_{rr} + \sin U = f \sin \Omega t + \frac{1}{r} U_r - \epsilon W(U, r), \tag{1}$$

where we have used l_J, the Josephson penetration depth, as the scale of length, and ω, the reciprocal plasma frequency, as scale of time. We then introduce the designations $\epsilon = I_\epsilon / I_0$, and $f = I_f / I_0$. Here I_0 is the critical Josephson current, and I_ϵ and I_f are the amplitudes of theinhomogeneous and external current respectivly. Let us also assume that the external current is being homogeneously applied and the inhomogenety is radially symmetric.

In order to apply perturbation theory it is necessary to renormalize and to separate a localised part in the perturbed terms [3]. For this we write the solution of (1) in the form

$$U(r,t) = U_s(r,t) + \phi(t),$$

where $U_s(r,t)$ is the localized part, $\phi(t)$ is induced field created by the external current. Solving linearized equatin for $\omega(t)$ we obtain the next equation for $U_s(r,t)$,

$$U_{stt} - U_{srr} + \sin U_s = f(1 - \cos U_s)\sin\Omega t - \epsilon W(U_s, r) - \frac{1}{r}U_{sr}. \qquad (1')$$

3. Formalism

Following [8] let as consider the dynamics of a kink-shaped bubble satisfying (1′) in which the energy is concentrated in a single well. We assume that the radius of such a bubble is large ($\rho \gg 1$) enough to be considered as a perturbation. At $\epsilon = 0$, $f = 0$,and $\rho \gg 1$, one obtains the fluxon solutin in the form

$$U(z) = 4\arctan\exp(z), \qquad z = \frac{r-\rho}{\sqrt{1-v^2}}, \qquad \rho = vt + \rho_0,$$

where v is the fluxon velocity and ρ is the radius.

We consider the fluxon dynamics according to perturbation theory by taking the right hand side in (1′) as a small perturbation. Then applying the equations of perturbation theory one obtains

$$\frac{dv}{dt} = \frac{1}{\rho} + \epsilon a(\rho) + \pi\frac{f\sin\Omega t}{8(1-\Omega^2)}, \qquad (2)$$

and

$$\frac{d\rho}{dt} = v(t), \qquad (3)$$

where

$$a(\rho) = \frac{1}{4}\int_0^\infty W(z,\rho)\,\mathrm{sech}\, z dz$$

In this equation we assume the nonrelativistic limit $v^2 \ll 1$ and take into account the renormalization of the external field following [3].

Let us investigate the case when $\epsilon > f$, and $1/\rho \sim f$. In this case the system (2), (3) can be studied with the account of radiality and external current as a perturbation. Let as consider the truncation of the system (2), (3)

$$\frac{dv}{dt} = \epsilon a(\rho), \qquad (2')$$

$$\frac{d\rho}{dt} = v. \qquad (3')$$

In general case the system can be integrated

$$t = \int_{\rho_0}^{\rho}\frac{d\rho}{H - \epsilon G(\rho)}, \qquad (4)$$

where $dG(\rho)/d\rho = a(\rho)$, and H is a constant of integration.

Let us consider two concrete models of the inhomogeneity. The first is

$$W_1(r) = \cos(ur). \tag{5}$$

This is a model of the periodically-inhomogeneous junction which can be created, for example, by periodic grading. It is known, that there exist librating, passing and sepratrix solutions. The separatrix solution takes the form

$$\begin{aligned} \rho_s &= (4\arctan[\exp(\omega_0 t)] - \pi)/u, \\ \omega_0 &= [\pi^2 \epsilon u/(4\sinh \pi u/2)]^{1/2} u. \end{aligned} \tag{6}$$

The second model is

$$W_2(r) = \delta(r - r_1) + \delta(r - r_2). \tag{7}$$

This is a model of the Josephson junction with micro short-circuits, the regions of strong supercondacting current, the sizes of which are much less than l_J. The separatrix solution given in non explicit form is found from the following equation [9]

$$\begin{aligned} 2H_s(t - t_0) &= \rho' - \frac{\alpha + y_0}{2\sqrt{y_0^2 - 1}} \ln \frac{y_0 - 1 + \sqrt{y_0^2 - 1}\tanh \rho'}{y_0 - 1 - \sqrt{y_0^2 - 1}\tanh \rho'} \\ y_0 &= \alpha - \frac{2}{\alpha}, \qquad \alpha = \cosh(r_2 - r_1), \\ \rho' &= \rho - \frac{(r_1 - r_2)}{2}, \\ H_s &= \frac{\epsilon \alpha^2}{4(\alpha^2 - 1)}. \end{aligned} \tag{8}$$

Taking into account the separatrix solution one can find the criteria of the chaotic motion near the separatrix with the help of the Melnikov method [10].

Let us consider a simple two dimensional autonomous system

$$\dot{X} = F_0(X) + \epsilon F_1(X, t),$$

where X, F_0, F_1 are two dimensional vectors. At $\epsilon = 0$ the system takes the exact solution

$$X = X_0.$$

Let us calculate the Melnikov integral $D(t_0)$

$$D_1(t_0) = -\int_{-\infty}^{\infty} F_0(X_0) \wedge F_1(X_0) dt,$$

where

$$X \wedge Y = X_1 Y_2 - X_2 Y_1.$$

It is known that if $D(t_0)$ changes a sign,then the separatrices is intersect and the motion becomes chaotic in this region. Applying this technique to the system (2), (3) one can obtain the following expression for $D(t_0)$. For the model (1) we have

$$\begin{aligned} D_2(t_0) &= \frac{\pi^2 f \epsilon \sin \Omega t_0}{2u\Omega \cosh(\pi\Omega/(2au))} - \ln \frac{n+1}{n}, \\ a &= \frac{\pi^2 \epsilon^2 u^2}{\sinh(\pi u/2)}, \end{aligned} \tag{10}$$

where n is defined by ρ_0 as

$$\frac{2\pi n}{u} < \rho_0 < \frac{2\pi(n+1)}{u}.$$

For the model (2)

$$D(t_0) = \frac{\epsilon a \sin \omega_1 t_0}{2(1-\Omega^2)} \int_0^L \cos(k + \ln \sinh x) dx - \ln \frac{r_1}{r_2},$$

where $\omega_1 = \Omega/2H_s$, $r_2 > r_1 \gg 1$, and $L = r_2 - r_1$. For $\rho_0 \gg 1$ we have

$$\ln \frac{r_1}{r_2} \sim \frac{1}{\rho_0},$$

and if $\omega_1 \gg 1$ we obtain

$$2D(t_0) \sim \frac{\epsilon a \sin(k+L) \sin \omega_1 t_0}{2\omega_1 * (1-\Omega^2)} - \frac{1}{\rho_0}.$$

4. Discussion

Having studied the results presented in formulae (10) and (11) we came to conclusion that in general case the fluxon dynamics in large-area Josephson junctions under the action of variable exter- nal field can have a stochastic character and the account of radial symmetry results in narrowing of the chaos region, i.e. it acts as damping. This conclusion should be taken into account while designing Josephson oscillators with variable frequency, in that case when we apply the alternating current to change the frequency.

References

1.D.W.McLaughlin, A.C.Scott. In Solitons in action.Acad.Press., N.Y., 1978.
2.K.K.Likharev.Introduction in dynamics of Josephson junctions.Mir, Moscow, 1986. (In Russian)
3.Y.S.Kivshar, B.A.Malomed.Reviews of Modern Physics, V.61, N.4, p. 763, 1989.
4.F.Kh.Abdullaev. Lebedev institute reports in physics.N.1, p.53.
5.f.Kh.Abdullaev, A.A.Abdumalikov, B.A.Umarov. Doklady Uzbek Academy of Sciences.N.3, p.25, 1989. (In Russian)
6.P.L.Christiansen, P.S.Lomdahl, N.J.Zabusky., Appl. Phys. Lett., v.39 N.2, p.170, 1981.
7.E.M.Maslov.Preprint IZMIRAN SSSR, N.15(769), 1989.
8.E.M.Maslov.Physica D15, p. 433, 1985.
9.F.Kh.Abdullaev, S.A.Darmanyan, B.A.Umarov. Phys.Rev.A, v.41, N.8,p.4498, 1990.
10.V.K.Melnikov. Trudy Moscovsk.Mat.Obshestva., v.12, p.3, 1963. (In Russian)

Soliton Dynamics in Tunnel-Coupled Fibers with Variable Coupling

S.A. Darmanyan

Thermal Physics Department of Uzbek Academy of Sciences,
700 135 Tashkent, Katartal Str. 28, Uzbekistan

Abstract. The problem of soliton propagation in a system of tunnel- coupled fibers in the case when a coupling parameter depend on coordinate along a fiber is considered. It is shown, that at periodic modulation of coupling both the parametric resonance of solitons and their dynamic stochastization are possible depending on the initial conditions. While in the case of a random coupling parameter the stochastic parametric resonance of solitons arises.

1. Introduction

The system consisting of two tunnel-coupled nonlinear waveguides has been first investigated in [1,2]. As has been shown by the authors of these works, the optical transistors, switchers, logical elements can be designed on the basis of tunnel-coupled optical waveguides. On the other hand the application of solitonic regimes of optical pulse propagation looks very promising to increase the ability of optical communication lines (see e.g.[3,4]). This makes the investigation of soliton propagation in the system of coupled fibers very actual both from the point of view of the general theory of solitons as well as its various applications. Such kind of investigation have been carried out in [4-9]. In particular, as shown in [7,8], the solitons propagating in different fibers under certain conditions can either form the bound states or repulse each other. In this paper in contrast to the previous ones we investigate the case when a parameter describing a coupling of two fibers is the function of the coordinate along a fibers.

Editors: F. Abdullaev · A.R. Bishop · S. Pnevmatikos

2. Solitons in Tunnel-Coupled Fibers

The system of equations describing the optical pulse propagation in coupled fibers is represented by the system of two nonlinear Schrodinger equation (NSE) [4]

$$\begin{aligned} iq_{1x} + \frac{1}{2} q_{1\tau\tau} + |q_1|^2 q_1 &= \varepsilon q_2 \\ iq_{2x} + \frac{1}{2} q_{2\tau\tau} + |q_2|^2 q_2 &= \varepsilon q_1 \, . \end{aligned} \tag{1}$$

Here q, x, τ are the dimensionless field amplitude, the coordinate along the propagation direction and the time in the "travelling" system of coordinates. The parameter ε characterizing the coupling between fibers and depending upon a distance between fiber cores and the difference of dielectric permeability of the fiber core and cladding is considered here to be a function of the coordinate x. It accounts the possibility of both the fluctuations of geometric factors of a system and the parameters of fiber core environment as well as their artificial modulation. A solitonic solution of unperturbed NSE has the form

$$q = 2\nu \,\mathrm{sech}[2\nu(\tau-\xi)] \exp[2i\mu(\tau-\xi) + i\delta], \tag{2}$$

where ν, ξ, μ and δ are the parameters characterizing the soliton amplitude, location of center, velocity and phase, correspondingly.

Assuming that $|\varepsilon| \ll 1$ and applying the perturbation theory in adiabatic approximation, analogously to [7] one can show that for the solitons with the equal amplitudes $\nu_1 = \nu_2 = \nu$, the potential of interaction is expressed by

$$V = 16\nu^2 \varepsilon(x) \cos(\psi) \, r/\sinh r \, , \tag{3}$$

where $r = 2\nu(\xi_2 - \xi_1)$, $\psi_0 = \delta_{02} - \delta_{01}$ the initial difference of soliton phases which equals either π or 0. The equation for the parameter r describing a normalized distance between the soliton centers has for small r the following form:

$$r_{xx} - \frac{16}{3} \nu^2 \cos(\psi) \, \varepsilon(x) \, r = 0. \tag{4}$$

As could be seen from (3) and (4) the attraction between solitons arises in the case of negative $\varepsilon\cos(\psi_0) < 0$. At ε = const the solution of equation (4) describing the bound states of solitons is $r = r_0\cos(kx)$, where $k = 4\nu(|\varepsilon|/3)^{1/2}$, with r_0 being the initial value of parameter r. Account of the dependence of coupling parameter ε on the coordinate x, can significantly change the solution of equation (4). Let us consider two particular cases of this dependence.

2.1 The Parametric Resonance

Let assume that $|\varepsilon| = \varepsilon_0 + \varepsilon'\cos(\text{æ}x)$. In this case the equation (4) is reduced to the Mathieu equation

$$r_{xx} + k^2\cdot(1 + \varepsilon_1\cos(\text{æ}x))r = 0, \quad (5)$$

where $k^2 = 16\nu^2\varepsilon_0/3$, $\varepsilon_1 = \varepsilon'/\varepsilon_0$.

The solutions of the above equation are well known and are expressed through the Mathieu functions. In the case of small ε_1 ($\varepsilon_1 \ll 1$), i.e. under a weak coupling modulation, the solution of equation (5) can be obtained by the perturbation theory (see e.g. [10]). It is necessary to note the possibility of the parametric resonance. In this case in the region of parameters $\text{æ} = 2k + \eta$, where η is the small deviation ($-k\varepsilon_1 < 2\eta < k\varepsilon_1$) the exponential growth of the oscillation amplitude of the distance between the soliton centers takes place:

$$r(x) = r_0 \exp(\lambda x)\cos[(k + \eta/2)x] ,$$

where $4\lambda = (k^2\varepsilon_1^2 - 4\eta^2)^{1/2}$.

At $\text{æ} \approx 2k/n$ ($n > 1$ being the integer number), the resonances of the higher order are provided. However in the case of a resonance of n-th order, the resonance region essentially narrows $\eta \sim \varepsilon_1^n$ and the corresponding increments of instabilities will also reduce [10].

2.2 The Stochastic Parametric Resonance

The second case corresponds to the fluctuations of the coupling parameter near its mean value ε_0: $|\varepsilon| = \varepsilon_0(1 + \varepsilon_1)$ where ε_1 is the random Gaussian function with the mean values defined as

$$\langle \varepsilon_1 \rangle = 0, \qquad \langle \varepsilon_1(x_1)\, \varepsilon_1(x_2) \rangle = B(x_1 - x_2).$$

In this case one can not write out the solution of the equation (4) in the explicit form, but this can be done for the mean values. As it is known [11] the stochastic parametric resonance is provided in the oscillating system under the presence of parameter fluctuations. The possibility of system parametric excitation is connected with the fact that the random function contains all the harmonics, some of which satisfy the conditions of parametric resonance. For the first time the existence of analog of phenomenon of stochastic parametric resonances for solitons has been reported in [12]. Here we present another example of this phenomenon for solitons.

For simplicity we assume that the correlation function has the form $B(x_1-x_2) = 2\sigma^2\delta(x_1-x_2)$ and omitting the intermediate calculations that can be found in [11] we write out an expression for average square of distances between solitons:

$$\langle r^2(x) \rangle = r_0^2 + \frac{1}{2k^2}\{e^{\sigma^2k^2x} - e^{-\sigma^2k^2x/2}[\cos 2kx + \frac{3}{4}\sigma^2k^2\sin 2kx]\}. \tag{6}$$

An analogous expression can be also written for mean value of soliton square velocity. As can be seen from (6), the mean value of square distance between solitons is exponentially increasing as soliton is propagating along a fiber, and the increment of this increase is $\lambda = \sigma^2k^2$. Generalization of these results on the processes with finite correlation radius can be found in [11]. The methods to construct the distribution function of random values r and r_x are also indicated there.

Thus, we have shown that in a system of tunnel-coupled fibers with variable coupling, the existence of parametric and stochastic parametric resonances of solitons is possible. These phenomena are expressed as exponential increase of the amplitude of the relative soliton oscillations. It is necessary to note, that at the distances of the order $\lambda x \sim |\ln r_0|$ (where $r_0 \ll 1$ is the initial distance between solitons) an approximation of harmonic interaction potential is not more valid and one has to solve the problem with the exact potential (3), but in this case the numerical calculations are required.

2.3 Dynamical Chaos of Solitons

There exists one more type of initial conditions, when the system of equations (1) can be investigated in the framework of the adiabatic perturbation theory. Let us assume that solitons propagating in different fibers have the equal velocities, their centers coincide and amplitudes slightly differ. In this case the system of equations describing the behavior of soliton phase difference $\psi = \delta_2 - \delta_1$ and their amplitude difference $\Delta = \nu_2 - \nu_1$ has the form:

$$\Delta_x = -2A\,\varepsilon(x)\sin\psi \tag{7a}$$

$$\psi_x = \left(2A + \frac{5}{A}\,\varepsilon(x)\right)\Delta\ , \tag{7b}$$

where $A = \nu + \nu = \mathrm{const}$, $\Delta/A \ll 1$.

From the system of equations (7) one can obtain a closed equation for the phase difference ψ. Neglecting the second term in brackets in the equation (7b) one obtains

$$\psi_{xx} + 4A^2\,\varepsilon(x)\sin\psi = 0\ . \tag{8}$$

This equation at $\varepsilon = \mathrm{const}$ coincides with one obtained in [9] and represents the well-studied equation of the non-linear pendulum (for ψ at $\varepsilon > 0$ and for $\Phi = \psi + \pi$ at $\varepsilon < 0$).

In the case when ε is a function of x the equation (8) describes the parametrically driven nonlinear pendulum. At $|\psi| \ll 1$ the equation reduces to the described above linear one and depending on the form of the $\varepsilon(x)$ function there can exist either parametric or stochastic parametric resonance. In result the soliton amplitude difference will increase exponentially, because in accordance with (7) it is expressed as $\Delta \approx \psi_x/2A$.

It is also important to emphasize the following circumstance: the nonlinear pendulum under action of regular perturbation has the parameters region where its behaviour becomes stochastic [13,14].

For example at $\varepsilon = \varepsilon_0 + \varepsilon'\cos(\text{æ}x)$, the width of stochastic layer wich arises near separatrix [14] is defined as:

$$|H - H_s| \leqslant 4A^2\varepsilon'æ/k \qquad \text{at } æ \leqslant k , \qquad (9a)$$

$$|H - H_s| \leqslant 4A^2\varepsilon'(æ/k)^3\exp[-\pi æ/2k] \qquad \text{at } æ >> k , \qquad (9b)$$

where $k^2=4A^2\varepsilon_0$, $H = \psi_x^2/2 - k^2\cos\psi$ is the hamiltonian of nonlinear pendulum, $H_s = k^2$ is the value of H on separatrix.

In our case $|H - H_s| = |\Delta^2/8A^2 - 2k^2\cos^2(\psi/2)|$, thus e.g. for $æ \sim k$ and initial phase difference being $\psi_0 = \pi$ soliton dynamics will be stochastic if initial soliton amplitudes satisfy the inequality: $|\nu_{01} - \nu_{02}| \leqslant 4A^2(2\varepsilon')^{1/2}$.

REFERENCES

1. S.Jensen, IEEE Quantum Electronics **QE-18**, 1580 (1982).
2. A.A.Mayer, Quantum Electronics **9**, 2296 (1982).
3. A.Hasegawa, F.Tappert, Appl. Phys. Lett. **23**, 142 (1973).
4. F.Kh.Abdullaev, S.A.Darmanyan, P.K.Khabibullaev, "Optical Solitons", Tashkent, FAN, 1987.
5. F.Kh.Abdullaev, P.M.Abrarov, S.A.Darmanyan, In "Technology and Properties of Fiber Waveguides", Tashkent, FAN, 1986.
6. L.M.Andrushko, K.S.Karilyuk, S.B.Ostrovskiy, Radiotekhnica i electronica, **32**, 427, (1987).
7. F.Abdullaev, R.Abrarav, S.Darmanyan, Opt.Lett **14**, 131(1989)
8. E.Wright, G.Stegeman,S.Wabnitz, Phys.Rev. **A40**, 4455 (1989).
9. Y.Kivshar, B.Malomed, Opt.Lett. **14**, 1365 (1989).
10. L.D.Landau, E.M.Lifshitz, "Mechanics", M.Nauka,1988.
11. V.I.Klyantskin "Stochastic Equations and Waves in Randomly Inhomogeneous Media", M.Nauka,1980. Editions de Physique, Besancon-Cedex,1985.
12. F.Kh.Abdullaev, S.A.Darmanyan, M.Djumaev. Uzvestia AN UzSSR Seriya Fis.Mat.Nauk, N6,1986. Phys.Lett.**141**, 423, (1989).
13. B.V.Chirikov, Phys.Rep.**52**, 263 (1979).
14. G.M.Zaslavsky, R.Z.Sagdeev, "Introduction in Nonlinear Physics", M.Nauka, 1988.

Instability of Solitons and Nonlinear Waves in Liquid Crystals

A.A. Abdumalikov[1] *and E.N. Tsoi*[2]

[1]Physics Department, Tashkent State University, 700 095 Tashkent, Uzbekistan
[2]Thermophysics Department of Uzbek Academy of Sciences, 700 135 Tashkent, Katartal Str. 28, Uzbekistan

The linear stability of solitons and nonlinear periodic waves for modified Boussinesq equation, which describe a various physical systems are investigated by Whitham method. It is shown, that fast solitons are stable. The stability region of wave parameters has been obtained.

1. Introduction

Nowadays the dynamics of nonlinear formations in various physical systems are intensively investigated. An interesting model, in which the nonlinear waves arise, is the following equation:

$$u_{tt} - u_{xx} + 6(u_x)^n u_{xx} + u_{xxxx} = 0 \tag{1}$$

At n = 1 we have the Boussinesq equation (BqE), exactly integrable by inverse scattering transform (IST) method [1]. The equation (1) at n =2, which we shall call "modified Boussinesq equation" (mBqE) is not , apparently, integrated by IST method.

In this work the linear stability of solitons and knoidal waves of mBqE is investigated by Whitham method. The given equation appears at a description of dynamics of liquid crystals director [2], and at a study of a nonlinear mechanics of monoatomic chain with a certain kind of interaction [3].

The stability of solitons and periodic solutions of BqE has been investigated by Berryman [5] and Falkovich et al.[5]. In particular in [5] it has been shown by Shabat method, that the BE soliton is stable at $v > 1/2$, and one is unstable at $v < 1/2$. The region of parameters for nonlinear waves has been also determined.

2. Main results

At first let us consider the stationary solutions of mBqE, that we will study here. For this purpose we substitute a function $u(\xi)$, where $\xi = x - vt$, in equation (1) and then twice integrating by ξ, we obtain:

$$(u_{\xi\xi})^2 = -(u_\xi)^4 + (1 - v^2)(u_\xi)^2 + C_2 \tag{2}$$

Here for simplicity we assume one constant of integration to be

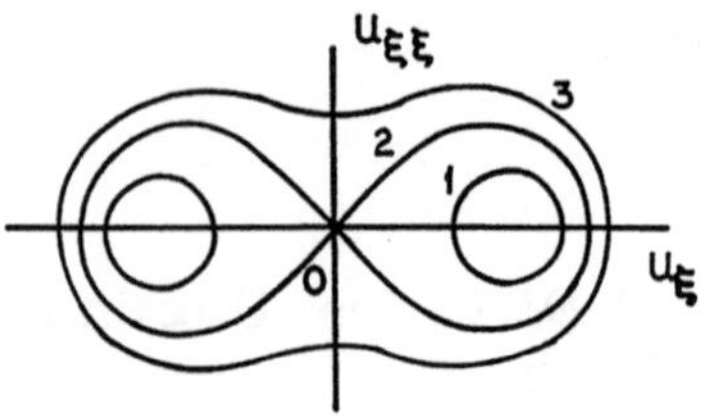

Fig.1.Phase plane for equation (1). (1 - "rotations wave", 2 - "oscilations wave", 3 - soliton).

equal to 0. The phase mapping of equation (2) is presented on Fig.1. One can see, that there exist 3 kinds of solutions, which we take down in following forms:

a) Fist type is a nonlinear periodic wave, corresponding to spiral director distribution (curve 1 on Fig.1):

$$u(x,t) = \mathrm{am}(2K\theta/\pi) + \pi/2, \qquad \theta = kx - \omega t \tag{3a}$$

with dispersion relation :

$$\omega^2 - k^2 + (2K/\pi)^2 \cdot k^4 (2 - m^2) = 0 \tag{4a}$$

We shall name such wave a "rotations wave" (RW).

b) Second type is a nonlinear periodic wave, corresponding to director distribution, in that the molecules is deviated from vertical line on an angle less than π (curve 2 on Fig.1):

$$u(x,t) = \sin^{-1}(m \cdot \mathrm{sn}(2K\theta/\pi)) + \pi/2, \tag{3b}$$

with dispersion relation :

$$\omega^2 - k^2 + (2K/\pi)^2 \cdot k^4 (2m^2 - 1) = 0 \tag{4b}$$

We shall name it "oscillations wave" (OW).

c) Third type is a solitary wave - soliton:

$$u(x,t) = 2\tan^{-1}[\ \exp(m_o\theta)\ ] \tag{3c}$$

with dispersion relation :

$$\omega^2 - k^2 + m_o^2 \cdot k^4 = 0 \tag{4c}$$

Here k is the wave number, ω is the frequency, m is the form-factor (elliptic function modulus), am and sn are the elliptic amplitude and sine respectively, $K \equiv K(m)$ is the complete elliptic integral of the first kind. To use the Whitham method let us introduce Lagrange function, corresponding to equation (1):

$$L = \frac{1}{2} \int_{-\infty}^{+\infty} [(u_t)^2 - (u_x)^2 + (u_x)^4 - (u_{xx})^2]\, dx \tag{5}$$

Then the set of equations, describing the evolution of stationary waves $u(\theta)$ modulations has the form [6]:

$$(\partial/\partial t)L_\omega - (\partial/\partial x)L_k = 0, \quad (\partial/\partial t)k + (\partial/\partial x)\omega = 0, \tag{6}$$

$$\text{where} \quad L(k,\omega,m) = \frac{1}{2\pi}\int_0^{2\pi} L\, dx \tag{7}$$

is the averaged Lagrangian (for soliton the integration limits must be changed for $-\infty$ and $+\infty$), indexes denote partial derivatives on corresponding variables. To close the set (5) it needs to append the dispersion relation (4), then we have complete system of equations on 3 parameters $k(x,t)$, $\omega(x,t)$, $m(x,t)$.

As to the stability of some stationary solutions one can judge from the type of modulation equations (6). Namely, if the type of the set (6) is hyperbolic, then the perturbations of initial wave are finite on time, but if the type of set (6) is elliptic, the perturbations grow exponentially. The type of modulation equations is determined by the sign of discriminant:

$$D = r^2 - pq \quad , \tag{8}$$

where

$$p = L_{\omega\omega}\cdot L_{mm} - L_{\omega m}^2, \quad q = L_{kk}\cdot L_{mm} - L_{km}^2, \quad r = L_{\omega k}\cdot L_{mm} - L_{\omega m}\cdot L_{km}.$$

At $D < 0$ the set (6) is hyperbolic, at $D > 0$ one is elliptic.

Now let us study the stability of periodic waves (3a,b), whose averaged Lagrangian has the form:

$$L(\omega,k,m) = (\omega^2 - k^2)\cdot f_1 + \alpha\, k^4\cdot f_2 \tag{9}$$

where $\alpha = (2/\pi)^2/3$, functions f_1 and f_2 for RW and OW are respectively equal to:

$$\begin{cases} f_1 \equiv f_1(m) = EK \\ f_2 \equiv f_2(m) = [\ (2 - m^2)E + (1 - m^2)K\ ]K^3 \end{cases} \tag{10a}$$

$$\begin{cases} f_1 \equiv f_1(m) = (E - m'^2K)K \\ f_2 \equiv f_2(m) = [\ (2m^2 - 1)E + m'^2(1 - 3m^2)K\ ]K^3 \end{cases} \tag{10b}$$

Here $E \equiv E(m)$ is the complete elliptic integral of the second kind, $m'^2 = 1 - m^2$. Then using the Lagrangian (9) one can show, that the waves (3a,b) are stable under a condition:

$$A(m) - B(m)k^2 > 0 \tag{11}$$

where

$$\begin{aligned} A(m) &= f_1{}^2 - 6(f_1)'f_1f_3/(f_3)' + 4(f_1)'f_2/(f_3)' \\ B(m) &= 6\alpha(\ f_1f_2 + 2(f_1)'f_3(\ f_2 - 2f_1f_3)/(f_3)'\ ) \end{aligned} \tag{12}$$

$(\ )' \equiv (d/dm)$ and function f_3 is defined respectively for each

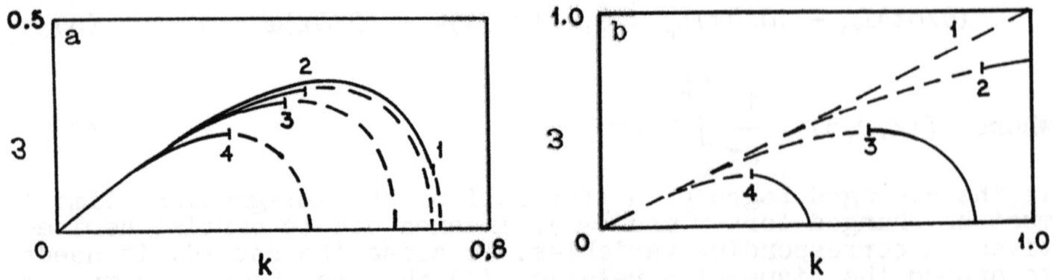

Fig.2a. Dispersion relations $\omega(k,m)$ for RW (curve 1 - m = =0.001, 2 - m=0.5, 3 - m=0.9, 4 - m=0.99). Solid lines correspond to stable modes, dashed lines correspond to unstable ones.

Fig.2b. Dispersion relations $\omega(k,m)$ for OW (curve 1 - m = $=(1/2)^{1/2}$, 2- m=0.8, 3- m=0.9, 4- m=0.99). Solid lines correspond to stable modes, dashed lines correspond to unstable ones.

waves:

$$f_3 \equiv f_3(m) = \begin{cases} (2 - m^2)K^2 & \text{(13a)} \\ (2m^2 - 1)K^2 & \text{(13b)} \end{cases}$$

In both cases the threshold wave number k_{th} is determined as $k_{th} = A(m)/B(m) \rightarrow [6\alpha \cdot \ln^2(4/m')]^{-1}$. On Fig.2 it has been presented the dispersion relations for RW and OW in dependence of form-factor m; a solid curves correspond to stable nonlinear modes, dashed lines correspond to unstable ones. As seen under preset m for RW the shorter wave lengths are stable, for OW the longer ones are stable.

For the mBqE soliton (3c) we obtain, that D = 0 , i.e. the set (6) is parabolically degenerate. Thus the Whitham method does not give an information on the mBqE soliton stability. However in [7] it has been investigated the stability condition for the solitary solutions of wide class of equations, whose special case is the equation (1). This condition has the form:

$$I(v^2) + 2v^2 \cdot [d/d(v^2)]I(v^2) < 0 \tag{14}$$

where $I(v^2) = \int_{-\infty}^{+\infty} (u_y)^2 \, dy$.

Using this criterion we find, that mBqE soliton is stable at $v^2 > 1/2$. Thus the fast solitons are stable, that coincides with a result for BqE.

3. Conclusion

It is interesting to note the following peculiarity. One can see that both the RW, which is the sequence of kinks and the OW, which is the sequence of kink-antikink pairs, transform to the soliton (3c) at m = 1. For this the threshold wave velocity $v_{th} = \omega_{th}/k_{th} \rightarrow (1/2)^{1/2}$ .But for RW the stability region is determined by the condition $v > v_{th}$ whereas for OW one is $v < v_{th}$. This fact can be apparently explained by the difference of a topological properties of solutions (3a,b).

References

1. V.E.Zakharov, Zh.Eksp.Teor.Fiz., **65** (1974) 219.
2. V.G.Kamensky,S.S.Rozhkov, Zh.Eksp.Teor.Fiz, **89** (1985) 106.
   F..Kh.Abdullaev, A.A.Abdumalikov, E.N.Tsoi, Phys.stat.sol. **146** (1988) 457.
3. A.M.Kosevich, Nonlinear mechanics of crystals, IEM, Sverdlovsk, preprint, 1975.
   N.Flytzanis, S.Pnevmatikos, M.Remoissenet, Physica D, **26** (1987) 311.
4. J.Berryman, Phys.Fluids, **19** (1976) 771.
5. G.E.Falkovich, M.D.Spector, S.K.Turitsyn, Phys.Lett.A **99** (1983) 271.
6. G.B.Whitham, Linear and nonlinear wave, N.Y., 1974.
7. A.F.Givental, Teor.Math.Fiz., **82** (1990) 28.

Phonons in Disordered Anharmonic Solids

M.I. Molina and D.C. Mattis

Department of Physics, University of Utah, Salt Lake City, UT 84112, USA

Abstract. A model of glass constructed entirely with phonons is introduced and solved in closed form. It yields $c_p(T) \sim T$ at low temperatures in the glassy phase. The fit to silica glass, requiring just a single adjustable parameter, is excellent. The model includes random scattering of normal modes and quartic anharmonicity (the latter required for stability,) but no two-level states.

1. INTRODUCTION. The theory of amorphous materials, i.e. glass, has interested physicists and chemists over a number of years.[1] Is there a genuine phase transition at T_g or merely a critical slowing down? Why is the low-temperature specific heat $c_p(T)$ in insulating glasses linear in T, rather than following the ubiquitous Debye T^3 law ? A number of other properties are equally anomalous.[2] For example, the inverse lifetime of phonons in a glass obeys $1/\tau \sim \omega^2$ at small ω, rather than the familiar Rayleigh law $1/\tau \sim \omega^4$.

The standard microscopic explanation, proposed by Anderson and by Phillips,[3] depends on the coexistence of hypothetical "tunnelling" or "two-level" states (TLS) with the normal modes of the material. (the energy of the two-level states is taken to be randomly distributed). However the TLS have proved hard to pin down. Some experimental support comes from Hunklinger and co-workers,[4] who have found saturation in the ultrasonic attenuation in glasses at low temperature, compatible with TLS. Even so, it is possible that TLS exist only in some glasses but not in others.

In this Letter we advance an alternative microscopic picture based *entirely* on phonons. Our theory explains a number of experimental data and has the additional virtue of being exactly soluble. We first observe that in a glass, just as in a slightly disordered crystal, a random potential H_1 scatters the normal modes. In the glass, however, an otherwise small quartic interaction H_2 enters into play because the scattering is so much greater, and causes the phonons to be overcoupled. This instability causes a symmetry to break and a new phase to become stable.

The glassy phase in our model exhibits a number of properties characteristic of all such glasses. Its properties are surprisingly independent of the spectrum of normal modes $\rho_0(\omega)$ in the reference crystal. In one application given below, we show that a fit to the measured $c_p(T)$ for vitreous silica, can be achieved *quantitatively* to a better than a fraction of a percent at low temperatures with just a single adjustable parameter. One important and general *qualitative* result, is that the Debye-Waller exponent in the glassy phase of the model is so large that the spatial correlations of the reference-crystal are lost.

Springer Proceedings in Physics, Vol. 67 Nonlinearity with Disorder
Editors: F. Abdullaev · A.R. Bishop · S. Pnevmatikos

2. THE MODEL. The Hamiltonian of the reference crystal is:

$$H_0 = \sum_{\mathbf{k}} \omega_{\mathbf{k}}(a^\dagger_{\mathbf{k}}a_{\mathbf{k}} + 1/2) \qquad (\hbar = 1). \tag{1}$$

where the label $\mathbf{k}$ includes wavevector and polarization indices. The $a^\dagger_{\mathbf{k}}(a_{\mathbf{k}})$ are the creation (destruction) operators of the crystal. In an Einstein model, we can take $\omega_{\mathbf{k}} = \omega_E$, a constant. In a Debye model, we assume $\omega_{\mathbf{k}} = (|\mathbf{k}|/k_D)\omega_D$. (However, we have found that most properties calculated in the glassy phase of our model are entirely insensitive to dispersion in the reference crystal.)

We introduce disorder just as in the exactly soluble model[5] we have previously studied, via

$$H_1 = (g/4)\sum_{\mathbf{k},\mathbf{k}'} M(\mathbf{k},\mathbf{k}')(\omega_{\mathbf{k}}\omega_{\mathbf{k}'})^{1/2}Q_{\mathbf{k}}Q_{\mathbf{k}'}. \tag{2}$$

where N is the number of normal modes, $Q_{\mathbf{k}} = a^\dagger_{\mathbf{k}} + a_{-\mathbf{k}}$, and the $M(\mathbf{k},\mathbf{k}')$ are random scattering amplitudes,

$$M(\mathbf{k},\mathbf{k}') = (1/\sqrt{N})\exp[i\Theta(\mathbf{k},\mathbf{k}')]. \tag{3}$$

with the $\Theta(\mathbf{k},\mathbf{k}')$ independently random parameters, lying in the interval $[-\pi,\pi]$ (subject only to $\Theta(\mathbf{k}',\mathbf{k}) = \Theta(\mathbf{k},\mathbf{k}') = -\Theta(-\mathbf{k},-\mathbf{k}')$ to insure that (2) is Hermitian.) g^2 is proportional to the concentration of defects introduced into the crystal. Finally, nonlinearity is introduced by means of the quartic Hamiltonian,

$$H_2 = \frac{\xi}{N\omega_D}\sum_{\mathbf{k}_1..\mathbf{k}_4} V(\mathbf{k}_1,\mathbf{k}_2,\mathbf{k}_3,\mathbf{k}_4)(\omega_{\mathbf{k}_1}\omega_{\mathbf{k}_2}\omega_{\mathbf{k}_3}\omega_{\mathbf{k}_4})^{1/2}Q_{\mathbf{k}_1}Q_{\mathbf{k}_2}Q_{\mathbf{k}_3}Q_{\mathbf{k}_4}. \tag{4}$$

with ω_D the characteristic frequency of the reference crystal, and ξ a dimensionless *anharmonicity parameter.* The small value of anharmonicity in almost all solids[6] gives a physical motivation to take ξ small. In fact, as we shall proceed to the limit $\xi \to 0$ in the present calculations, the precise value of ξ will be irrelevant.

It is convenient to define $H(\lambda)$ by $H_0 + \lambda(H_1 + H_2)$, setting $\lambda = 1$ at the end of the calculation. The free energy, entropy and specific heat are:

$$F = F_0 + \int_0^1 d\lambda < (H_1 + H_2) >_\lambda \;,\; S(T) = -(\partial F/\partial T) \;,\; C_p(T) = T(\partial S/\partial T). \tag{5}$$

To evaluate the thermal averages we resort to the double-time Green functions.[7] All observables (Green functions, F, etc.,) are averaged over the set of random variables $\{\Theta(\mathbf{k},\mathbf{k}')\}$.

3. THE TRANSFORMATION. For an explicit closed-form solution, we specialize to $V(\mathbf{k}_1,\mathbf{k}_2,\mathbf{k}_3,\mathbf{k}_4) = \delta(\mathbf{k}_1,-\mathbf{k}_2)\delta(\mathbf{k}_3,-\mathbf{k}_4)$, i.e.

$$H_2(\lambda) = \frac{\lambda\xi}{N\omega_D}\sum_{\mathbf{k},\mathbf{k}'} \omega_{\mathbf{k}}\omega_{\mathbf{k}'}Q_{\mathbf{k}}Q_{-\mathbf{k}}Q_{\mathbf{k}'}Q_{-\mathbf{k}'}. \tag{6}$$

Now let us perform a symmetry-breaking transformation:

$$a_{\mathbf{k}} \to a_{\mathbf{k}} + f_{\mathbf{k}}/\sqrt{\omega_{\mathbf{k}}}\ ,\ a_{\mathbf{k}}^{\dagger} \to a_{\mathbf{k}}^{\dagger} + f_{\mathbf{k}}^{*}/\sqrt{\omega_{\mathbf{k}}}\ ,\ \text{with } f_{\mathbf{k}}^{*} = f_{-\mathbf{k}}. \tag{7}$$

This introduces a number of new terms into $H(\lambda)$, enumerated below.
(i) A shift in the vacuum energy,

$$\delta H_0 = (N\omega_D/32)\Delta + \frac{\lambda g}{\sqrt{N}} \sum_{\mathbf{k},\mathbf{k}'} f_{-\mathbf{k}} f_{\mathbf{k}'} \exp\left[i\Theta(\mathbf{k},-\mathbf{k}')\right] + (N\lambda\xi\omega_D/64)\Delta^2. \tag{8}$$

where

$$\Delta \equiv (32/N\omega_D) \sum_{\mathbf{k}} |f_{\mathbf{k}}|^2. \tag{9}$$

Δ provides the norm for $f_{\mathbf{k}}$, as well as the correction to the Debye-Waller factor.
(ii) Terms linear in the $Q_{\mathbf{k}}$,

$$-\sum_{\mathbf{k}} \{ f_{\mathbf{k}} + \frac{\lambda g}{\sqrt{N}} \sum_{\mathbf{k}'} f_{\mathbf{k}'} \exp\left[i\Theta(\mathbf{k},-\mathbf{k}')\right] + \lambda\xi\Delta f_{\mathbf{k}} \} \sqrt{\omega_{\mathbf{k}}} Q_{\mathbf{k}}. \tag{10}$$

(iii) Additional terms quadratic in the $Q_{\mathbf{k}}$'s,

$$\frac{\lambda\xi\Delta}{4} \sum_{\mathbf{k}} \omega_{\mathbf{k}} Q_{\mathbf{k}} Q_{-\mathbf{k}} + \frac{24\lambda\xi}{N\omega_D} \Big(\sum_{\mathbf{k}} f_{\mathbf{k}} \omega_{\mathbf{k}}{}^{1/2} Q_{\mathbf{k}}\Big)^2. \tag{11}$$

(iv) Cubic terms,

$$\frac{-8\lambda\xi}{N\omega_D} \Big(\sum_{\mathbf{k}} \omega_{\mathbf{k}}{}^{1/2} f_{\mathbf{k}} Q_{\mathbf{k}} \Big(\sum_{\mathbf{k}'} \omega_{\mathbf{k}'} Q_{\mathbf{k}'} Q_{-\mathbf{k}'}\Big), \tag{12}$$

with only the quartic terms remaining unaffected. We determine $f_{\mathbf{k}}$ by the stability requirement that all linear contributions, (10), vanish. A nontrivial solution is achieved by solving the eigenvalue problem defined by the integral equation,

$$f_{\mathbf{k}}(1 + \lambda\xi\Delta) + \lambda(g/\sqrt{N}) \sum_{\mathbf{k}'} \exp\left[i\Theta(\mathbf{k},-\mathbf{k}')\right] f_{\mathbf{k}'} = 0 \tag{13}$$

for which $f_{\mathbf{k}}$ is the eigenfunction, normalized according to Δ, Eq. (9). Let us define μ to be the corresponding eigenvalue of the random kernel, $M(\mathbf{k},\mathbf{k}')$, of Eq. (3). Then, (13) reduces to:

$$\Delta = -(1 + \mu\lambda g)/\lambda\xi. \tag{14}$$

which, by (9), implies that $f_{\mathbf{k}}$, a function of the random $\Theta(\mathbf{k},\mathbf{k}')$'s, has a magnitude which diverge as $1/\sqrt{\xi}$ in the limit $\xi \to 0$. The shift in vacuum energy can now be expressed in terms of Δ as: $\delta H_0 = -N\lambda\xi\omega_D\Delta^2/64$. The lowest vacuum energy is obtained by choosing the most negative μ, known from previous studies of random matrices[8,5] to be $\mu = -2$ in the thermodynamic limit $N \to \infty$. Thus, (14) becomes:

$$\Delta = (2\lambda g - 1)/\lambda\xi \quad (\text{for } \lambda g \geq 1/2). \tag{15}$$

The divergence of Δ in the limit $\xi \to 0$ implies a divergent Debye-Waller exponent, hence a loss of atomic correlations in the glassy phase. Thus, in our model all the Bragg peaks disappear discontinuously as g is increased through $g_c = 1/2$.

For $\lambda g < 1/2$, (14) yields $\Delta < 0$, incompatible with the definition (9). Therefore, in this instance we must adopt the trivial solution,

$$\Delta = 0 \quad (\text{for } \lambda g < 1/2). \tag{16}$$

and the theory reduces to the previously solved disordered crystal.[5] Note that for thermodynamic purposes it is *not* required to know $f_{\mathbf{k}}$, the eigenfunction of the random kernel, explicitly. Only its eigenvalue is explicitly required.

According to (8), the shift in vacuum energy at a given λ is:

$$\delta E_0(\lambda) = \frac{-N\omega_D(\lambda g - 1/2)^2}{16\lambda\xi} \quad (\text{for } \lambda g \geq 1/2)\ , \quad \text{or } \delta E_0 = 0 \quad (\text{for } \lambda g < 1/2). \tag{17}$$

Of the new quadratic contributions (10), the first term is O(N), whereas the second is a random walk (we can safely assume $< f_{\mathbf{k}} f_{-\mathbf{k}'} >= O(1/\sqrt{N})$ for $\mathbf{k} \neq \mathbf{k}'$) and being O(1), becomes negligible in the thermodynamic limit. Both are independent of ξ in the limit $\xi \to 0$. The cubic terms and quartic terms in (12) and (6) vanish as powers of ξ in the limit $\xi \to 0$, and can be ignored henceforth.

Although, for $\lambda g < 1/2$ the quartic terms are irrelevant, for $\lambda g \geq 1/2$, they serve to renormalize the vacuum energy and the quadratic terms. The vacuum energy is highly discontinuous at $\lambda g = 1/2$, indicating that the crystal-to-glass transition is first-order at g_c. In both phases, the *dynamic* part of the Hamiltonian, $H'(\lambda)$, is quadratic in the $Q_{\mathbf{k}}$'s and is exactly soluble. With the help of a step function ($u(x) = 1$ for $x > 0$, $u(x) = 0$ for $x \leq 0$,) we write a single expression for both phases, of the form:

$$\begin{aligned} H'(\lambda) &= \sum_{\mathbf{k}} \omega_{\mathbf{k}}(a^\dagger_{\mathbf{k}} a_{\mathbf{k}} + 1/2) + \frac{\lambda g}{4\sqrt{N}} \sum_{\mathbf{k},\mathbf{k}'} (\omega_{\mathbf{k}}\omega_{\mathbf{k}'})^{1/2} \exp[i\Theta(\mathbf{k},\mathbf{k}')] Q_{\mathbf{k}} Q_{\mathbf{k}'} + \\ & \frac{1}{2} u(\lambda g - 1/2)(\lambda g - 1/2) \sum_{\mathbf{k}} \omega_{\mathbf{k}} Q_{\mathbf{k}} Q_{-\mathbf{k}}. \end{aligned} \tag{18}$$

$H'(\lambda)$ is explicitely independent of ξ, and the limit $\xi \to 0$ can be taken trivially. However, the parameter $0 \leq \lambda \leq 1$ remains to be integrated over. The sole adjustable coupling constant in the theory is g, with $g_c = 1/2$ signalling onset of the glassy phase.

4. THE GREEN FUNCTIONS. For $\lambda g < 1/2$, the results of ref. 5 remain unchanged. For $\lambda g \geq 1/2$, we obtain a modification:

$$< G_{\mathbf{k},\mathbf{k}'}(\omega) >= \frac{(\omega_{\mathbf{k}}/\pi)\delta_{\mathbf{k},\mathbf{k}'}}{\omega^2 - \omega_{\mathbf{k}}{}^2(1 + \Phi + Z(\omega))}. \tag{19}$$

where $Z(\omega)$ is:

$$Z(\omega) = \frac{(\lambda g)^2}{N} \sum_{\mathbf{q}} \frac{\omega_{\mathbf{q}}{}^2}{\omega^2 - \omega_{\mathbf{q}}{}^2(1 + \Phi + Z(\omega))}. \tag{20}$$

with

$$\Phi = (1/2)(\lambda g - 1/2)u(\lambda g - 1/2). \tag{21}$$

One can also obtain such other averages over random configurations as may be useful, including $< G_{\mathbf{k},\mathbf{k}'} \exp[i\Theta(\mathbf{k},\mathbf{k}')] >$.

With the help of the above one calculates $F = F_0 + \delta F_0 + F_1 + F_2$. Here F_0 is the free energy of the reference crystal, δF_0 is $\int_0^1 d\lambda < \delta H_0 >_\lambda$,

$$\delta F_0 = -(\omega_D/64\xi)[4g(g - \log(2g)) - 1]u(g - 1/2). \tag{22}$$

$$F_1 = k_B T \int_0^\infty d\omega\ \rho_1(\omega) \log(2\sinh(\beta\omega/2)) \qquad \beta \equiv 1/k_B T. \tag{23}$$

$$F_2 = k_B T \int_0^\infty d\omega\ \rho_2(\omega) \log(2\sinh(\beta\omega/2)). \tag{24}$$

The spectral densities $\rho_1(\omega)$ *and* $\rho_2(\omega)$ are given by,

$$\rho_1(\omega) = \frac{2}{\pi g^2} \int_0^1 \frac{d\lambda}{\lambda^3} \frac{\partial}{\partial\omega}[-I(\omega,\lambda)]. \tag{25}$$

$$\rho_2(\omega) = \frac{u(g - 1/2)}{\pi g^2} \int_{(2g)^{-1}}^1 \frac{d\lambda}{\lambda^3}(2\lambda g - 1)\frac{\partial}{\partial\omega}[-I(\omega,\lambda)]. \tag{26}$$

5. SOME RESULTS.

Einstein Model

Define $x \equiv \omega/\omega_E$ and $Z(x) = R(x) + iI(x)$, and $x_1 \equiv \sqrt{1 - 2\lambda g}$, $x_2 \equiv \sqrt{1 + 2\lambda g}$.
In the range $\lambda g < 1/2$:

$$I(x) = \begin{cases} (1/2)[(2\lambda g)^2 - (x^2 - 1)^2]^{1/2} & x_1 < x < x_2 \\ 0 & 0 < x < x_1 \text{ or } x > x_2 \end{cases} \tag{27}$$

$$R(x) = \begin{cases} (1/2)(x^2 - 1) & x_1 < x < x_2 \\ (1/2)(x^2 - 1) - (1/2)[(x^2 - 1)^2 - (2\lambda g)^2]^{1/2} & 0 < x < x_1 \text{ or } x > x_2 \end{cases} \tag{28}$$

In the range $\lambda g \geq 1/2$:

$$I(x) = \begin{cases} (1/2)[(2\lambda g)^2 - (2\lambda g - x^2)^2]^{1/2} & 0 < x < \sqrt{4\lambda g} \\ 0 & x > \sqrt{4\lambda g} \end{cases} \tag{29}$$

$$R(x) = \begin{cases} (x^2 - 2\lambda g)/2 & 0 < x < \sqrt{4\lambda g} \\ (1/2)(x^2 - 2\lambda g) - (1/2)[(2\lambda g - x^2)^2 - (2\lambda g)^2]^{1/2} & x > \sqrt{4\lambda g} \end{cases} \tag{30}$$

For $g > 1/2$, the low-frequency gap in the spectral density is completely eliminated and the low-temperature specific heat becomes linear in T in the Einstein model, just as in all glasses. At higher temperatures,

$$c_p \approx 3k_B - O(T_E/T)^\delta. \tag{31}$$

recovering the Dulong-Petit law with correction exponent $\delta = 3$ for $g < 1/2$, $\delta = 2$ for $g \geq 1/2$.

Debye Model.

In the Debye approximation, set $x = \omega/\omega_D$ and obtain:

$$Z(x) = \frac{-(\lambda g)^2}{(1 + \Phi + Z(x))}\left[1 + 3\Omega^2 - (3/2)\Omega^3 \log\frac{(\Omega + 1)}{(\Omega - 1)}\right] \tag{32}$$

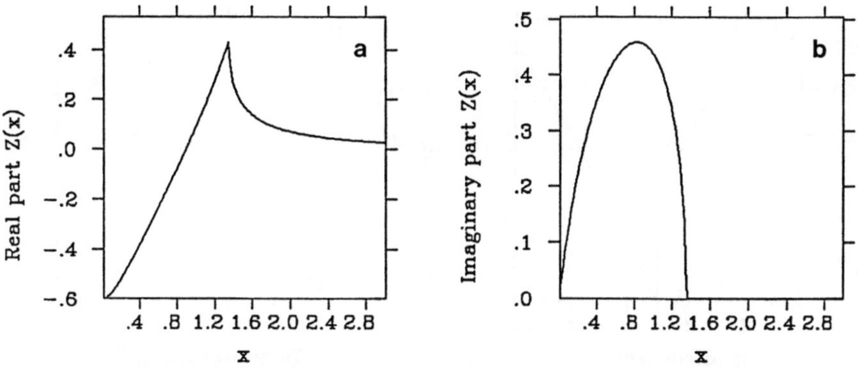

Fig. 1a: Real part of self-energy $Z(x)$ at $g = 0.6$ (above critical disorder) for the Debye model, where $x \equiv \omega/\omega_D$.

Fig. 1b: Imaginary part of self-energy $Z(x)$ at $g = 0.6$ (above critical disorder) for the Debye model.

where $\Omega^2 \equiv x^2/(1 + \Phi + Z(x))$. One requires $I(x) \geq 0$, $I(0) = 0$, $Z(\infty) \to 0$, and lim. $g \to 0\{R(x) = 0, I(x) = 0^+\}$. Figs. $1a, b$ shows the calculated R, I for g=0.6. The speed of sound V_s at long wavelength, as function of g, is given by the expression:

$$V_s(g) = \begin{cases} V_0[1/2 + \sqrt{(1/4 - g^2)}]^{1/2} & g < 1/2 \\ V_0\sqrt{g} & g \geq 1/2 \end{cases} \tag{33}$$

as shown in fig. 2, with $V_0 \equiv \omega_D/k_D$. There is a cusp at $g = 1/2$.

The specific heat can be obtained numerically from Eq. (5). At low-temperatures, it is well approximated by a low-frequency expansion of Eqs. (28), (29):

$$g < 1/2: \qquad c_p(T) = c_3(T/\Theta_D)^3 + O(T/\Theta_D)^5. \tag{34}$$

where $c_3 = (12\pi^4/5)(1/2 + \sqrt{(1/4 - g^2)})^{-3/2}$.

$$g \geq 1/2: \qquad c_p(T) = \bar{c}_1(T/\Theta_D) + \bar{c}_3(T/\Theta_D)^3 + O(T/\Theta_D)^5. \tag{35}$$

where:

$$\begin{aligned} \bar{c}_1 &= (2\pi/3\sqrt{3})[(2g)^{3/2} - 1]/g^{3/2}. \\ \bar{c}_3 &= \{661.23 - \frac{491.55}{g^{3/2}}[(2g)^{3/2} - 1] + \frac{50.22}{g^{5/2}}[(2g)^{5/2} - 1]\}. \end{aligned} \tag{36}$$

As in the Einstein model glass, it is the anomalous increase in the spectral width of the phonons which produces the linear specific heat at low temperatures. Now consider silica glass, SiO_2. We have fit two parameters: the theoretical $\bar{c}_1$ to the experimental $\bar{c}_1$, the theoretical $\bar{c}_3$ to the experimental $\bar{c}_3$, with a single adjustable coupling constant $g = 0.5001382$ determined by least squares. This result in the comparison illustrated over two orders of magnitude of T in Fig. 3. At $T \gg T_D$, c_p monotonically approaches the Dulong-Petit limit.

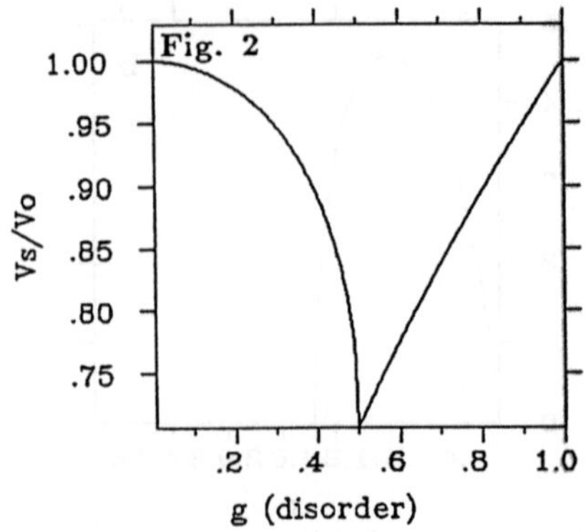

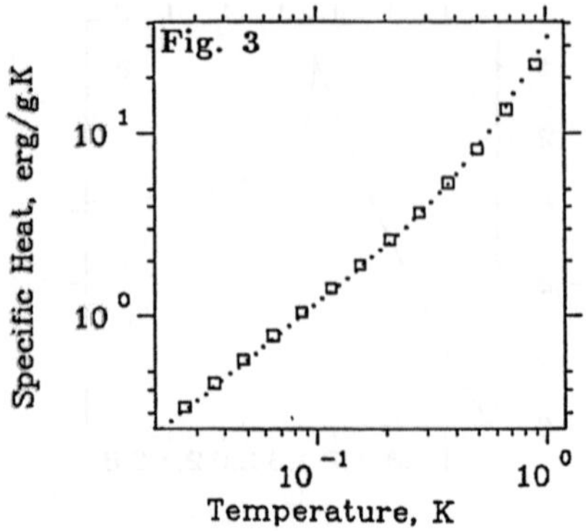

Fig. 2: speed of sound V_s/V_0 of long wavelength phonons, versus disorder.

Fig. 3: specific heat of SiO_2 at low temperatures. Dotted line: theoretical fit. Squares: experimental values.

6. CONCLUSIONS. Our model is based on the twin pillars of the glassy phase, randomness and anharmonicity. It yields many features common to all glasses. We find the glass transition to be first-order in the disorder parameter g,[10] while at fixed g, $F(T)$ is analytic in T. Although the model permits explicit evaluation of the Kubo formulas for thermal transport, the resulting expressions are too involved to display here. Thermodynamics equilibrium requires the use of the most negative eigenvalue, $\mu = -2$, of the random matrix $M(\mathbf{k}, \mathbf{k}')$. The role of the higher eigenvalues of $M(\mathbf{k}, \mathbf{k}')$ in the transport, metastable, and hysteretic properties of glass remains to be understood. Further details concerning the model will be published elsewhere, as well as in the forthcoming Ph.D. dissertation of one of us (M.M.).

FOOTNOTES AND REFERENCES

1. A delightful introduction to the subject of the amorphous phase is:
R. Zallen in " *The Physics of Amorphous Solids* ", eds. John Wiley & sons, 1983.

2. R.O. Pohl, J. J. de Yoreo, M. Meissner and W. Knaak, "Are we beginning to understand the vibrational anomalies of glass ? " in: " *Physics of disordered materials* ", eds. D. Adler, H. Fritzche and S. Ovshinsky (Plenum, New York, 1985) p. 529.
W. A. Phillips (ed.), " Amorphous Solids. Low Temperature Properties ",(Berlin: Springer, 1981).

3. P.W. Anderson, B. Halperin and C. Varma, Phil. Mag. **25**, 1468 (1972).
W. A. Phillips, J. Low Temp. Phys. **7**, 351 (1972).

4. S. Hunklinger, W. Arnold, S. Stein, R. Nova and D. Dransfeld, Phys. Lett. **42**A, 253 (1972).
S. Hunklinger, W. Arnold and S. Stein, Phys. Lett. **45**A, 311 (1973).

5. D.C. Mattis, Phys. Lett. **A, 117**, 6 (1986).
D.C. Mattis, Phys. Lett. A, **120**, 7 (1987).

6. N. W. Ashcroft and N. D. Mermin in " *Solid State Physics* ", eds. Holt, Rinehart and Winston, (1976), p. 494.

7. D. N. Zubarev, Sov. Phys. Usp.**3**, 320 (1960).

8. E. P. Wigner, Proc. 4th. Can. Math. Congr., Toronto, 1959, p. 174.
D. C. Mattis and R. Raghavan, Phys. Lett. , **75**A, 4 (1980).
A review of the theory of random matrices appear in:
M. L. Mehta, " *Random Matrices* ", (Acad. Press, NY, 1967).

9. R. C. Zeller and R. O. Pohl, Phys. Rev. **B4**, 2029 (1971).

10. for example, in the amount of radiation damage to a solid.

Two-Sublattice Solitons in Hydrogen-Bonded Chains with Dynamical Disorder

A.V. Savin[1] and *A.V. Zolotaryuk*[2]

[1]Institute for Physico-Technical Problems, 119 034 Moscow, Russia
[2]Institute for Theoretical Physics, Ukrainian Academy of Sciences, 252 120 Kiev, Ukraine

Abstract. *In the framework of the two-sublattice model the dynamics of ionic defects in a thermalized hydrogen-bonded chain has been numerically studied. The mobilities of a positive and a negative defects have been evaluated that are in good agreement with the experimental data. The dynamics of non-topological supersonic acoustic solitons and their interaction with ionic defects have been also investigated.*

1. Introduction

Macromolecular quasi-one-dimensional structures that are long periodic chains of cells coupled by hydrogen bonds have been studied extensively during the last decades. These studies were essentially intensified since the basic transport processes in bioenergetics, i.e. the transfer of energy and charge in biomolecular systems, came to be associated with the motion of nonlinear collective localized excitations (solitary waves) along these systems (see, for instance, the book [1] and further references cited therein). One of the aspects of this problem is the investigation of the soliton models for the proton transport in hydrogen-bonded systems [2-5].

The transfer of protons along a hydrogen-bonded or ice-like network proceeds via the motion of the ionic and rotational (Bjerrum or bonding) defects [6]. The former involve translational motions of the hydrogen-bonded protons, whereas the latter are results of rotations of the water molecules. The transport of protons may begin either with the passage of an ionic defect or the passage of a rotational defect, but thereafter the motion of these defects must strictly alternate.

In the soliton models studied up to now [2-5] the proton-proton coupling or the characteristic velocity of the small-amplitude waves in the protonic subsystem, and therefore the soliton width, was supposed to be sufficiently large for the continuum approximation to be applied. In this paper we shall consider the two-sublattice model [7] of ionic defects

Springer Proceedings in Physics, Vol. 67 Nonlinearity with Disorder
Editors: F. Abdullaev · A.R. Bishop · S. Pnevmatikos

in a hydrogen-bonded chain with a weak intersite interaction when the continuum limit can not be considered (in our case the width of an ionic defect does not exceed two periods of the chain). In this case the pinning of the defect occurs and it can move only in the thermalized chain when an external electric field is applied.

2. The Model

We consider a typical one-dimensional network $\cdots O-H\cdots O-H\cdots O-H\cdots$ of hydrogen-bonded water molecules where O denotes an oxygen and H is a hydrogen forming a covalent bond with an adjacent oxygen ion (line segment) and a hydrogen bond with the other ion (dotted segment). This chain can be found in the two degenerate ground states $\cdots O-H\cdots O-H\cdots$ and $\cdots H-O\cdots H-O\cdots$. The passage of the chain from one of the ground states into the other corresponds to the motion of a positive (I^+) or negative (I^-) ionic defect. Under this passage the covalent and hydrogen bonds change their places, i.e. the proton in a hydrogen bridge $O-H\cdots O$ tunnels between two equilibrium positions that are energetically equivalent. Taking into account the correlative motion of the protons in adjacent hydrogen bonds it may be assumed in the model that the protons are connected by springs and moving between each pair of oxygen ions in a symmetrical double-well potential [2-4]. This potential may be constructed as the sum of two pair ion-proton potentials (for example, the Morse or Lennard-Jones potentials [7-9]).

It is convenient to describe the interaction of a proton with an oxygen ion by the Morse potential $U(r) = D_0\{\exp[-b(r-r_0)]-1\}^2$ where D_0 and r_0 are the energy and the length of the $O-H$ bond in an isolated water molecule and b is the phenomenological parameter determined by the "bare" $O-H$ frequency. To construct correctly the on-site double-well proton potential we consider a pair of adjacent oxygens that are allowed to move. The distance between them is given by $l+\rho$ where l is the lattice spacing and ρ is the relative lengthening of the $O-H\cdots O$ bond. Then the on-site proton potential may be defined as the sum [7]

$$\begin{aligned} V(u,\rho) &= U(\frac{l+\rho}{2}+u)+U(\frac{l+\rho}{2}-u)-U(\frac{l}{2}+q_0)-U(\frac{l}{2}-q_0) = \\ &= \varepsilon_I(\alpha-1)^{-2}\{[\alpha-\cosh(bu)e^{-b\rho/2}]^2+\frac{1}{2}(1-e^{-b\rho})\} \qquad (1) \end{aligned}$$

where u is the displacement of the proton from the midpoint of the distance between the oxygens, the points $u=\pm q_0$ are the minima of the function (1) when $\rho=0$ and they are determined by the equation

$$\alpha = \cosh(bq_0) = \frac{1}{2}\exp[b(\frac{l}{2} - r_0)] \, , \tag{2}$$

$\varepsilon_I = D_0(1-\alpha^{-1})^2$ is the height of the barrier at $\rho = 0$. When $\rho \neq 0$ the positions of the minima $\pm u_0(\rho)$ are determined by the equation $\cosh(bu_0) = \alpha\exp(b\rho/2)$ while the height of the barrier is $\varepsilon(\rho) = D_0[1 - \alpha^{-1}\exp(-b\rho/2)]^2$. At the critical value $\rho = -2b^{-1}\ln\alpha$ the potential (1) becomes a function with the single minimum at $u = 0$.

The energy of the hydrogen bond, when the proton is located at the ground state, may be also given by the Morse potential

$$V_{OO}(\rho) = \varepsilon_H[\exp(-\beta\rho) - 1]^2 \tag{3}$$

where ε_H is the energy of the hydrogen bond and β is the phenomenological parameter. Note that the last term in (1) can be absorbed by the potential $V_{OO}(\rho)$, so that the total energy of the bond $O - H \cdots H$ can be written as the sum $W(u,\rho) = V_{OH}(u,\rho) + V_{OO}(\rho)$ where

$$V_{OH}(u,\rho) = D_0[1 - \alpha^{-1}\cosh(bu)e^{-b\rho/2}]^2. \tag{4}$$

Consider now the proton-proton interactions. In the cyclic hydrogen-bonded chain of N hydroxyl groups the total energy of the proton-proton Coulomb interactions, when only the nearest neighbors are considered, can be represented as the sum [10]

$$\frac{e^2}{4\pi\varepsilon_0}\sum_{n=1}^{N}\left(\frac{1}{l + y_{n+1} - y_n} - \frac{1}{l}\right) = \sum_{n=1}^{N} V_{HH}(y_{n+1} - y_n) \tag{5}$$

where the potential V_{HH} is given by the function

$$V_{HH}(x) = \kappa x^2/2(1 + x/l) \, . \tag{6}$$

Here y_n denotes the displacement of the proton from the midpoint of the n-th and $(n+1)$-th oxygens located in the equilibrium positions, e the protonic charge, ε_0 the dielectric constant and $\kappa = e^2/2\pi\varepsilon_0 l^3$.

Thus, the total energy of the chain can be written in the form

$$\begin{aligned} H &= \sum_n [\frac{1}{2}m\dot{y}_n^2 + \frac{1}{2}M\dot{Y}_n^2 + \\ &+ V_{HH}(y_{n+1} - y_n) + V_{OO}(\rho_n) + V_{OH}(u_n,\rho_n) + V_O(Y_n)] \end{aligned} \tag{7}$$

where Y_n measures the displacement of the n-th oxygen with mass M from its equilibrium position, m is the proton mass, the dots over the variables y_n and Y_n denote the differentiation with respect to time t, the potentials V_{HH}, V_{OO}, V_{OH} and V_O (substrate) are given by (6), (3), (4), and $V_O(Y) = K_o Y^2/2$ (with stiffness K_o), respectively, the variables u_n and ρ_n are defined by

$$u_n = y_n - \frac{1}{2}(Y_n + Y_{n+1}), \qquad \rho_n = Y_{n+1} - Y_n. \tag{8}$$

In fact, the Hamiltonian (7) is a small generalization of the model proposed in Ref. [7].

3. Parameter Values for the Hydrogen-Bonded Chain

For the chain of hydrogen-bonded water molecules in ice the lattice constant $l = 2.76\mathring{A}$, the $O - H$ bond distance $r_1 = 1.01\mathring{A}, q_0 = \frac{l}{2} - r_1 = 0.37\mathring{A}$, $m = m_p$ and $M = 17m_p$ (m_p is the proton mass), the elastic stiffness of the $O - H$ bond $K_{OH} = 704N/m$ [11]. For the height of the barrier in the double-well potential we choose $\varepsilon_I = 0.31eV$. The phenomenological parameter b can be found from the equation

$$K_{OH} = \frac{\partial^2}{\partial u^2} V_{OH}(u = \pm q_0,\ \rho = 0) = 2\varepsilon_I b^2 \frac{\alpha + 1}{\alpha - 1}. \tag{9}$$

The solution of Eq. (9) is $b = 7.393\mathring{A}^{-1}$.

For the vibrations of the oxygen sublattice in ice there are two translational modes (symmetric and antisymmetric) with the frequences $\omega_s = (K_o/M)^{1/2} = 60cm^{-1}$ and $\omega_a = [(K_o + 4K_{OO})/M]^{1/2} = 229cm^{-1}$ where K_{OO} is the elastic stiffness of the hydrogen bond. From these equations we find $K_o = 4N/m$ and $K_{OO} = 13N/m$.

The other phenomenological parameter β can be found from Eq. (3) by using the value $\varepsilon_H = 0.22eV$ for the energy of the hydrogen bond. As result, we find $\beta = (K_{OO}/2\varepsilon_H)^{1/2} = 1.358\mathring{A}^{-1}$. The calculation of the stiffness for the proton sublattice according to the definition gives the value $\kappa = 22N/m$. Finally, we give the values for the sound velocities in the proton and oxygen sublattices : $v_H = (\kappa/m_p)^{1/2} l = 31.7 \times 10^3 m/s$ and $v_O = (K_{OO}/17m_p)^{1/2} l = 5.9 \times 10^3 m/s$.

4. Stationary States of the Defects

In oder to find the stationary states of the ionic defects we have to solve the following minimization problem (see (7)) :

$$E = \sum_{n=2}^{N-1} [V_{HH}(y_{n+1} - y_n) \; + \; V_{OO}(\rho_n) + V_{OH}(u_n, \rho_n) + \frac{K_\rho}{2} Y_n'^2] \rightarrow$$

$$\rightarrow \; \min : \; u_3, ..., u_{N-2}, \; \rho_3, ..., \rho_{N-2} \qquad (10)$$

with the conditions $u_1 = u_2 = \pm q_0$, $u_{N-1} = u_N = \mp q_0$, $\rho_1 = \rho_2 = \rho_{N-1} = \rho_N = 0$ for a defect $I^{\pm}$. Here $Y_1 = 0$, $Y_n = \sum_{k=1}^{n-1} \rho_k$, $n = 2, ..., N$; $y_n = u_n + \frac{1}{2}(Y_n + Y_{n+1})$, $n = 1, ..., N-1$. The problem (10) with $N = 50$ was solved by the steepest descent method.

It is convenient to describe a defect by the position of its center $n_c = \sum_{n=2}^{N-1} nP_n$ and its root-mean-square width $L = 2[\sum_{n=2}^{N-1}(n - n_c)^2 P_n]^{1/2}$ given in the units of the lattice constant l. Here $P_n = \frac{1}{C}|\tilde{u}_{n+1} - \tilde{u}_{n-1}|$ where $\tilde{u}_n = u_n$ if $|u_n| < q_0$ and $\tilde{u}_n = (\mathrm{sign} u_n) u_0$ if $|u_n| \geq q_0$, the constant C is determined from the condition $\sum_{n=2}^{N-1} P_n = 1$, i.e. $C = 4q_0$. The problem (10) admits solutions with two types of symmetry: a solution with an *integral* center of symmetry when $u_{\frac{N}{2}+k} = -u_{\frac{N}{2}-k}$, $\rho_{\frac{N}{2}+k} = \rho_{\frac{N}{2}-k}$ ($u_{\frac{N}{2}} = 0$, $n_c = \frac{N}{2}$) and a solution with a *half-integral* center of symmetry for which $u_{\frac{N}{2}+k} = -u_{\frac{N}{2}-k+1}$, $\rho_{\frac{N}{2}+k} = \rho_{\frac{N}{2}-k+1}$ ($n_c = \frac{N+1}{2}$), $k = 0, 1, 2, ...$.

First we consider ionic defects in the isolated chain (when $K_\rho = 0$). The profiles of these defects are plotted in Figs. 1 and 2. The positive defect I^+ with the half-integral center $n_c = 25.5$ has the energy $E_1^+ = 0.444eV$ and the width $L_1^+ = 1.19$. When this defect is located at $n_c = 25$. its energy $E_2^+ = 0.464eV$ and width $L_2^+ = 1.44$. The negative defect I^- in the state with $n_c = 25.5$ has the energy $E_1^- = 0.104eV$ and the width $L_1^- = 1.58$. When it is centered at $n_c = 25$, its energy and width are $E_2^- = 0.206eV$ and $L_2^- = 1.75$. The chain is stretched out in the region of localization of a positive defect whereas it is shrinked in the region of a negative defect. It should be emphasized that the energy of a defect in the half-integral-centered state is always less than its energy in the integral-centered state. Similarly, in the case with $K_0 = 4N/m$ we have $E_1^+ = 0.445eV$, $L_1^+ = 1.53$, $E_2^+ = 0.488eV$, $L_2^+ = 2.65$ and $E_1^- = 0.144eV$, $L_1^- = 1.46$, $E_2^- = 0.272eV$, $L_2^- = 1.50$.

Now we proceed to get the dependence of the energy of a stationary defect on the position of its center, i.e. the Peierls potential relief. To this end we fix in the problem (10) any point $u_{N/2}$ between $-q_0$ and q_0 and accomplish the minimization procedure with respect to the rest of the variables u_n. Changing monotonically $u_{N/2}$ we obtain a monotonic varying of the defect center. From here we can find the dependence of the defect energy on the position of its center. For both the cases with $K_\rho = 0$ and $K_\rho = 4N/m$ the Peierls profiles $E^{\pm}(n_c) - E_1^{\pm}$ are plotted in Fig. 3 . It follows from these plots that a defect I^+ is stable in both the states with integral and half-integral centers while a defect I^- is sta-

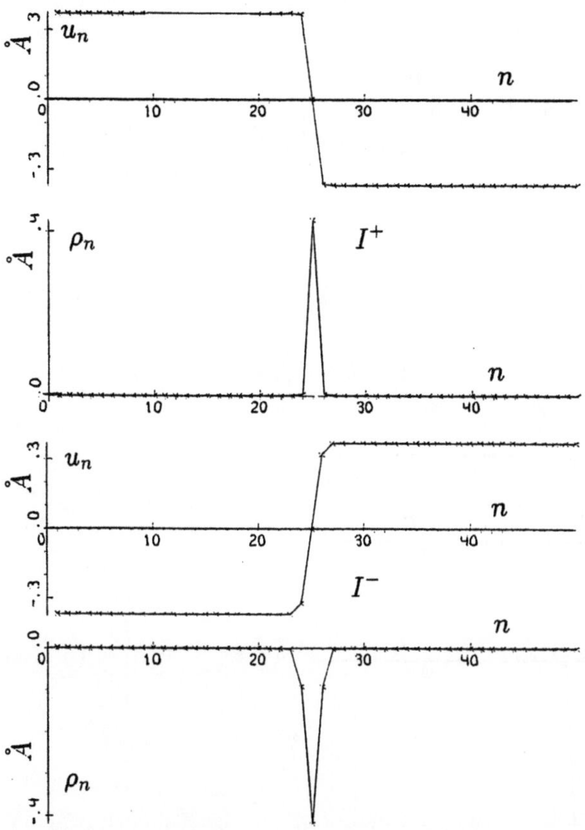

Figure 1: The profiles of the stationary ionic defects with integral center $n_c = 25$ in the chain with $K_o = 0$.

ble only in the state with a half-integral center. For the transition of the stationary defect I^+ from the state with a half-integral center into the state with an integral center it is necessary to get over the barrier with the height $\Delta E_1^+ = 0.045eV$ if $K_o = 0$ and $\Delta E_1^+ = 0.050eV$ if $K_o = 4N/m$. The reverse transition of the defect I^+ from the state with an integral center into the state with a half-integral center requires the energy $\Delta E_2^+ = 0.025eV$ if $K_o = 0$ and $\Delta E_2^+ = 0.008eV$ if $K_o = 4N/m$. For the passage of one chain cell the defect I^- must get over the barrier with the height $\Delta E_1^- = 0.102eV$ if $K_o = 0$ and $\Delta E_1^- = 0.128eV$ if $K_o = 4N/m$. The energy required to create a pair of defects (dissociation energy) can be defined as $E_I = E_1^+ + \Delta E_1^+ + E_2^-$. Then $E_I = 0.695eV$ in the isolated chain and $E_I = 0.767eV$ in the chain with the substrate. These values are in good agreement with the energy 0.76 ± 0.06 eV of the formation of a pair of ionic defects in ice crystal [12].

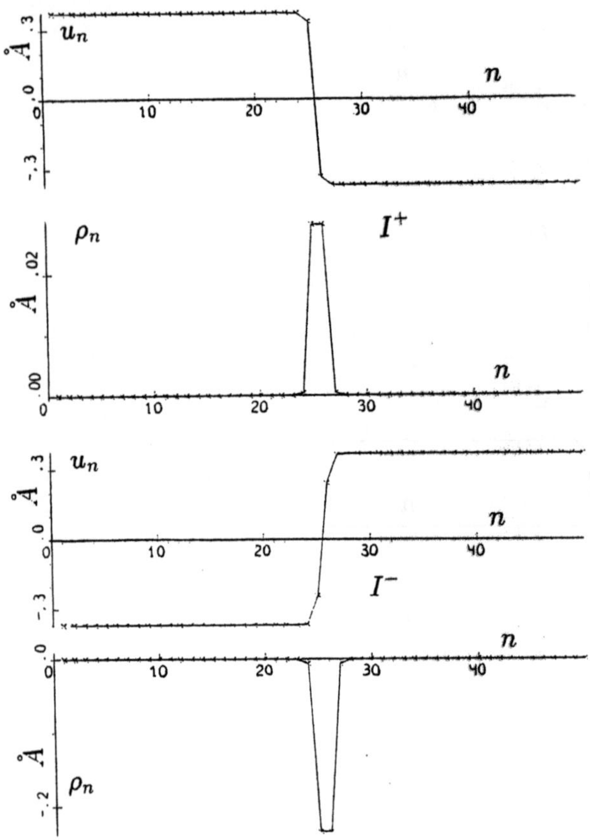

Figure 2: The profiles of the stationary ionic defects with half-integral center $n_c = 25.5$ in the chain with $K_o = 0$.

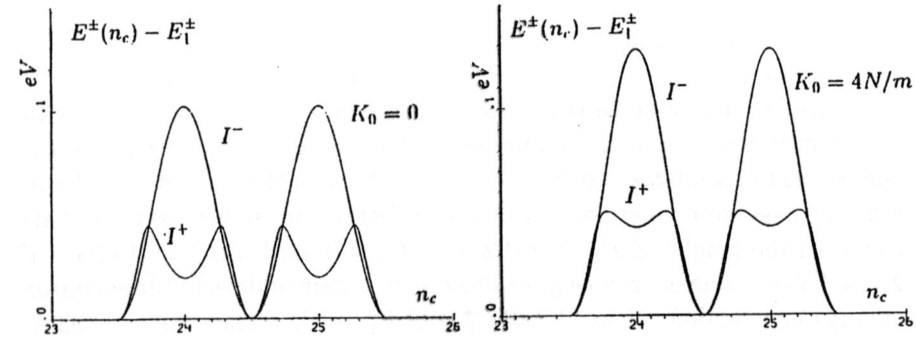

Figure 3: The Peierls potential reliefs for the ionic defects $I^{\pm}$.

5. Dynamics of Ionic Defects in a Thermalized Chain

The transfer of a defect $I^{\pm}$ along the chain requires overcoming the barrier $\Delta E_1^{\pm}$ in the Peierls potential relief. Therefore at temperatures $T < T_0 = \Delta E_1^{\pm}/k_B$ (k_B is the Boltzmann constant) a defect diffuses along the chain randomly jumping over the Peierls barriers. These jumpings happen at those instants of time when the energy occasionally taken by the defect from the heat bath is enough for overcoming the barrier. In the chain with $K_o = 0$ $(4N/m)$ we have $T_0 = 552K$ $(580K)$ for I^+ and $T_0 = 1184K$ $(1485K)$ for I^-.

Consider a chain of N water molecules with fixed ends. The dynamics of this chain at temperature T and an external electric field with intensity E is described by the Langevin-type equations of motion

$$m\ddot{y}_n = -\frac{\partial}{\partial y_n}H + eE, \qquad (11)$$

$$M\ddot{Y}_n = -\frac{\partial}{\partial Y_n}H - \Gamma M\dot{Y}_n + \xi_n(t) - eE \qquad (12)$$

where the Hamiltonian H is given by (7), Γ is the damping coefficient for the oxygen sublattice and $\xi_n(t)$ is the normal random external force satisfying the conditions

$$\langle\xi_n(t)\rangle = 0, \ \langle\xi_k(s)\xi_n(t)\rangle = 2\Gamma M k_B T\delta_{kn}\Delta(s-t). \qquad (13)$$

Here $k, n = 2, ...N-1$ and the function $\Delta(s-t)$ is defined as $\Delta(s-t) = 0$ if $|s-t| > t_0'/2$ and $\Delta(s-t) = 1/t_0'$ if $|s-t| \le t_0'/2$. The correlation time t_0' gives the duration of the interaction of an oxygen atom with a molecule of the heat bath. It should be noticed that the oxygen size considerably exceeds the proton size and therefore we neglect the interaction of molecules from the heat bath directly with chain protons. The damping coefficient is defined by $\Gamma = 1/t_0''$ where t_0'' is the relaxation time of the velocity of an oxygen atom (when Langevin-type equations are used, it is always assumed that $t_0' \ll t_0'' \ll t$ where t is time of the dynamics of the system).

For numerical simulations of the set of Eqs. (11) and (12) it is convenient to use the dimensionless time $\tau = (\kappa/m)^{1/2}t$. Throughout this paper the damping coefficient is taken to be $\Gamma = 0.1(K_{OO}/M)^{1/2}t$. This value corresponds to the relaxation time $\tau_0'' = 53.64$ ($t_0'' = 4.68 \times 10^{-13}s$). We consider the dynamics of the system at the correlation times $\tau_0' = 0.1$ and $\tau_0' = 1$ ($t_0' = 8.72 \times 10^{-16}s$ and $t_0' = 8.72 \times 10^{-15}s$) for which the condition $\tau_0' \ll \tau_0''$ is satisfied.

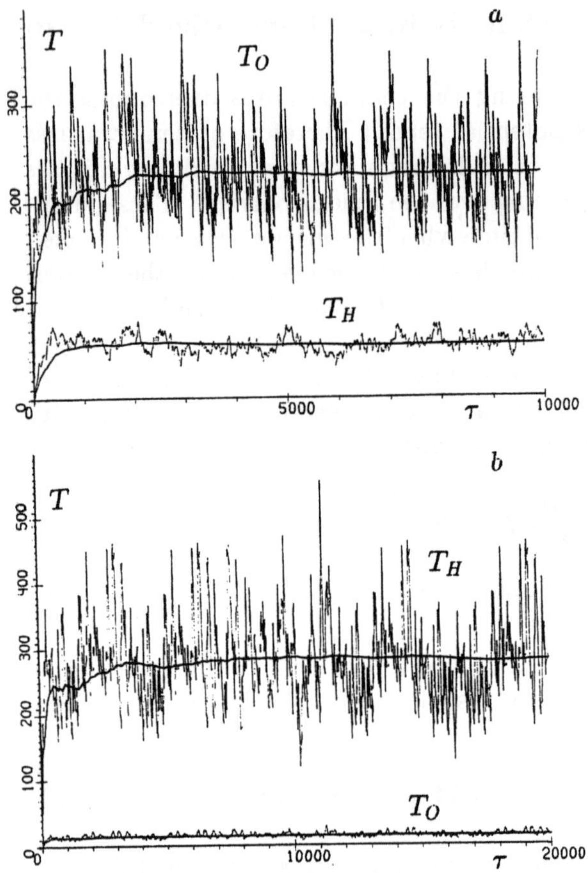

Figure 4: The thermalization of the chain consisting of $N = 20$ molecules at $\tau_0' = 0.1$ (a) and $\tau_0' = 1$ (b), $T = 300K$.

At any instant of time the chain can be described by the instantaneous temperature of the oxygen sublattice $T_O(t) = (M/(N-2)k_BT)\sum_{n=2}^{N-1} \dot{Y}_n^2$, the instantaneous temperature of the proton sublattice $T_H(t) = (m/(N-2)k_BT)\sum_{n=2}^{N-1} \dot{y}_n^2$, the root-mean-square amplitude of the $O-H\cdots O$ bond distance vibrations $R_{OO}(t) = (\frac{1}{N-1}\sum_{n=1}^{N-1} \rho_n^2)^{1/2}$, and the root-mean-square amplitude of the $O-H$ bond distance vibrations $R_{OH}(t) = [\frac{1}{N-2}\sum_{n=2}^{N-1}(q_0 + y_n - Y_n)^2]^{1/2}$. Consider first the dynamics of the chain at $T = 300K$ when it is found in one of its equilibrium positions (say $Y_n \equiv 0$ and $u_n \equiv -q_0$). The dynamics of the chain can be described by the mean quantities $\bar{T}_O = \frac{1}{t}\int_0^t T_O(s)ds$, $\bar{T}_H = \frac{1}{t}\int_0^t T_H(s)ds$, $\bar{R}_{OO} = \frac{1}{t}\int_0^t R_{OO}(s)ds$, and $\bar{R}_{OH} = \frac{1}{t}\int_0^t R_{OH}(s)ds$. The time dependence of the instantaneous temperatures of the sublattices $T_O(t), T_H(t)$ and their

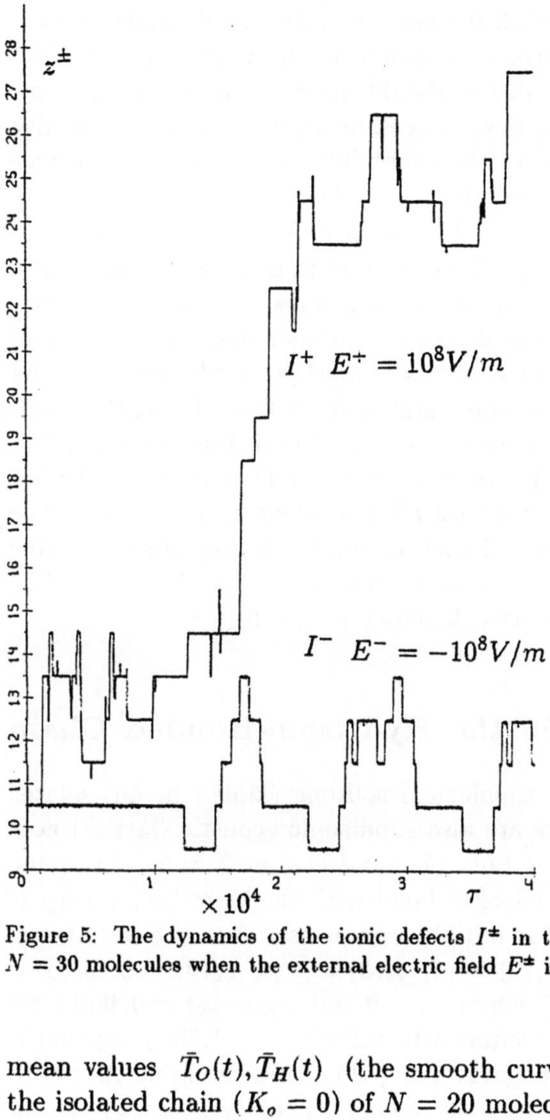

Figure 5: The dynamics of the ionic defects $I^{\pm}$ in the thermalized chain cosisting of $N = 30$ molecules when the external electric field $E^{\pm}$ is applied.

mean values $\bar{T}_O(t), \bar{T}_H(t)$ (the smooth curves) has been calculated for the isolated chain ($K_o = 0$) of $N = 20$ molecules at $E = 0$ with $\tau_0' = 0.1$ (Fig. 4a) and $\tau_0' = 1$ (Fig. 4b). When $\tau_0' = 0.1$ we obtain $\bar{T}_O = 228K$, $\bar{T}_H = 54K$, $\bar{R}_{OO} = 0.0947\AA$, $\bar{R}_{OH} = 0.0098\AA$ while for $\tau_0' = 1$ we have $\bar{T}_O = 283K$, $\bar{T}_H = 15K$, $\bar{R}_{OO} = 0.1058\AA$, $\bar{R}_{OH} = 0.0039\AA$. At the correlation time of the random force $\tau_0' = 1$ the thermalization of the $O - H$ vibrations practically does not occur which is clearly shown in the equality $\bar{T}_O/\bar{T}_H = 18.9 \simeq M/m = 17$. Indeed, the frequences of the covalent vibrations $\omega \geq (K_{OH}/m)^{1/2} = 6.5 \times 10^{14} s^{-1}$ while the frequency spectrum of the random force is located below the frequency $\omega_\xi = 2\pi/t_0'$ equal to $7.2 \times 10^{14} s^{-1}$ if $\tau_0' = 1$ and $7.2 \times 10^{15} s^{-1}$ if $\tau_0' = 0.1$.

At $\tau_0' = 0.1$ the excitation of high-frequency covalent oscillations appears which results in increasing the mean temperature of the proton sublattice.

The dynamics of an ionic defect should mainly depend on the temperature of the proton sublattice. Therefore at $\tau_0' = 0.1$ the mobility of defects should be larger than their mobility at $\tau_0' = 1$. The dynamics of the centers $z^{\pm}$ of the defects $I^{\pm}$ in the thermalized chain with $N = 30$, $T = 300K$, $\tau_0' = 1$ and $E = \pm 10^8 V/m$ ($\bar{T}_O = 285K$ and $\bar{T}_H = 18K$) is shown in Fig. 5 . The motion of these defects is thermally activated, so that they diffuse in the chain by means of jumps from site to site. For time $\tau = 40000$ the defect I^+ travels along the direction of the electric field at the distance $R^+ = \frac{2}{\tau}\int_0^\tau [z^+(u) - z^+(0)]du = 17.84$ while the defect I^- passes only the distance $R^- = 0.67$. Using these values we find the mobilities of the defects according to the formula $\mu^{\pm} = (R^{\pm} l/\tau E)(\kappa/m)^{1/2}$: $\mu^+ = 1.4 \times 10^{-3} cm^2/V \cdot s$ for I^+ and $\mu^- = 0.037\mu^+$ for I^-. At $\tau_0' = 0.1$ the mobilities $\mu^{\pm}$ are larger by one order. The value for the mobility μ^+ at $\tau_0' = 1$ corresponds to the experimental value $\mu^+ = (2.4 \pm 1.6) \times 10^{-3} cm^2/V \cdot s$ in ice crystal at $T = 263K$ (-10^oC) [12] better than the value for μ^+ calculated at $\tau_0' = 0.1$.

6. Lattice Solitons in the Hydrogen-Bonded Chain

In addition to the pinned topological solitons (kinks) in an isolated hydrogen-bonded chain there are also supersonic acoustic (lattice) nontopological solitons. Consider Eqs. (11) and (12) at $T = 0$, $\Gamma = 0$ and $E = 0$. When the left-end hydrogen bond with the hydrolysis energy of one molecule ATP ($0.48eV$) is initially compressed, i.e. the initial conditions $y_1(0) = -q_0 + y_0$, $Y_1(0) = y_0$, $y_n(0) = -q_0$, $Y_n(0) = 0$, $\dot{y}_n(0) = 0$, $\dot{Y}_n(0) = 0$, $n = 2, 3, ..., N$, where $y_0 > 0$ and $V_{OO}(-y_0) = 0.48eV$, are chosen, a supersonic lattice soliton with velocity $v = 1.79 v_O$ appears in the chain. This soliton propagates along the chain retaining its profile and velocity and carrying almost all the ATP hydrolysis energy. In the chain with a substrate (for example, when $K_o = 4N/m$) no lattice solitons are formed. In this case the energy propagates along the chain in the form of a spreading wave packet.

The lattice solitons in our two-sublattice model have all the properties of the acoustic solitons in one chain of the Morse potentials. They travel with steady supersonic velocities and their interaction with each other is elastic. Consider the interaction of a lattice soliton with ionic defects. As is shown in Fig. 6a , the acoustic soliton passes through the defect I^+ almost without energy losses retaining its velocity and shape. When this

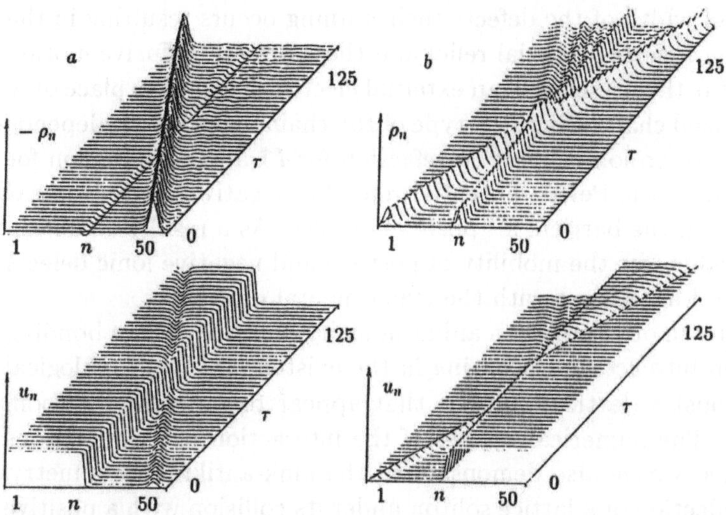

Figure 6: The passage of the positive (a) and negative (b) ionic defects by the acoustic soliton.

soliton passes through the defect I^-, the reflections of some part of the soliton energy occurs (see Fig. 6b). As a result, the soliton amplitude and velocity have been decreased (before the collision $v = 1.79v_0$ while after it $v = 1.72v_0$). The numerical simulation of the dynamics of an acoustic soliton in the thermalized chain has shown its thermal stability at $T = 300K$.

7. Conclusions and Discussion

We have presented here some numerical studies for a one-dimensional hydrogen-bonded chain focusing our attention on the dynamics of ionic defects in the case when this chain is thermalized and the defects are very narrow (their width is no more than two periods of the chain). In our studies the oxygen sublattice has been allowed to move resulting in the dependence of the on-site double-well for the hydrogen-bonded protons on the displacements of the oxygen atoms. Therefore, the resulting two-sublattice model [7] includes the set of the two (proton and oxygen) displacement fields on a one-dimensional lattice. In the framework of this model the profiles of the stationary ionic defects accompanied by local deformation of the chain have been found by using the steepest descent method applied before to other models [10].

Due to small width of the defects their pinning occurs resulting in the existence of the Peierls potential relief and therefore the diffusive motion of the defects (at the presence of an external electric field) takes place only in the thermalized chain. Since the type of the chain deformation depends on the polarity of an ionic defect (rarefaction for I^+ and compression for I^-) the barriers of the Peierls potential relief for negative defects appear to be higher than the barriers for positive defects. As a result, we obtain the different values for the mobility of positive and negative ionic defects that are in good agreement with the experimental data [12].

Furthermore, in our studies an anharmonicity of the hydrogen bonding has been taken into account resulting in the existence of non-topological supersonic acoustic (lattice) solitons that appear to be stable at room temperatures. The numerical studies of the interaction of these solitons with ionic defects have also demonstrated the kink-antikink asymmetry. There is no reflection of a lattice soliton under its collision with a positive ionic defect while this is not the case when the lattice soliton collides with a negative defect.

References

[1] A.S.Davydov, *Solitons in Molecular Systems* (Reidel, Dordrecht, 1985).

[2] J.H.Weiner and A.Askar, Nature **226**, 842 (1970).

[3] Y.Ya.Antonchenko, A.S.Davydov, and A.V.Zolotaryuk, phys. status solidi (b) **115**, 631 (1983); E.W.Laedke, K.H.Spatschek, M.Wilkens, Jr. and A.V.Zolotaryuk, Phys. Rev. A **32**, 1161 (1985); M.Peyrard, St.Pnevmatikos, and N.Flytzanis, Phys. Rev. A **36**, 903 (1987).

[4] A.V.Zolotaryuk, K.H.Spatschek, and E.W.Laedke, Phys. Lett. A **101**, 517 (1984); A.V.Zolotaryuk, Theor. Mat. Fiz. **68**, 415 (1986) [Sov. Theor. Math. Phys. **68**, 916 (1987)]; J.Halding and P.S.Lomdahl, Phys. Rev. A **37**, 2608 (1988); D.Hochstrasser, H.Büttner, H.Desfontaines, and M. Peyrard, Phys. Rev. A **38**, 5332 (1988).

[5] St.Pnevmatikos, Phys. Rev. Lett. **60**, 1534 (1988); G.P.Tsironis and St.Pnevmatikos, Phys. Rev B **39**, 7161 (1989); St.Pnevmatikos, G.P.Tsironis and A.V.Zolotaryuk, J. Molec. Liquids **41**, 85 (1989).

[6] J.F.Nagle and H.J.Morowitz, Proc. Natl. Acad. Sci. (USA) **75**, 298 (1987); J.F.Nagle, M.Mille, and H.J.Morowitz, J. Chem. Phys. **72**, 3959 (1980); J.F.Nagle and S.Tristram-Nagle, J. Membrane Biology **74**, 1 (1983).

[7] A.V.Zolotaryuk, St.Pnevmatikos, and A.V.Savin, Physica D (in press).

[8] M.C.Lawrence and G.N.Robertson, Ferroelectrics **25**, 363 (1980); M.C.Lawrence and G.N.Robertson, J. Phys. C **13**, 1053 (1980).

[9] A.V.Zolotaryuk and A.V.Savin, in *Proceedings of the International Symposium on Solitons in Non-Integrable Systems (Theory and Applications), JINR, Dubna, 1988*, edited by V.G.Makhankov (JINR, Dubna, 1990), p.429; E.S.Kryachko, Chem. Phys., in press, (1990).

[10] St.Pnevmatikos, A.V.Savin, A.V.Zolotaryuk, Yu.S.Kivshar, and M.J.Velgakis, Phys. Rev. A (in press).

[11] M.G.Sceats and S.A.Rice, in *Water: a Comprehensive Treatise*, edited by F.Franks (vol. 7, *Water and Aqueous Solutions at Subzero Temperatures*, Plenum Press, New York).

[12] B.Bullemer, H.Engelhard, and N.Riehl, in Physics of Ice (Plenum Press, New York, 1969), p.416 .

Phase Transition in a Kink and Dynamical Structure Factor of Quasi-One-Dimensional Antiferromagnets

B.A. Ivanov, A.K. Kolezhuk, and G.K. Oksyuk

Institute for Metal Physics, 252 142 Kiev, Vernadsky Str. 36, Ukraine

The dynamics of nonlinear excitations in quasi-one-dimensional magnets has been widely investigated both from the theoretical point of view and experimentally [1]. It is known that existence of solitons (kinks) in the one-dimensional systems leads to the so-called central peak (CP) in the dynamic structure factor (DSF), which determines the cross section of inelastic neutron scattering [2]. A characteristic feature of antiferromagnets (AFM), as it was shown in [3-5], is that the external magnetic field can transform the kink structure. Authors of [3-6] have considered only the stability of static kink solutions depending on the value of the external field. However, in recent works [7,8] it was found that the structure and the symmetry of moving kink and kink at rest can differ sharply. In particular this leads to kink instability at finite velocity and to phase transitions between different type of kinks [8,9]. In this paper we will show that the effects of phase transition essentially change the DSF for the case of the easy- plane (or rhombic) AFM in an external magnetic field (i. e. in the standard situation [10,11]).

We consider a model of one-dimensional two-sublattice AFM which can be described in terms of magnetization and sublattice magnetization $\vec{m}$ and $\vec{l}$ vectors, $\vec{m}^2+\vec{l}^2=1$. It is known [12,13] that equations for the vectors $\vec{m}$ and $\vec{l}$ can be reduced to one equation for $\vec{l}$, and $\vec{m}$ can be expressed in terms of $\vec{l}$, $\partial\vec{l}/\partial t$:

$$2\delta\vec{m} = (1/g)\left[\vec{l}\times\partial\vec{l}/\partial t\right] + \vec{h} - \vec{l}\left(\vec{h},\vec{l}\right), \qquad (1)$$

where $g = 2\mu_0/\hbar$, and μ_0 is the Bohr magneton. The effective equations for the unit vector $\vec{l}$, $|\vec{l}|=1$, can be obtained from the following Lagrange function:

Springer Proceedings in Physics, Vol. 67 Nonlinearity with Disorder
Editors: F. Abdullaev · A.R. Bishop · S. Pnevmatikos

$$L=M_0^2 a^2\int dx\left\{\frac{\alpha}{2}\left[\frac{1}{c^2}\left(\frac{\partial\vec{l}}{\partial t}\right)^2-\left(\frac{\partial\vec{l}}{\partial x}\right)^2\right]+\frac{4}{g\delta M_0}\vec{h}\left[\frac{\partial\vec{l}}{\partial t}\times\vec{l}\right]-w_\alpha(\vec{l})-\frac{2}{\delta}\left(\vec{l},\vec{h}\right)^2\right\},$$

$$w_\alpha=\beta_1 l_x^2/2+\beta_2 l_y^2/2 \qquad (2)$$

Here the anisotropy constants $\beta_1>\beta_2>0$ determine the anisotropy energy w_a, the z axis being the easiest one; $c=gM_0\sqrt{\alpha\delta}/2$ is the terminal velocity, M_0 is an equilibrium sublattice magnetization, δ and α are uniform and nonuniform exchange constants respectively, $\vec{h}=\vec{H}/M_0$, $\vec{H}$ is an external magnetic field, and a is the lattice constant. For the easy-plane TMMC-type AFM the constant β_2 determines the weak anisotropy in the basal plane, $\beta_2\ll\beta_1$, but in the case of brightly rhombic AFM such as $CsMnCl_3\times 2H_2O$ we have $\beta_2\simeq\beta_1$. For $w_a=0$, $\vec{h}=0$ this Lagrangian corresponds to the well-known Lorentz-invariant chiral σ-model of $\vec{n}$-field. If one takes w_a into account, then (2) may be considered as an anisotropic generalization of this model, but for $\vec{h}=0$ the Lorentz invariance is hold. It means that for $\vec{h}=0$ all effects of kink dynamics are reduced to Lorentz contraction of its thickness, and therefore if a kink is stable for $v=0$, then it remains stable for all velocities $v<c$. But for the case of nonzero $\vec{h}$ the Lagrangian will contain the term linear in $\partial\vec{l}/\partial t$, which will break Lorentz invariance.

The effects of the magnetic field are determined by the field orientation. Let's firstly consider the case, when far from the kink we have $\vec{l}=\pm\hat{z}$, where $\hat{z}$ is a unit vector along the z-axis (the behaviour in the spin-flop phase we shall discuss below). In angular variables θ,φ for vector $\vec{l}$, $l_z=\cos\theta$, $l_x+il_y=\sin\theta\exp(i\varphi)$ we have the following equations of motion:

$$\alpha\partial_\mu^2\theta-\sin\theta\cos\theta\left[\alpha(\partial_\mu\varphi)^2+\bar{\beta}_2\sin^2\varphi+\bar{\beta}_1\cos^2\varphi\right]-\frac{8h}{g\delta M_0}F(\theta,\varphi)\frac{\partial\varphi}{\partial t}=0 \qquad (3)$$

$$\alpha\partial_\mu(\sin^2\theta\,\partial_\mu\varphi)-(\bar{\beta}_2-\bar{\beta}_1)\sin^2\theta\sin\varphi\cos\varphi+\frac{8h}{g\delta M_0}F(\theta,\varphi)\frac{\partial\theta}{\partial t}=0 \qquad (4)$$

Here $\partial_\mu=\left[\partial/\partial x,\ \partial/\partial(ict)\right]$, the function $F(\theta,\varphi)$ is determined by the field direction; $\bar{\beta}_1$ and $\bar{\beta}_2$ are the effective anisotropy constants determined by the renormalization of w_a, $w_a\to\bar{w}_a=w_a+2(\vec{h}\vec{l})^2/\delta$.

We suppose that soliton solutions of (4) have the form of a wave with constant velocity v: $\theta=\theta(\xi)$, $\varphi=\varphi(\xi)$, where $\xi=\gamma(x-vt)$, $\gamma=(1-v^2/c^2)^{-1/2}$. In this case $\partial_\mu^2\theta\to\theta''$, $\partial\theta/\partial t\to -v\theta'\gamma$ and so on, where the prime denotes a differentiation by ξ. Further consideration requires to choose the direction of magnetic field $\vec{H}$, to find $\bar{\beta}_1$, $\bar{\beta}_2$, and $F(\theta,\varphi)$ and then to solve the system (3,4). Let's suppose that $\vec{H} \parallel \vec{e}_y$, and so $\bar{\beta}_1=\beta_1$, $\bar{\beta}_2=\beta_2+4h^2/\delta$, $F(\theta,\varphi)=\sin^2\theta\sin\varphi$. In the static case $v=0$ there are solutions describing two types of kinks. In one of them (in zy-kink, namely) the vector $\vec{l}$ rotates in the easy plane ($\varphi=\pm\pi/2$), and in another (in zx-kink) $\vec{l}$ turns through the hard axis Ox ($\varphi=0,\pi$). Both kinks are described by the solution

$$\cos\theta_0 = \pm\ \mathrm{th}(\xi/x_0), \tag{5}$$

where $x_0=(\alpha/\bar{\beta}_2)^{1/2}$ for (zy)-kink and $x_0=(\alpha/\bar{\beta}_1)^{1/2}$ for (zx)-kink. If $\bar{\beta}_2<\bar{\beta}_1$, i.e. $H<H_c=(1/2)M_0[\delta(\beta_1-\beta_2)]^{1/2}$, then zy-kink is lower in energy, but if $H>H_c$ then zx-kink is preferable.

At $v\neq 0$, however, only the solution with $\varphi=0,\pi$ survives without any changes (except of the Lorentzian contraction), while the solution with $\varphi=\pm\pi/2$ must change its form, because $F(\theta,\pm\pi/2)\neq 0$ and so $\vec{l}$ deviates from the (zy)-plane. In this case we can find an approximate solution for a zy-kink using the technics of [8] (we use the notation "zy-kink" for simplicity although it would be more correct to denote such solution as a zxy-kink). Let us suppose that $\bar{\beta}_2$ is close to $\bar{\beta}_1$, i.e. H is close to H_c (as we shall see below, it is the most interesting case). Then from (4) follows that the characteristic length on which the angle φ changes considerably, is $\left[\alpha/|\bar{\beta}_1-\bar{\beta}_2|\right]^{1/2}$, what is much more than kink thickness $(\alpha/\bar{\beta}_1)^{1/2}$. So "in a kink", we can put $\varphi=\varphi_0(v)=\mathrm{const}$. The eq.(3) will give for $\theta(\xi)$ the solution (5) in which x_0 must be replaced by $x_0(\varphi_0)=[\alpha/\beta(\varphi_0)]^{1/2}$, $\beta(\varphi_0)=\bar{\beta}_2+(\bar{\beta}_1-\bar{\beta}_2)\cos^2\varphi_0(v)$. Integrating the eq.(4) in infinite limits and using (5), we obtain the following form of $\varphi_0(v)$:

$$\cos\varphi_0(v) = (v/v_c)\left[(c^2-v_c^2)/(c^2-v^2)\right]^{1/2}\mathrm{sign}(h\theta_0') \tag{6}$$

where the critical velocity v_c is given by

$$v_c=c|\bar{\beta}_2-\beta_1|\delta^{1/2}\left[(\pi h)^2\beta_1+(\bar{\beta}_2-\beta_1)^2\delta\right]^{-1/2}\simeq(4c|H-H_c|/H_c)\ll c. \quad (7)$$

It's easy to see that for $v\to\pm v_c$ the (zy)-type solution transforms continually into the (zx)-type one with $\varphi=0$ and $\varphi=\pi$, respectively. In particular, for $H<H_c$ we come to simple picture, where (zy)- and (zx)-kinks can be treated as one unified nonlinear excitation, which undergoes the second-order "phase transition" at $|v|=v_c$. Values $(\pi-\varphi_0)$ and φ_0 have the meaning of the order parameter (near transition point they behave as $(v_c-v)^{1/2}$), and the (zy)- kink with $\varphi\neq0,\pi$ has lower symmetry that the (zx)-kink. For all kinks their energy $E(v)$ is equal to $E_0(\varphi_0)\gamma$, $E_0(\varphi_0)=2M_0^2[\alpha\beta(\varphi_0)]^{1/2}$.

Let's discuss the stability of kink solutions in respect to small perturbations. Putting $\varphi=\varphi_0+\left[\mu(\xi)/\sin\theta_0\right]\exp(i\omega\tau)$, $\theta=\theta_0+\vartheta(\xi)\exp(i\omega\tau)$, where $\tau=\gamma(t-xv/c^2)$, we obtain the following eigenvalue problem (omitted terms will be discussed below):

$$\begin{aligned}&[\hat{H}_0-\Omega^2-\Delta\beta(\varphi_0)]\mu=2i\varepsilon\Omega\vartheta\sin\theta_0\sin\varphi_0-2\varepsilon(v/c)\operatorname{sign}\theta_0'\cos\varphi_0\mu,\\&(\hat{H}_0-\Omega^2)\vartheta=-2i\varepsilon\Omega\mu\sin\theta_0\sin\varphi_0\end{aligned} \quad (8)$$

Here $\Delta\beta(\varphi_0)=\left[(\beta_1-\bar{\beta}_2)/\beta(\varphi_0)\right]\cos2\varphi_0$, $\Omega=\omega x_0(\varphi_0)/c$, $\varepsilon=2h/\sqrt{\beta(\varphi_0)\delta}$, and $\hat{H}_0=-x_0^2d^2/d\xi^2+1-2\operatorname{ch}^{-2}(\xi/x_0)$ is the well- known Shroedinger-like operator. We are interested in the region near the instability point, where Ω^2 tend to zero, so only the lowest eigenvalue of (8) is to be considered. Supposing that $|\beta_1-\bar{\beta}_2|/\beta_1\ll1$ and $v\ll c$, we can expand ϑ and μ in eigenfunctions of $\hat{H}_0$: $\vartheta=A\psi_0+\sum_k a_k\psi_k$, $\mu=B\psi_0+\sum_k b_k\psi_k$, and construct the perturbation theory. The coefficients a_k, b_k appear to be small as compared to A, B (because $\Omega\ll1$), and so they can be approximately expressed in terms of A,B. Thus the system of eqs for A,B can be closed and we obtain the minimal eigenvalue of (8):

$$\Omega^2=(\pi^2\varepsilon^2\sin^2\varphi_0/4G^2)-[\Delta\beta(\varphi_0)/G]-(\pi\varepsilon v\cos\varphi_0\operatorname{sign}\theta'/2cG), \quad (9)$$

where $G=G(\varepsilon,\varphi_0)=1+\varepsilon^2D\sin^2\varphi_0$, the factor $D\simeq0.390$. (The terms omitted in (7), although they formally are of the same order in small parameters $\Delta\beta(\varphi_0)$ and v/c as the remained ones, give higher order corrections to the final result (9) because they contribute only through the terms containing a_k,b_k; for more detailed discussion see [8]). The expression (9) determines the stability of kink solutions (kinks are stable if $\Omega^2>0$ and unstable if $\Omega^2<0$). Let's discuss this formula.

For $\varphi=0,\pi$ (what corresponds to $\varkappa x$-kinks) eq.(9) coincides with the result of Wysin et al [9]. We obtain that $\varkappa x$-kinks with different signs of the value $\sigma=\text{sign}(v\theta' \cos\varphi_0)$ have different behaviour. If $H<H_c$, i,e. $\bar{\beta}_2<\beta_1$ then $\varkappa x$-kinks with $\sigma>0$ ($\sigma<0$) are stable in the velocity intervals $v_c\leq v\leq c$ and $-c\leq v\leq -v_c$, respectively, and if $H>H_c$ then stability regions become $-v_c\leq v\leq c$ (for $\sigma>0$) and $-c\leq v\leq v_c$ (for $\sigma<0$). So, the presence of Lorentz- noninvariant terms in (2-4) leads to the fact that a kink "feels" the velocity direction. Physically it is because the component of magnetization $\vec{m}$ proportional to $\left[\partial\vec{l}/\partial t \times \vec{l}\right]$, can be directed both parallel and antiparallel to the external magnetic field $\vec{H}$.

For $\varphi=\varphi_0(v)$ eq.(9) describes the stability of $\varkappa y$-kinks. It can be easily shown that $\Omega^2>0$ in the interval $-v_c\leq v\leq v_c$, and so the $\varkappa y$-kink is stable in the whole region where the solution (6) has a sense, both for $H>H_c$ and $H<H_c$. It is very interesting that although $\varkappa y$-kinks at $H>H_c$ have greater energy that $\varkappa x$-kinks, but they are stable in respect to the small linear perturbations. This fact confirms the numerical results of Wysin et al [9], who had revealed that $\varkappa y$-kinks at $H>H_c$ does not tend to decay into energetically preferable $\varkappa x$-kinks.

Now we proceed to the calculation of the dynamical structure factor. As $v_c \ll c$, the effects in question are important only in the region of small velocities. The DSF can be easily calculated with the help of the standard soliton ideal-gas phenomenology [14]; it yields

$$S^{ik}(q,\omega) = \left(\frac{2\pi}{q^2 v_T^2}\right)^{1/2} \frac{\pi^2 x_0^2 n_s}{\text{ch}^2(\pi q x_0/2)} \exp\left[-\frac{\omega^2}{2q^2 v_T^2}\right] A_{ik}(q,\omega). \quad (10)$$

Here $i,k = (x,y)$; $v_T=c\sqrt{T/E(0)}$ is the thermal velocity of a soliton, $x_0\simeq\sqrt{\alpha/\beta}$, n_s is the soliton gas density (for our purposes we needn't calculate the dependence $n_s(T)$), and factor $A_{ik}= A_{ii}\delta_{ik}$ is due to phase transition,

$$A_{yy} = \begin{cases} (1-\omega^2/q^2 v_c^2), & |\omega/q|<v_c \\ 0, & |\omega/q|>v_c \end{cases}; \quad A_{xx}= \begin{cases} (\omega/q v_c)^2, & |\omega/q|<v_c \\ 1, & |\omega/q|>v_c \end{cases} \quad (11)$$

It is easy to see, that the phase transition leads to considerable narrowing of the central peak in the longitudinal DSF $S^{yy}(q,\omega)$ when the external magnetic field H tend to the critical value H_c (i. e. when $v_c \rightarrow 0$; really the effect becomes observable for $v_c \leq v_T$). Secondly, the component $S^{xx}(q,\omega)$ becomes nonzero due to the appearance of the (zx)-kink contribution. This component has the form of two symmetrically placed peaks with maxima at finite frequencies; if v_c decreases (i. e. the field increases) then the peaks increases and become closer one to another.

In conclusion of this section we also would like to mention the interesting effects which occur at $H > H_c$: the instability of one among the two (zx)-type kinks causes the reduction of $S^{xx}(q,\omega)$ in the region $|\omega/q|>v_c$.

The analysis of other geometries shows that for $\vec{H} \parallel \vec{e}_x$ the value of $v_c \ll c$, and the DSF in fact will not be modified. In the spin-flop geometry ($\vec{H} \parallel \vec{e}_z$) for $H<H_{SF}=(1/2)M_0\sqrt{\beta_2\delta}$ the value of v_c is also close to c, and for $H>H_{SF}$ the situation is the same as in the case $\vec{H} \parallel \vec{e}_y$. More concrete, for $H_{sf}<H<H_c'$, and $H>H_c'$, where $H_c'= (1/2)M_0\sqrt{\beta_1\delta}$, the situation is same as for $H_y<H_c$ and for $H_y>H_c$, respectively. Corresponding components $S^{xx}(q,\omega)$ and $S^{zz}(q,\omega)$ of the DSF have the same behaviour as the components S^{xx} and S^{yy} for the case of $\vec{H} \parallel \vec{e}_y$.

So, the dynamical phase transitions lead to essential anomalies in the soliton-induced central peak in the DSF of one-dimensional antiferromagnets. Non-Gaussian forms of the CP appear, and even two-hump form occurs. Maximal effects can be observed close to the critical field value ($H=H_c$ for $\vec{H} \parallel \vec{e}_y$, and $H=H_c'$ in the spin-flop phase at $\vec{H} \parallel \vec{e}_z$, for TMMC $H_c \simeq 70$ kOe [5]).

REFERENCES

1. Physics in One Dimension, ed. by J.Bernasconi and T.Schneider.- Springer,Berlin,1981.
2. Izyumov Yu.A.-Usp.Fiz.Nauk, 155,553 (1988)
3. Fluggen N., Mikeska H.J.-Solid State Comm., 48, 293 (1983).
4. Mikeska H.J.-J.Phys.C, 13, 2913 (1980).
5. Gouvea M.E., Pires A.S.T.- Phys.Rev.B,34, 306(1986).
6. Lemmens L.F.,Kimura I.,de Jonge W.J.M.-J.Phys.C 19, 139 (1986)
7. V.G.Bar'yakhtar, V.A.Lvov, D.A.Yablonskij Zh. Eksp. Teor. Fiz. 87,1863 (1984).

8. E.V.Gomonaj, B.A.Ivanov, V.A.Lvov, G.K.Oksyuk. Zh. Eksp. Teor. Fiz.,97, 307 (1990) [Sov Phys.JETP 70, 174 (1990)].
9. Wysin G.M., Bishop A.R., Oitmaa J.-J.Phys.C, 19, 221 (1986).
10. Regnault L.P., Boucher J.P., Rossat-Mignod J. et.al.-J.Phys.C, 15,1261 (1982).
11. Boucher J.P., Regnault L.P., Pynn R. et. al. -Europhys.Lett., 1, 415 (1986).
12. I.V.Bar'yakhtar, B.A.Ivanov. Fiz. Nizk. Temp., 5,759, 1979 [Sov. J. Low Temp. Phys.5, 361, 1979].
13. Bar'yakhtar I.V., Ivanov B.A.-Sol.St.Comm., 34,545(1980).
14. Currie J.F., Krumhansl J.A., Bishop A.R., and Trullinger S.E.-Phys.Rev.B, 22, 477(1980).

Layer Disordering and Optical Properties of α-Si/SiO_2 Superlattices

E.A. Vinogradov, V.N. Denisov, A.V. Zayats, B.N. Mavrin, D.N. Nikogosyan, Yu.A. Repeyev, and F.A. Pudonin

Institute of Spectroscopy, The Russian Academy of Sciences, 142 092 Troitsk, Moscow region, Russia

Abstract. Amorphous silicon / quartz superlattices (SL) with ultrathin layers have been investigated. In SL Raman spectra the interface modes and the redistribution of bands intensities and at the SL period variation due to the Brillouin zone folding were found. Disordering of Si layers is stronger in SL than in a single layer. Electron states caused by the quantum size effect in SL have been investigated. Localized states due to Si layers disordering influence the intensity, band shape and peak position of intersubband radiative recombination in SL

1. Introduction

Semiconductor superlattices attract great attention today in view of both scientific problems and their technical applications in micro- and optoelectronics. Fast optical switches and bistable elements based on superlattices are most perspective for practical applications [1]. These effects are due to a new (compared to bulk materials) structure of phonon and electron states arising in SL.

In this paper we investigate the phonon and electron states of short period amorphous Si/SiO_2 superlattices. The one-dimensional Kronig-Penney potential formed artificially in the amorphous SL determines the energy states in such superlattices. The depth of this potential ΔE depending on the components band gaps ($\Delta E = E_g(SiO_2) - E_g(Si) \sim 7$ eV) is larger than the modulation of potential in SL based on A^3B^5 or A^2B^6 semiconductor compounds.

In addition to the effects determined by the size quantization, different kinds of disorders of layers creating the superlattice influence strongly the optical and vibrational properties of SL. The SL potential can influence also the layers disorders.

Springer Proceedings in Physics, Vol. 67 Nonlinearity with Disorder
Editors: F. Abdullaev · A.R. Bishop · S. Pnevmatikos

Amorphous Si/SiO_2 superlattices have were prepared by RF magnetron sputtering of Si and SiO_2 targets in turn in spectrally pure Ar at 10^{-3} Torr. The deposition rates of Si and SiO_2 were 0.09 and 0.04 nm/s, respectively. The temperature of substrates (Si or SiO_2 crystals) did not exceed $70^{\circ}C$. To avoid the influence of the crystalline field of the substrate on the superlattice an amorphous Ni layer 0.1μm -thick was deposited on the substrate. The first and the last layers of the superlattices were a-SiO_2 layers. The layers thicknesses and regularity were verified by the profile Auger spectroscopy [2, 3] and the transmission electron microscopy.The abruptness of the layers interfaces is good. The mutual diffusion and mixing of layers were not observed that confirms the regularity of the one dimensional potential. Therefore, the quantum size effects should be observed in amorphous SL as well as in crystalline superlattices.

2. Raman Scattering in SL

Let us consider quantum size effects such as the folding of the Brillouin zone leading to Raman activity of acoustic modes, the confinement of optical modes and the appearance of interface optical modes which should be observed in the Raman spectra of the amorphous SL [4]. Raman spectra of the amorphous Si/SiO_2 superlattices were excited by light of an Ar^+ laser in back scattering geometry at the room temperature and were recorded by a triple stage multichannel spectrometer [5].

As the Raman cross-section of a-Si is essentially larger than that of a-SiO_2 the Raman spectra of a-Si were actually registered. Fig.1a shows the Raman spectra of amorphous Si/SiO_2 superlattices with the parameters presented in Table 1. These spectra look like the Raman spectrum of a-Si thick film (~1μm) that testifies to the formation of amorphous layers of Si at the preparation technique used. Low and high frequency bands correspond to TA- and TO- phonons, respectively. It should be noted that the ratio of the TA- and TO- bands integral intensities in SL differs from that in the bulk a-Si. The relative intensity of the TO- mode decreases with the thickness of the a-Si layers possibly due to the further localization of the TO-mode in SL [6]. The shift of the TO- band to lower frequency (6-9 cm^{-1}) can be attributed to both the confinement of the TO-

Table 1. Parameters of a-(Si/SiO_2) SL.

N^o	a-Si layer thickness	a-SiO_2 layer thickness	total a-SL thickness
1	2.7 Å	10 Å	667 Å
2	5.5 Å	10 Å	889 Å
3	8.2 Å	10 Å	737 Å
4	11 Å	10 Å	629 Å

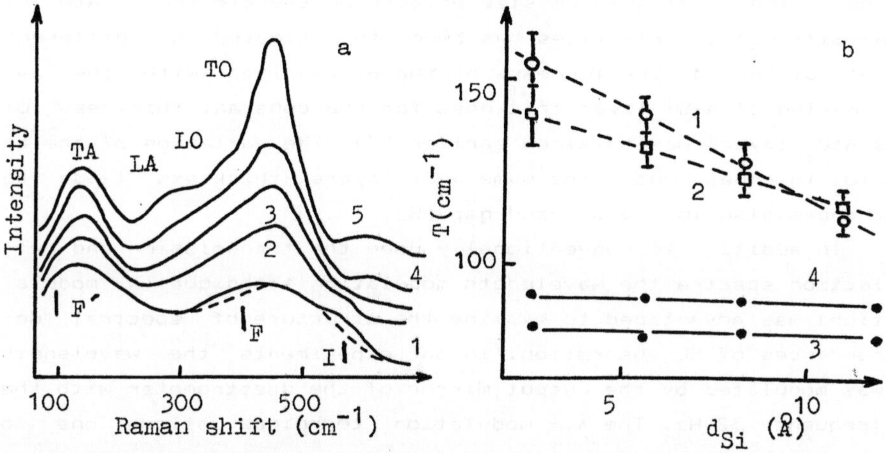

Fig.1. (a) Raman spectra of a-SL (1-4) and 3D a-Si (5). Numbers of the curves correspond to SL samples from the Table 1.
(b) TO- (1, 3) and TA- (2, 4) bandwidth dependences on a-Si layers thickness in SL (1, 2) and in a single layer (3, 4).

mode and strain in the $Si\text{-}SiO_2$ interfaces. The increase of the Si-Si bond length in thin a-Si layers can lead to the TO- band shift as well. An essential increase of the halfwidths of both TA- and TO- bands is observed (Fig.1b).The width of the a-Si TO-band characterizes the bond angle distortion $\Delta\theta$ ($\Gamma_{TO} = 15 + 6\Delta\theta$) [6]. The widths of TO- bands observed indicate that ÂË is 8-10° larger in a-Si layers of SL compared to Âë in single a-Si layers. Hence, the disorder of amorphous silicon layers is higher in SL than in ordinary a-Si.

AS the thickness of a-Si layers decreases the peculiarities F, F' and I are observed in the SL Raman spectra (Fig.1a). F and F′ peculiarities on the low frequency side of the TO- and TA- bands are associated with the folding of the optical and acoustical phonon bands of a-Si, respectively. The interface mode (I) is observed in SL with a small number of Si monolayers when the contribution of $SiSi_3O$ interface tetrahedrons increases compared to the bulk $SiSi_4$ tetrahedrons.

3. Optical Absorption in SL

The influence of quantum size effects on the electron states of a-(Si/SiO_2) superlattices has been investigated by different optical methods.The increase of the SL band gap with the decreasing of a-Si layer thickness for the constant thickness of a-SiO_2 layers was obtained earlier [7]. The variation of the a-SiO_2 thickness under the same a-Si layers thickness leads to changes also in the SL band gap [8].

In addition to conventionally used the transmission and reflection spectra the wavelength modulation technique (λ- modulation) was advantaged to examine the structure of spectral dependences of SL absorption. In our experiments the wavelength was modulated by the output mirror of the spectrometer with the frequency 32 Hz. The λ - modulation technique allows one to obtain the first derivative of the transmission or reflection spectra when the signal is amplified at the modulation frequency [3]. The spectral dependence of the absorption coefficient and its derivative can be obtained from these spectra.

When the thickness of the semiconductors layer restricting the electron motion is small, the quantum size effects arise and conduction and valence bands split into a number of subbands (k = 0, 1 ...). Transitions between these subbands of valence and conduction bands obey the rule $\Delta k = 0$, and the absorption coefficient $\alpha(\omega)$ is given by the equation [9]

$$\alpha(\omega)\ n(\omega)\ h\omega \propto (h\omega-E_k)\ \Theta(h\omega-E_k), \qquad (1)$$

where E_k is the energy difference between k -th subbands, $n(\omega)$ is the refractivity. The structure of the absorption spectrum (1) is weak and the subbands edges E_k are not visible from the spectra (fig.2a). The λ-modulation technique allows one to take the spectrum of the derivative of the absorption coefficient

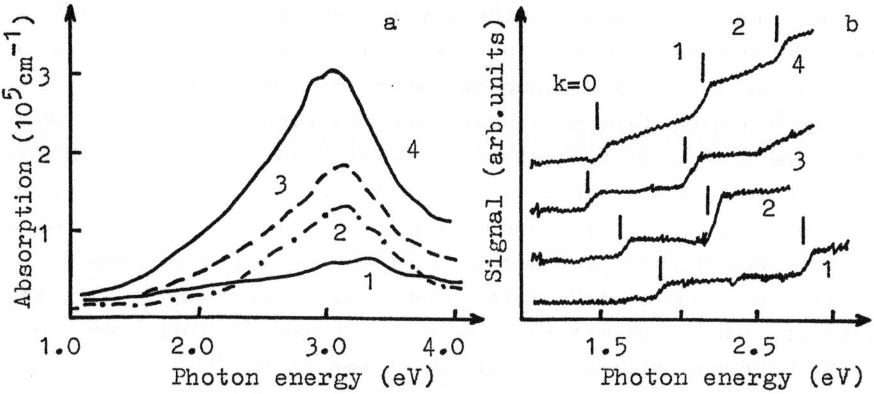

Fig.2. Spectral dependences of the absorption (a) and the first derivative of the absorption coefficient (b) of a-SL. Curve numbers correspond to SL samples from the Table 1. The arrows show the transitions between k -th subbands. The measurement temperature is 300 K.

$$d\alpha / d\lambda \propto \sum_k E_k \ \theta(h\omega - E_k) \qquad (2)$$

in which the band peculiarities are revealed as some "stair".

The spectra of the absorption and of λ- absorption of the SL with the parameters from the Table 1 are shown in fig.2. The λ-absorption spectra observed confirm that the subbands caused by the size quantization are present in the electron states spectra of SL. The energy distances between the subbands with the same k- quantum number correspond to the energy positions of the appropriate steps in the λ- modulation spectra. The transition between the nearest (k = 0) subbands of valence and conduction bands determines the SL band gap. Shorter wavelength steps are associated with the transitions between the k = 1 and higher subbands. The band gap of SL examined exceeds considerably the band gap of 3D amorphous silicon (0.9 eV). The SL band gap depends nonmonotonously on the SL period (see fig.2b) Analogous results were obtained for SL with longer periods [8]. Such nonmonotonous dependence can not be explained by changes of barrier and well widths of the Kronig-Penney potential. In SL with ultrathin layers the potential may be not right-angle for a number of reasons [8] (e.g. due to the charged Si - SiO_2 interfaces). The deformation of the Si - Si bonds discussed earlier can lead to changes in the effective mass of charge carriers when the layers thickness is changed (analogously to the polaron effect in a thin film [10]). These reasons can give rise to the band gap dependence observed. The band gap of SL with 10 Å a-SiO_2 layers changes over the range 1.46-2.0 eV when a-Si layers thickness decreases from 11 Å to 2.7 Å (see fig.2b)

The comparison of absorption coefficients of different SL in the low energy spectral range shows that the relative density of distributed localized states at the band tails increases for SL with thinning of a-Si layers, and this corresponds to the Raman scattering investigations which showed an increase in the Si-Si bonds disordering in short period SL.

4. Photoluminescence of SL

Photoluminescence (PL) spectra of the amorphous Si/SiO_2 superlattices was investigated under the continuous and pulse excitation. Ar^+ or He-Cd continuous lasers were used for excitation of PL. Pulse excitation was produced by 20 ps- light pulse of parametric generator pumped by second harmonic of Nd:YAG laser with passive mode synchronization. Parametric generator pulse energy was 0.13 and 0.09 mJ for 450 and 580 nm light.

Continuously excited PL spectral dependences of all SL are analogous to a-SiO_2 PL spectra in this spectral range (fig.3). But the intensity of SL photoluminescence is much stronger (by ~ 10^4 times) than that of a-SiO_2 single layer. The PL intensity depends on the excitation light wavelength. The intensity of SL PL excited by 514 nm- laser emission is 2 - 3 times lower than the PL excited by the 488 nm- light. There is no PL under short wavelength excitation 330 nm. The SL absorption differs slightly for these three wavelengths, so the sharp PL intensity dependence cannot be explained by changes in SL absorption.

The luminescence bands observed in the a-SiO_2 spectra are caused by Ge impurities in the fused quartz [11]. The luminescence excitation efficiency of these bands is very low in a-SiO_2 because the a-SiO_2 band gap is about 3 times greater than the exciting photons. Absorption of the exciting light is stronger in SL since there are a-Si layers. Therefore, the electrons excited are created in a-Si layers of SL and after that they go to the impurities states in the adjacent a-SiO_2 barriers layers where recombination occurs. (This effect is the

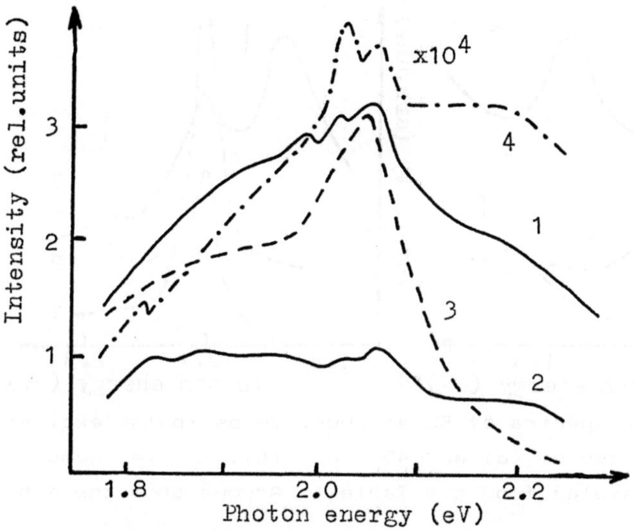

Fig.3. Luminescence spectra of the sample 4 at 4.2K (1) and 300K (2), of the sample 1 at 4.2K (3) and of the single a-SiO_2 layer at 300K (4). Continious excitation wavelength is 488 nm.

reverse of the resonant transfer of carriers between the states of different layers in a GaAs / AlGaAs superlattice [12].) The transfer probability depends on the mutual arrangement of the energy states in different layers and, therefore, this probability changes due to the quantum size effect when the layers thicknesses are changed. For the superlattices investigated the variation of the a-Si thickness does not take out the carriers states from the resonant transfer condition, so that the PL spectra of SL with different periods do not differ. The initial energy of the carriers in the subbands dependent on the excitation wavelength strongly influences the transfer condition [12] and the PL intensity drops.

The short lifetime of photoexcited carriers due to nonradiative recombination with participation of localized states in the band tail caused by disordering results in the absence of luminescence in a-Si layers of SL. To eleminate partially the influence of fast nonradiative relaxation picosecond excitation was used. There are previously observed PL and new luminescence bands due to carriers transition between the k -th subbands of conduction and valence bands of Si layers in the

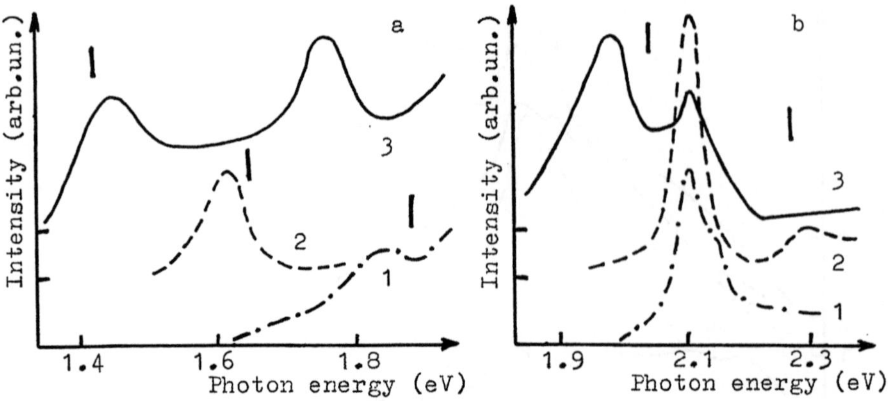

Fig.4. Luminescence spectra of SL at 300K. 20 ps -pulse excitation wavelength is 580 nm (a) and 450 nm (b). Curve numbers correspond to SL samples from the Table 1. Arrows show the subband edges.

spectra (fig.4a, b). The luminescence caused by transitions between the nearest (k=0) subbands is excited by light which is absorbed due to transitions between the same (k=0) subbands. When next subbands (k=1) start to participate in absorption the luminescence related to these subbands appears but the previously observed PL band (due to shallower k=0 subbands) disappears. Such PL dependence on the energy of exciting photon points to a low probability of intersubband carriers relaxations (relaxation between the subbands of conduction or valence bands with a different quantum number k). The fact that subbands radiative recombination is observed under the pulse excitation only shows that this recombination occurs faster than that due to impurities in a-SiO_2 (including time needed for carriers transfer between Si and SiO_2 layers).

All PL bands observed have a longwavelength tail and large width ~ 0.1 eV compared to Boltzman width ~k_BT. which is expected for "ideal" intersubbands transitions. These band features prove the participation of localized states of the subbands tails in the recombination. The longwavelength tail of PL band becomes more intensive with the SL period decreases. At the same time the band shifts to the longwavelength range compared to the subbands edge. This can be caused by increasing density of localized states due to larger disorder of a-Si layers in SL with shorter periods.

As to additional luminescence bands observed in the spectra, they can be caused by transitions with participation of both states of subbands of Si well layers and impurity states of SiO_2 barrier layers. PL bands intensities are mainly a linear function of the excitation intensity, but intensities of bands related with the impurity in a-SiO_2 layers of SL are proportional to second power of the excitation intensity, that power for PL band due to k=0 intersubband transitions is less than 1 (from 0.5 to 0.8). These intensity nonlinearities confirm strong interaction between different carriers states of SL.

5. Conclusions

Our investigations of ultrashort period a-Si/SiO_2 superlattices show considerable disordering of amorphous silicon layers depending on the SL period: disordering increases for shorter SL period. The distortion of Si-Si bond length and deflection of the angle between these bonds from the tetrahedral angle $\Delta\theta = 8\text{-}10^{\circ}$, leads to a shift and the width increase of TO- and TA- band in Raman spectra. The distortion of Si atom coordination near the surface of layers results in the appearance of an interface mode in SL investigated. The Si layers disordering is higher in SL than in isolated Si layers.

The disordering of Si layers manifests itself also in optical absorption. It leads to an increase in the density of localized states in the bands tails but does not influence quantum size effects. These effects cause the arising of subbands of conduction and valence bands of a-Si.

Fast nonradiative relaxation of the excited carriers in Si layers probably due to its disordering results in the absence of luminescence from these SL layers under continuous excitation. At such condition impurity luminescence is observed only from a-SiO_2 layers. The intersubband radiative transitions appear under the 20 ps- pulse excitation. The shape and spectral shift of PL bands may be determined to a considerable extent by the density of band tail states depending on the SL period. The strong interaction between electronic states of different SL layers gives rise to the nonlinearity of intensity dependences of PL bands.

References

1. H.M.Gibbs, Optical Bistability: Controlling Light with Light (Academic Press, Ney York, 1985).
2. A.F.Plotnikov, F.A.Pudonin, V.B.Stopachinsky, Pisma JETP 46, 443 (1987).
3. E.A.Vinogradov, A.V.Zayats, F.A.Pudonin, Fiz.Tverd.Tela, in press.
4. E.A.Vinogradov, V.N.Denisov, B.N.Mavrin, F.A.Pudonin, Fiz. Tverd.Tela, in press.
5. A.F.Goncharov, Sov.Phys.: JETP 94 321 (1988).
6. Z.Vardeny, O.R.Bratman, Phys.Rev. B19, 3290 (1979).
7. S.A.Kryukov, A.F.Plotnikov, F.A.Pudonin, V.B.Stopachinsky, Short Commun. P.N.Lebedev Physical Inst. 5, 34 (1986).
8. E.A.Vinogradov, G.I.Makarov, F.A.Pudonin, Fiz. Tverd. Tela 31, 111 (1989).
9. K.Hattori, T.Mori, H.Okamoto, Y.Hamakawa, Phys.Rev.Lett. 60, 825 (1988).
10. E.P.Pokatilov, V.M.Fomin, S.I.Beril, Vibrational excitations, polaritons and excitons in multilayer systems and superlattices, Kishinev, Shtiinza, 1990, 278p.
11. A.A.Gordeev, A.P.Gorchakov, Optika i spectroskopiya, 66, 662 (1989).
12. Y.Murayama, Phys.Rev. B36, 2500 (1986).

Modulational Polarization Instabilities and Disorder in Birefringent Optical Fibers

S. Trillo and S. Wabnitz

Fondazione Ugo Bordoni, Via B. Castiglione 59, I-00142 Rome, Italy

Abstract. We study the nonlinear dynamics of modulated waves in a nonlinear birefringent fiber. We show that, in analogy with the scalar case that is described by the nonlinear Schrödinger equation, a close correspondence may exist between the unstable behavior of two coupled periodic waves and the simple phase-space description of an equivalent nonlinear oscillator that is associated with a truncated Fourier expansion of the fields. The regular motion along these one-dimensional trajectories may break up into disordered spatio-temporal patterns.

1. Introduction

Among the nonlinear optical processes that may be described in terms of a Taylor expansion of the polarizability in powers of the electric field, parametric scattering involves the local and istantaneous interaction between the optical fields and the medium. Therefore the propagation of the electromagnetic field is completely determined by a conserved nonlinear field Hamiltonian [1]. It appears that important and new insights in classical and perhaps quantum nonlinear optics may be obtained by employing the powerful techniques that have been developed for the study of the dynamics and chaos of finite-dimensional conservative systems [2].

In fact, pioneering work on the nonlinear dynamics of parametric three-wave mixing was done in the sixties by Akhmanov and Khokhlov [3]. In recent years, considerable attention was dedicated to the Hamiltonian instabilities and chaos that may arise in the propagation of polarized waves through cubic nonlinearity media [4-13]. These instability phenomena are readily observable in fiber optics experiments involving intense light pulses and are relevant when using fibers for all-optical signal generation and processing [14-16]. Up-to-date reviews on the experimental and theoretical developments of the research on polarization instabilities have appeared in the literature [17-18].

In this work we provide a nonlinear optics example where only a few modes of the pde system are linearly unstable, whereas all the remaining higher order modes are essentially slaved to these few driving modes. In fact, a one-degree-of-freedom nonlinear oscillator model may well describe the modulational and polarization instabilities that occur in a nonlinear and dispersive birefringent optical fiber. This is interesting because, in the presence of dispersion, the nonlinear propagation of the optical field is described by two coupled nonlinear Schrödinger (NLS) equations that are not integrable by means of the inverse scattering transform. We will also point out that these regular one-dimensional curves separate wide regions of chaotic spatio-temporal evolutions, that may arise from the beating between several unstable modes.

2. Two-mode truncations

The coupled NLS equations that describe the evolution of a modulated wave in the birefringent fiber read, in dimensionless units,

$$\begin{aligned} i\frac{\partial u}{\partial \xi} \pm \frac{1}{2}\frac{\partial^2 u}{\partial t^2} + \frac{\Delta}{2}u + (|u|^2 + \frac{2}{3}|v|^2)u + \frac{1}{3}u^2 v^* = 0, \\ i\frac{\partial v}{\partial \xi} \pm \frac{1}{2}\frac{\partial^2 v}{\partial t^2} - \frac{\Delta}{2}v + (|v|^2 + \frac{2}{3}|u|^2)v + \frac{1}{3}u^2 v^* = 0, \end{aligned} \tag{1}$$

where Δ is the linear birefringence, the upper (lower) sign holds in the anomalous (normal) dispersion regime, whereas ξ and t are the longitudinal and temporal coordinates. The above equations may be written

$$\begin{aligned} -i\frac{\partial u}{\partial \xi} = \frac{\delta H}{\delta u^*}, \\ -i\frac{\partial v}{\partial \xi} = \frac{\delta H}{\delta v^*}, \end{aligned} \tag{2}$$

where the field Hamiltonian reads $H = \int_{-\infty}^{+\infty} \mathcal{H} dt$, and

$$\begin{aligned} \mathcal{H} = \pm \frac{1}{2}(| u_t |^2 + | v_t |^2) + \frac{1}{2} | u |^4 + \frac{1}{2} | v |^4 + \\ \frac{\Delta}{2}(| u |^2 - | v |^2) + \frac{2}{3} | u |^2 | v |^2 + \frac{1}{6}(u^{*2}v^2 + c.c.). \end{aligned} \tag{3}$$

Consider first, as a reference case, the stability of the single mode continuous wave solution of eqs.(1). In other words, we may set an initial condition

$$u(\xi = 0, t) = 1 + b\, cos(\Omega t), v(\xi = 0, t) = 0, \tag{4}$$

where $b = \epsilon exp(i\phi_0/2)$, and $|\epsilon| \ll 1$. In this case the propagation is governed by the scalar NLS equation for the u component. By inserting eq.(4) in eqs.(1) and linearizing it with respect to b, one obtains the usual condition for modulational instability (MI) of the continuous wave solution $\Omega \leq 2$, whenever the propagation is in the regime of anomalous dispersion of the fiber [19-20].

Clearly the linearized analysis does not give informations on the evolution of the field, past the early stage of exponential growth of the energy in the sidebands with distance ξ. On the other hand, it is known from theory and experiments [21] that in pde systems energy may flow back and forth in a periodic or quasiperiodic fashion (Fermi-Pasta-Ulam recurrence [22]) between the carrier and an infinite number of harmonics. In recent years, rigorous investigations of the nonlinear dynamics of MI's in integrable model nonlinear evolution equations were carried out [23-24]. In particular, analytical and numerical studies of the NLS equation have revealed that the linearized MI may be prolonged into a separatrix trajectory in the phase space of the solutions [23,25-27].

However, a global understanding of the homoclinic structure of these nonlinear modulation problems is somewhat obscured by the fact that one has to deal with infinite dimensional Hamiltonian systems. Therefore the possibility of finding an approximate ode model whose solutions would still mimic the behavior of the solutions of the full pde system is very appealing, especially for the immediate physical transparency of the results. Furthermore, such a perturbative approach to the problem appears to be unavoidable in cases (as with eqs.(1)) where a pde system is not integrable.

In fact, it was demonstrated by Infeld [28] that the period of depletion of a modulated continuous wave that obeys the NLS equation could be very accurately estimated by a one-degree-of-freedom oscillator model involving only three Fourier modes of the field (i.e., the pump and two symmetric sidebands). In ref.[29], a general bifurcation and stability analysis of the nonlinear eigenmodes of this three-wave system was carried out. Whereas in refs.[30-32] it was shown how close the homoclinic structure and homoclinic crossings (in the presence of dissipative and periodic perturbations) of this truncated Hamiltonian model may be to the actual behavior of the solutions of the NLS equation.

Here we point out the close analogy between modulational and polarization instabilities. In fact, as we shall see, in some significant cases it is possible to describe the nonlinear dynamics of these phenomena by means of an one-degree-of-freedom Hamiltonian function of the same form.

Let us consider the ansatz

$$u(\xi,t) = e^{\mp i\Omega^2\xi/4}(A_0(\xi) + \sqrt{2}A_1(\xi)\,cos(\Omega t)). \qquad (5)$$

In other words, we want to study the evolution of the initial conditions

(4) for distances ξ where the approximation (4) fails. By inserting eq.(5) into eqs.(1), one obtains

$$
\begin{aligned}
-i\frac{dA_0}{d\xi} &= \frac{\mu}{2}A_0 + (\mid A_0 \mid^2 + \sigma \mid A_1 \mid^2)A_0 + \varrho A_1^2 A_0^* \\
i\frac{dA_1}{d\xi} &= -\frac{\mu}{2}A_1 + (\gamma \mid A_1 \mid^2 + \sigma \mid A_0 \mid^2)A_1 + \varrho A_0^2 A_1^*,
\end{aligned}
\tag{6}
$$

where $\mu = \pm\Omega^2/2$, $\gamma = 3/2$, $\sigma = 2$, and $\varrho = 1$. The system (6) may be reduced to a single nonlinear oscillator equation in terms of the new variables $\eta \doteq \mid A_0 \mid^2 / S_0, \phi \doteq 2\phi_1 - 2\phi_0$, where $A_j \doteq \mid A_j \mid exp\{i\phi_j\}, j = 0, 1$, and $S_0 = \mid A_0 \mid^2 + \mid A_1 \mid^2$. Note that, from eq.(4) one has $\eta(\xi = 0) = 1/(1 + \epsilon^2/2)$, and $\phi(\xi = 0) = \phi_0$. The evolution of the variables η and ϕ obeys

$$
\begin{aligned}
\dot{\eta} &= \frac{\partial \tilde{H}}{\partial \phi}, \\
\dot{\phi} &= -\frac{\partial \tilde{H}}{\partial \eta}.
\end{aligned}
\tag{7}
$$

Here the dots denotes derivatives with respect to $\zeta \doteq \xi S_0$, and $\tilde{H}$ reads

$$\tilde{H}(\eta, \phi) = 2\eta(1 - \eta)\, cos(\phi) + (1 + \kappa)\eta - \frac{3}{2}\eta^2, \tag{8}$$

where $\kappa = \mu/S_0$ is a normalized phase mismatch between the pump and the sidebands. We may also rewrite eqs.(6) in terms of the real parameters

$$
\begin{aligned}
&S_0 = |A_0|^2 + |A_1|^2; S_1 = |A_0|^2 - |A_1|^2; \\
&S_2 = A_0^* A_1 + c.c.; S_3 = iA_0^* A_1 + c.c..
\end{aligned}
$$

Note that S_0 is a conserved quantity of eqs.(4), whereas S_0 is generally not conserved in the original system of eqs.(1). The evolution of the optical field in the fiber may be graphically displayed as the motion of the tip of the *modulation vector* $\mathbf{S} = (S_1, S_2, S_3)$ on the sphere $\{\mathbf{S} : S_1^2 + S_2^2 + S_3^2 = S_0^2\}$. We obtain

$$\frac{d\mathbf{S}}{d\xi} = (\mathbf{\Omega}_L + \mathbf{\Omega}_{NL}(\mathbf{S})) \times \mathbf{S}, \tag{9}$$

where $\mathbf{\Omega}_L = (\mu, 0, 0)$, and $\mathbf{\Omega}_{NL} = (10S_1 - S_0, 17S_2, 9S_3)/4$.
Eq.(9) is formally identical to the equation of motion for the Stokes vector, that describes the evolution of a *polarization* state in nonlinear anisotropic crystals [7,10]. Clearly here the origin of the light-induced

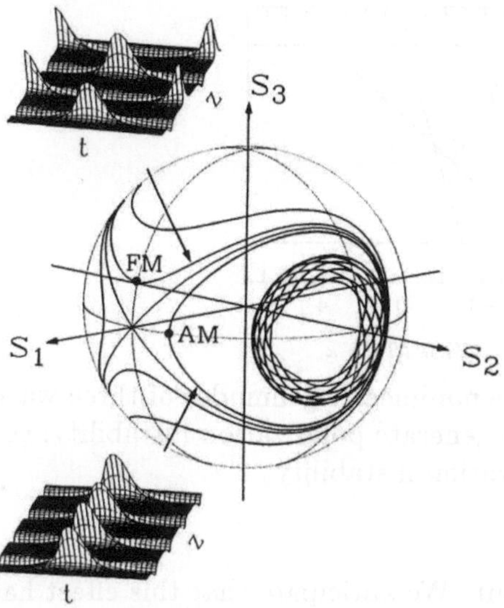

Figure [1]: Modulational sphere and AM-FM switching in a dispersive nonlinear medium based on modulational instability.

anisotropy resides in the nondegenerate mixing between the pump and the two sideband modes.

Figure (1) shows the projection on the sphere of the solutions of eqs.(1), with initial conditions (4), $\kappa = 1$, and the choice of anomalous dispersion. Each point on the sphere represents a different modulation state. For example, the points $\mathbf{S}_{p,s} = (\pm S_0, 0, 0)$ represent the unmodulated pump and the sidebands, respectively. Points on the equator (i.e., on the $S_3 = 0$ plane) represent pure amplitude modulations (i.e., with $\phi = 0$), whereas points on the meridian with $S_2 = 0$ correspond to pure frequency modulations (i.e., $\phi = \pi$). All the remaining points represent mixed AM-FM modulation states. The vector $\mathbf{S}$ was evaluated from the first three Fourier components of the full numerical solution. Note the agreement with the prediction of eq.(9), in particular in the proximity of the separatrix trajectory that emanates from the unstable pump mode.

The insets in figure (1) show the intensities in one temporal period T that correspond to either an initial AM or FM wave ($\phi(z = 0) = 0, \pi$). As can be seen, in both cases a strong AM modulation periodically develops in the fiber. However, whereas the trajectory *inside* the separatrix evolves around the same stable AM eigenmodulation, the trajectory that lies *outside* the separatrix switches between the two stable AM eigenmodulations that are time-shifted by half a period. As a result, a change from FM to AM of a weak initial modulation may result into orthogonal

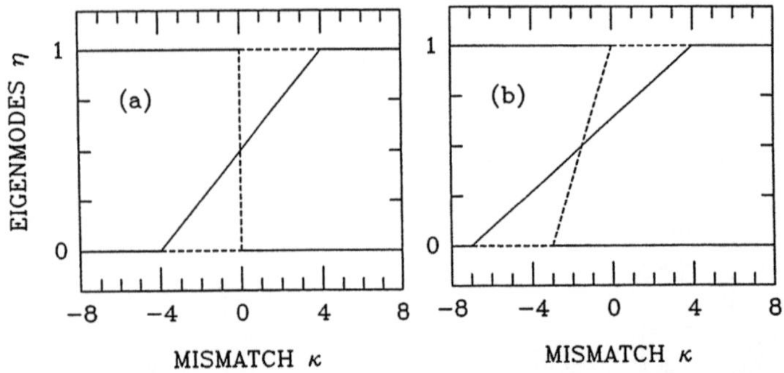

Figure 2. Bifurcation diagram of the nonlinear eigenmodes of three-wave mixing in a birefringent fiber: (a) degenerate polarization instability; (b) nondegenerate modulational polarization instability.

modulation states at the fiber output. We anticipate that this effect has potential applications to all-optical information processing devices.

Consider now the case of sidebands that are orthogonally polarized with respect to the pump. We may set

$$\begin{aligned} u(\xi,t) =& A_0(\xi)e^{\mp i\Omega^2\xi/4}, \\ v(\xi,t) =& \sqrt{2}A_1(\xi)e^{\mp i\Omega^2\xi/4}\, cos(\Omega t). \end{aligned} \tag{10}$$

By inserting the above expression in eqs.(1), one obtains eqs.(6), with $\mu = \Delta \pm \Omega^2/2$, $\sigma = 2/3$, $\varrho = 1/3$, and $\gamma = 3/2$ whenever $\Omega \neq 0$, whereas $\gamma = 1$ in the degenerate $\Omega = 0$ case.
In terms of the variables η and ϕ, and the coordinate $\zeta \doteq \xi S_0/3$, the reduced Hamiltonian, say, $\hat{H}$, reads

$$\hat{H}(\eta,\phi) = 2\eta(1-\eta)\cos\phi - (\kappa+5)\eta + \frac{7}{2}\eta^2, \tag{11}$$

where now $\kappa = -12\mu/S_0$ is the normalized phase mismatch. Let us consider the stability of the fixed points of the above dynamical system. In the degenerate case [6-7], one obtains the symmetric bifurcation diagram of figure (2a), where $\eta = (S_1+1)/2$ determines the fractional content of power in the eigenmodes. Dashed lineas indicate the unstable points: figure (2a) shows that mode A_0 is unstable for $0 \leq \kappa \leq 4$ whenever $\Delta \leq 0$, whereas the mode A_1 is unstable for $0 \geq \kappa \geq -4$ for $\Delta \geq 0$ (fast mode instability [6-8]). The oblique line gives the location of the two stable elliptical eigenpolarizations on the sphere (with opposite handedness).

In the nondegenerate case, one obtains the diagram of figure (2b). As can be seen, the nonlinear anisotropy due to the mutual phase shifts

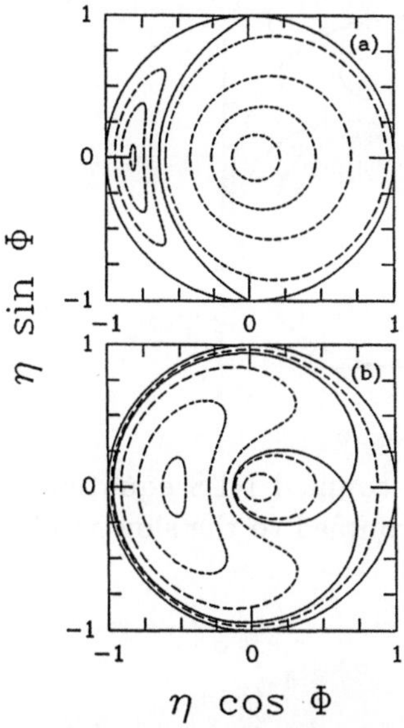

Figure 3: Phase-space portraits of the modulational polarization instability, with a detuning:(a) $\kappa = 2$; (b) $\kappa = -1$.

between the two sidebands breaks the symmetry with respect to the $\kappa = 0$ axis. The unstable region of the pump mode for $\kappa \geq 0$ corresponds to the predictions of the linearized stability analysis in ref.[33]. Figure (2b) shows that, for a certain set of negative values of κ, also the sideband mode is linearly unstable. Note that the dashed oblique line is associated with two unstable modes on the modulational sphere that are a certain linear combination of the pump and sideband mode.

The bifurcation diagram in figure (2) serves as a guide for predicting the topology of the motion in the phase space of the system (11). Figure (3) shows the trajectories and the homoclinic orbits (solid lines) in the plane (η, ϕ) for two different values of κ. As can be seen, varying κ may have a profound influence on the motion. Note that in the framework of the finite dimensional description the same value of κ may be obtained both in the normal and in the anomalous dispersion regime, as long as the sign of the birefringence is exchanged.

From a physical point of view, we may point out that for $0 > \kappa > -3$ a double loop separatrix divides the motion on the plane in two regions. This means that a period-doubling results in the spatial variation of the exchange of power between the pump and the sidebands. This important property is absent from the system (8) that is associated with the scalar case.

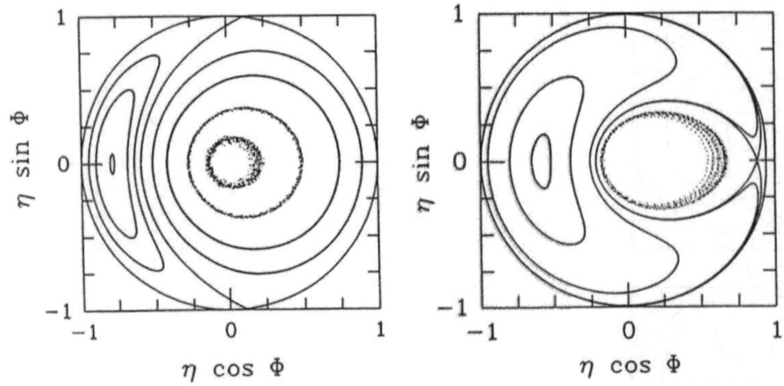

Figure 4: Projections from the solution of the coupled NLS equations: (left) $\kappa = 2$; (right) $\kappa = -1$. Here the pump is coupled to the slow mode and the dispersion is normal.

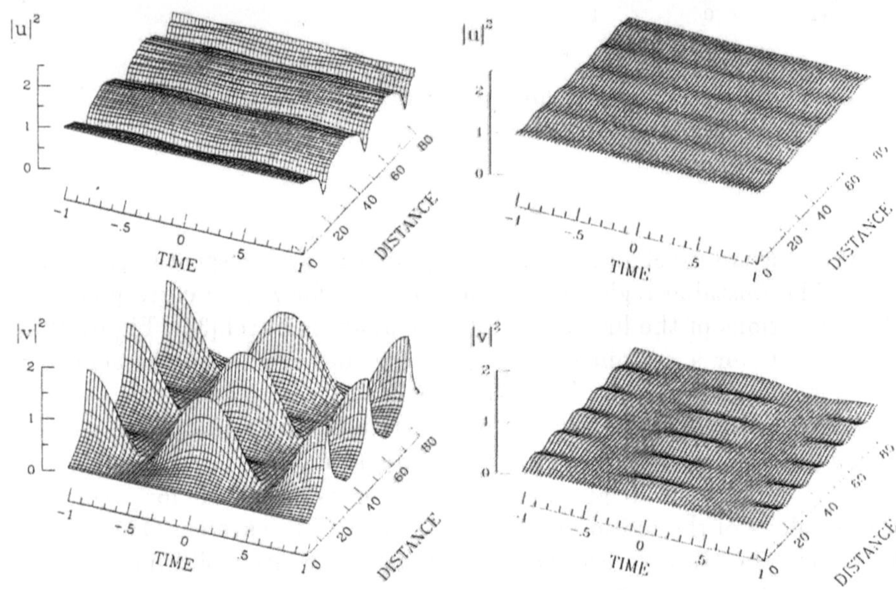

Figure 5: Evolution of one period of the time-periodic pattern of the two components of the field. Left: evolution inside the separatrix. Right: evolution outside the separatrix.

Figure (4) demonstrates the validity of the ode description of modulational polarization instability: here the trajectories have been computed from a numerical solution of eqs.(1), with the same κ's as in the previous figure, normal dispersion, and the pump mode was coupled to the slow fiber axis. The physical meaning of the separatrix crossing effect in the case of figure 3(b) is illustrated by figure 5. Here the motion occurs inside and outside the double-loop separatrix, respectively, and the separatrix crossing is forced by a slight change of the depth of the initial modulation. As can be seen, this small variation of the initial conditions leads to a dramatic difference in the subsequent exchange of power between the waves.

3. Modulational Chaos

As we have seen, in many cases the nonlinear interaction of modulated waves in a fiber may be described in relatively simple terms by means of modal truncations. Clearly this is not always the case: we want to point out in this final section that the general behavior of the periodic solutions of eqs.(1) may be very complex. In fact, in analogy with the completely degenerate interaction of polarized waves in a fiber [5,9-13,18], we may expect that the sensitivity of the output state of polarization with respect to small changes of the initial conditions will lead to spatial chaos in the presence of periodic modulations of the parameters of the fiber or whenever several modes are linearly unstable. Moreover, because of the infinite dimensionality of the underlying pde's (1), we would expect and indeed we find a rich unstable behavior.

In the following two numerical examples we will see that, although the spatio temporal disorder of modulated waves in a birefringent fiber may result from the competition between just a few unstable Fourier modes, in other cases one may observe a highly multimodal chaos. Figure (6) shows a near-separatrix evolution as it is computed from eqs.(1) with the same value of $\kappa = 2$ as in the case of figure (3a), but with a pump on the fast axis of the fiber and in the anomalous dispersion regime. As can be seen, after some distance sideband modes are generated also in the polarization that is parallel to the pump. The result of this few-mode interaction is an apparently irregular behavior of the time pattern in the pump mode. The origin of the growth of the additional modes in the input axis is the presence of the scalar MI which occurs in the anomalous dispersion regime. Still, at least over the relatively long distance that is shown here, the number of unstable modes is limited which would suggest the possibility of studying the observed chaotic behavior in terms of a finite dimensional truncation [31].

On the other hand, figure (7) has been obtained by keeping $\kappa = -1$ and inverting the pump axis and the sign of dispersion with respect to the case in figure (3b). Here the input condition was set close to the pump mode (weak initial modulation). As can be seen from figure (2), for $\kappa = -1$ the polarization of the carrier is linearly stable. In fact, figure (7) shows that only small polarization coupling occurs, whereas

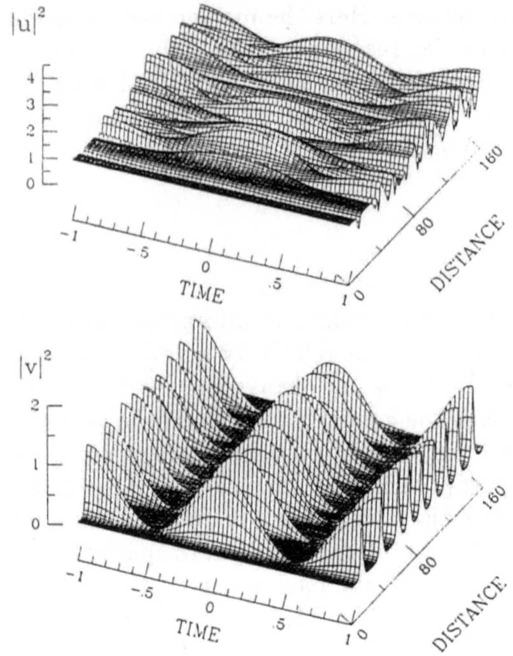

Figure 6: As in fig.(5), with a pump on the fast axis in the anomalous dispersion regime and $\kappa = 2$.

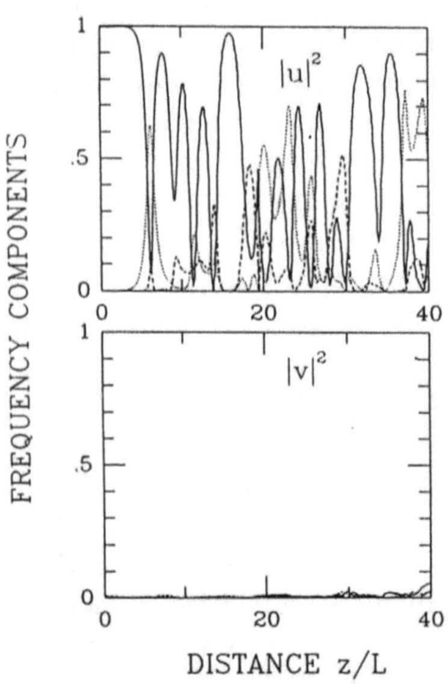

Figure 7: Intensities of spectral components with a pump on the slow axis in the anomalous dispersion regime and $\kappa = -1$.

the pump mode rapidly decays in an erratic fashion into the sidebands with parallel polarization. The gain for the highly multimodal instability that is observed here is essentially provided by the scalar MI, whereas the chaotic behavior is triggered by the small perturbation owing polarization coupling through the cross-phase modulation terms in eqs.(1).

4. Conclusions

We have presented integrable two mode truncations that may well describe the unstable manifolds that are observed in the nonlinear modulation of polarized waves in birefringent fibers. We have provided a connection with the scalar MI and the degenerate polarization instability. We have also illustrated the possibility that the present instabilities may lead to highly disordered multimode wave patterns.

5. Acknowledgements

We acknowledge stimulating discussions with N.N. Akhmediev, B. Daino, and G. Forest, and the valuable help of R. Chisari in the preparation of the figures. This work was carried out in the framework of the agreement between the Fondazione Ugo Bordoni and the Istituto Superiore Poste e Telecomunicazioni.

References

[1] A. Yariv, *Quantum Electronics* (Wiley, New York, 1975) Chapt. 16-18.

[2] A.J. Lichtenberg, and M.A. Lieberman, *Regular and Stochastic Motion* (Springer, New York, 1983); S. Wiggins, *Global Bifurcations and Chaos*, (Springer, New York, 1988).

[3] S.A. Akhmanov, and R.V. Khokhlov, *Problems of Nonlinear Optics* (Gordon and Breach, New York, 1972).

[4] M.I. Dykman, and G.G. Tarasov, Sov. Phys. Solid State **24**, 1361 (1982).

[5] J. Yumoto, and K. Otsuka, Phys. Rev. Lett. **54**, 1806 (1985).

[6] B. Daino, G. Gregori, and S. Wabnitz, J. Appl. Phys. **58**, 4512 (1985); Opt. Lett. **11**, 42 (1986).

[7] G. Gregori, and S. Wabnitz, in *Optical Bistability III*, H.M. Gibbs, *et al.*, eds. (Springer, Berlin, 1986) pp. 359-362; Phys. Rev. Lett. **56**, 600 (1986).

[8] H.G. Winful, Opt. Lett. **11**, 33 (1986).

[9] S. Wabnitz, Phys. Rev. Lett. **58**, 1415 (1987).

[10] M.V. Tratnik, and J.E. Sipe, Phys. Rev. A **35**, 2965 (1987), *ibid.* 2976.

[11] E. Caglioti, S. Trillo, and S. Wabnitz, Opt. Lett. **12**, 1044 (1987); in *Advances in Nonlinear Dynamics and Stochastic Processes II*, G. Paladin, and A. Vulpiani, eds., (World Scientific, Singapore, 1987) pp. 225.

[12] S. Trillo, and S. Wabnitz, Phys. Rev. A **36**, 3881 (1987); S. Trillo, and S. Wabnitz, J. Opt. Soc. of Am. B **5**, 483 (1988).

[13] D. David, D.D. Holm, and M.V. Tratnik, Physics Letters A **137**, 355 (1989); *ibidem*, **138**, 29 (1989).

[14] S. Trillo, S. Wabnitz, R.H. Stolen, G. Assanto, C.T. Seaton, and G.I. Stegeman, Appl. Phys. Lett. **49**, 1224 (1986).

[15] S. Trillo, S. Wabnitz, N. Finlayson, W.C. Banyai, C.T. Seaton, G.I. Stegeman, and R.H. Stolen, Appl. Phys. Lett. **53**, 837 (1988); IEEE J. Quantum Electron. **QE-25**, 104 (1989).

[16] S.F. Feldman, D.A. Weinberger, and H.G. Winful, Opt. Lett **15**, 311 (1990).

[17] N.I. Zheludev, Sov. Phys. Usp. **32**, 357 (1989).

[18] D. David, D.D. Holm, and M.V. Tratnik, Physics Reports **187**, 281 (1990).

[19] V.I.Bespalov, and V.I.Talanov, JETP Lett. **3**, 307 (1966).

[20] T.J. Benjamin, and J.E. Feir, J. Fluid Mech. **27**, 417 (1967).

[21] H.C. Yuen, and W.E. Ferguson, Jr., Phys. Fluids **21**, 1275 (1978).

[22] E. Fermi, J. Pasta, and S. Ulam, in *Collected Papers of Enrico Fermi*, E. Segrè, ed. (Univ. of Chicago, Chicago, 1965), Vol. 2, p.978.

[23] N. Ercolani, M.G. Forest, and D.W. McLaughlin, Physica **18D**, 472 (1986); *Geometry of the Modulational Instability. Part III: Homoclinic Orbits for the periodic sine-Gordon equation*, University of Arizona Preprint (1987), and Physica D (1990), and refs. therein.

[24] E.R. Tracy, H.H. Chen, and Y.C. Lee, Phys. Rev. Lett. **53**, 218 (1984); E.R. Tracy, H.H. Chen, Phys. Rev. A **37**, 815 (1988).

[25] N.N. Akhmediev, and V.I. Korneev, Theor. and Math. Physics **69**, 1089 (1986).

[26] H.T. Moon, Phys. Rev. Lett. **64**, 412 (1990).

[27] M.J. Ablowitz, and B.M. Herbst, SIAM J. Appl. Math. (April 1990).

[28] E. Infeld, Phys. Rev. Lett. **47**, 717 (1981).

[29] G. Cappellini, and S. Trillo, J. Opt. Soc. Am. B (April 1991).

[30] A.R. Bishop, in *Partially Integrable Evolution Equations in Physics*, R. Conte and N. Boccara, eds. (Kluver Academic, Amsterdam, 1990) pp. 1-38.
[31] A.R. Bishop, M.G. Forest, D.W. McLaughlin, and A.E. Overman II, Phys. Lett. A **144**, 17 (1990).
[32] S. Trillo, and S. Wabnitz, Optics Letters, in press (1991).
[33] S.Wabnitz, Phys. Rev. A **38**, 2018 (1988); S.Trillo, and S. Wabnitz, J. Opt. Soc. Am. B **6**, 238 (1989).

Pattern Formation in Nonlinear Physical Systems with Characteristic Electric Properties

F.-J. Niedernostheide[1], *R. Dohmen*[1], *H. Willebrand*[1], *H.-J. Schulze*[2], *and H.-G. Purwins*[1]

[1]Institut für Angewandte Physik, Universität Münster, W-4400 Münster, Fed. Rep. of Germany
[2]Siemens AG, ZFE, W-8000 München, Fed. Rep. of Germany

Systems are considered which consist of two layers, one of them characterized by a high ohmic, approximately linear and one by an S-shaped current-voltage characteristic. We describe such systems mathematically in terms of two-component reaction-diffusion equations. Analytical and numerical investigations of these equations reveal that the systems can self-organize different stable stationary inhomogeneous nonequilibrium structures as well as spatio-temporal irregular behaviour. These results are in good qualitative agreement with the current density patterns in two experimentally investigated systems, namely a one-dimensional npnp-semiconductor structure and a one-dimensional gas discharge system.

1. Introduction

The self-organization of spatial and spatio-temporal patterns in various types of spatially extended systems is subject of intensive studies in many branches of chemistry, biology and physics. In particular stationary spatial patterns are observed in condensed, liquid and gaseous matter. In gas discharges the spontaneous generation of striations in the positive column is well-known [1]. Radehaus et al. [2], Purwins et al. [3] and Willebrand et al. [4] report upon the observation of solitary current density filaments in a special one-dimensional dc-driven gas discharge system. Müller [5] investigates a two-dimensional discharge system consisting of a hollow steel cathode and a spherical steel anode at which the discharge forms spots of typical size. In an ac-driven two-dimensional helium discharge between two dielectric glass-plates the current flows in small discharge channels or stripes depending on the system parameters [6]. As example for pattern formation in liquids one may refer to

Springer Proceedings in Physics, Vol. 67 Nonlinearity with Disorder
Editors: F. Abdullaev · A.R. Bishop · S. Pnevmatikos

liquid cristals which form cells characterized by a definite orientation of the director due to electrohydrodynamic convection or due to mutual effects of the molecule dipole moments with the electric field vector of incident light [7,8]. Semiconductors and semiconductor devices are classical materials for pattern formation of the current density or of the electric field in the condensed state. Solitary current density filaments are observed in thin GaAs films by detecting the emitted radiation [9] or using the technique of scanning electron microscopy [10]. In gold compensated silicon pin-diodes an inhomogeneous spatial distribution of the surface potential, the temperature and the emitted infrared radiation is measured [11]. In many cases the observed patterns are the result of two or more competing components which are mathematically describable in terms of reaction-diffusion equations [12,13].

Besides spatial structures also spatio-temporal structures are of great interest. In semiconductors under certain conditions oscillating boundaries of current filaments are observed. The interaction of such oscillating walls may cause complex oscillations of the total current [14,15]. In gas discharge systems the generation process of current density filaments which is accompanied by regular and irregular current oscillations has been investigated and can be explained by periodically repeated division and subsequent extinction processes of filaments [16]. Such oscillating filament walls and division processes can be described by using reaction-diffusion equations if the time and diffusion constants of the competing components fulfill certain conditions [12].

In this article in chapter 2 we shortly review a model for systems consisting of two layers with different electric properties. Analytical and numerical solutions of the model equations are presented. In chapters 3 and 4 we report on experiments concerning pnpn-semiconductor structures and a gas discharge system. The observed spatial and spatio-temporal patterns are compared with those predicted by the model.

2. Two-layer model

In the following we describe the structure and electronic behaviour of an idealized system, which may be divided into two

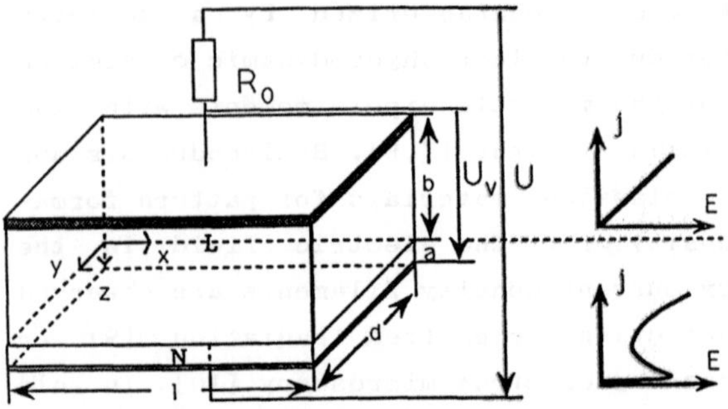

Fig. 1: Schematic diagram of the two-layer model and typical j(E)-characteristics of the linear (L) and nonlinear (N) layer

parts with different electric properties. One layer (N) shall be characterized by a nonlinear S-shaped current-voltage characteristic, the other layer (L) is assumed to have ohmic electric properties. The whole sample is provided with ideal metallic contacts. In addition it is required that the resistivity of the nonlinear layer is larger than that of the linear layer in a wide current range. A possible physical realization of this model is a gas discharge system in which parts of the gas gap cause an S-shaped current-voltage characteristic while other parts of the gap and the electrodes can be regarded as nearly ohmic. Other realizations are semiconductors and semiconductor devices, e. g. pnpn-diodes. Even homogeneous materials may exhibit different electric properties due to sample preparation or carrier injection leading, e. g., to a region of larger conductivity near the contacts with respect to the bulk material. Fig. 1 depicts the two-layer device which is connected to a voltage source U via an external resistor R_0.

Taking into account convection and dielectric displacement currents, the diffusion of charge carriers and the nonlinear electric behaviour for the layer N and the ohmic properties for the linear layer L and using a suitable potential ansatz one can deduce a two-component system of reaction-diffusion equations defined on the interface Ω between the two layers. Details concerning this derivation can be found in [13]. The two components are the current density $j(x,y,t)$ through this interface and the potential drop $U_L(x,y,t)$ across the linear layer. In the vicinity of the inflection point of the S-shaped characteristic the current density j acts as activator according to the negative differential resistivity (NDR) of the nonlinear j(E)-characteristic of the layer N. Physically this leads to an autocatalytic increase of j in the NDR range. The voltage U_L works as inhibitor and stabilizes current filaments because it prevents the spreading of existing filaments as well as the generation of new filaments in the environment of the latter. The mathematical description of the model leads finally to the following system of coupled differential equations in one-dimensional space [17]:

$$\begin{aligned} \partial v/\partial \tau &= \sigma \Delta v + f(v) - w + \mu(\nabla v)(\nabla w) + \varkappa_1 - (\varkappa_2/|\Omega|)\int v d\Omega, \\ \delta\, \partial w/\partial \tau &= \Delta w + v - w, \end{aligned} \tag{1}$$

$$v = (j-j_o)/j^*,\quad w = (U_L-U_{Lo})/U_L^*,\quad U_L^*/j^* = b\varrho,$$
$$\tau = (b\varrho/\ell)t,\quad x/d_w^{1/2} \longrightarrow x$$

and

$$d_v = D_0\ell/(\varrho b),\qquad d_w = (5/26)b^2,\qquad \sigma = d_v/d_w,$$
$$\mu = 26\mu_0 j^*\ell/(5b^2),\qquad \delta = \varepsilon_0\varepsilon_L\varrho^2 b/\ell,$$
$$\varkappa_1 = U/U_L^*,\qquad \varkappa_2 = R_0 1/(\varrho b),\ 1/(d_w)^{1/2} \longrightarrow 1.$$

One of the quantities U_L^* or j^* can be chosen arbitrarily. (j_o, U_{Lo}) represents a typical homogeneous reference state of the system. ℓ is a distributed inductance, D_0 and μ_0 are an ambipolar diffusion and mobility, respectively, ϱ is the specific resistivity of the linear layer, 1, a and b are geometrical quantities according to Fig. 1. ε_0 and ε_L denote the dielectric constant and the dielectric number of the linear layer, respectively. The integral term results from the external resistor and causes a global inhibition. The function f(v) is determined by the S-shaped $j(U_N)$-characteristic of the nonlinear layer. Depending on the concrete form of the function f(v) the system may have one or three stationary homogeneous states. Obviously the component v activates the component w and itself if $f' \equiv \partial f/\partial v > 0$ whereas the component w tends to suppress itself and the activator. The competition of these two components allows a wide variety of spatial and spatio-temporal patterns which are presented subsequently.

We first consider Eqs. (1) with $\sigma, \delta < 1$ and ask for the stability properties of a spatially homogeneous state (v_s, w_s), i. e. a low or high current state of the system. From linear stability analysis one knows that such a stationary solution (v_s, w_s) is stable if the eigenvalues of the matrix of the linearized Eqs. (1) have negative real parts and a bifurcation occurs, if by changing a parameter one of the eigenvalues crosses the imaginary axis. To calculate the eigenvalues one has to expand v and w with respect to the eigenfunctions of the Laplace operator with suitable boundary conditions. In a one-dimensional case with Neumann boundary conditions we obtain the eigenfunctions $\varphi_0 = 1/l$ and $\varphi_n(x) = (2/l)\cos(k_n x)$ with $k_n = n\pi/l$ and $n \geq 1$, where l is the length of the system. In this orthonormal basis Eqs. (1) can be written as an infinite system of nonlinear differential equations determining the temporal development of the amplitudes (v_n, w_n) which occur as factors of φ_n when expanding v and w. Writing Eqs. (1) in the orthonormal basis φ_n the linear part corresponding to the mode k_n is given by

$$C_n = \begin{pmatrix} -\sigma k_n^2 + f' - \delta_{0n}\varkappa_2 & -1 \\ 1/(\delta(1+k_n^2)) & -1/\delta \end{pmatrix}, \tag{2a}$$

where $f' \equiv f'(v_s)$ and δ_{0n} denotes the Kronecker-symbol. The eigenvalues $\omega(k_n)$ of C_n are determined by

$$\omega^2(k_n) - \omega(k_n)\mathrm{Tr}(C_n) + \mathrm{Det}(C_n) = 0. \tag{2b}$$

The system is stable if $\mathrm{Det}(C_n) > 0$ and $\mathrm{Tr}(C_n) < 0$ and may become linearly unstable in two different ways, $\mathrm{Tr}(C_n)=0$ or $\mathrm{Det}(C_n)=0$, called hardmode and softmode instability, respectively. In the following we restrict ourselves to the softmode instability in which e. g. a variation of the external voltage U or of the load resistor R_0 causes the real part of one of the eigenvalues to become positive at a critical point called Turing point. This eigenvalue refers to a critical mode k_c which is given by

$$k_c = (\sigma^{-1/2} - 1)^{1/2}. \tag{3}$$

To investigate the development of the arising spatially periodic structure near the instability point one can use the centre manifold approach [18,19]. As a result one obtains a critical value of f' where the amplitude p of the critical mode k_c may develop according to a super- or a subcritical bifurcation depending on the system parameters (cf. Ref. [19]). The subcriti-

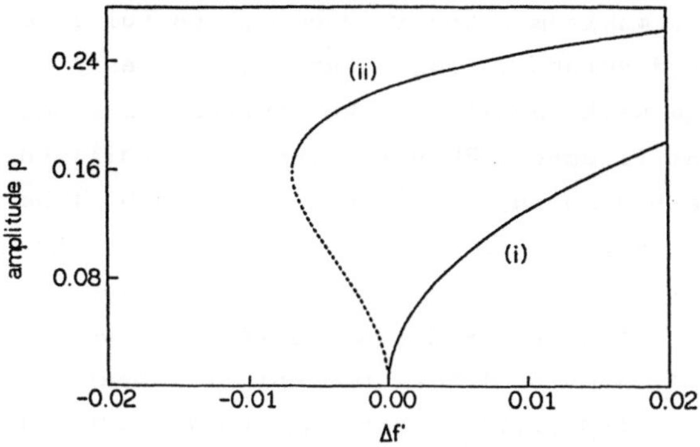

Fig. 2: Amplitude p of the developing critical mode k_c near the bifurcation point f'_c as a function of $\Delta f' = f' - f'_c$ for the supercritical (i) and the subcritical (ii) case. Solid and dashed lines mark stable and unstable branches, respectively. The parameters are $\sigma = 0.071$, $\mu = 0$, $\delta = 1$, $\varkappa_2 = 0$, and $l = 14.14$. The function $f(v)$ is defined as $f(v) = -v^3 + 0.48v$ and $f(v) = -v^3 + 0.6v$ for the cases (i) and (ii), respectively.

cal case can be found in a monostable system with one or three stationary states and has the property that the bifurcating mode k_c is not stable in the vicinity of the critical point. The related bifurcation therefore implies a jump in the amplitude of the critical mode. The supercritical bifurcation occurs only in a monostable system with one stationary state and is characterized by a square root like increase of the amplitude of the mode k_c with increasing bifurcation parameter f'. For two different parameter sets Fig. 2 shows the calculated amplitude of the mode k_c as function of $\Delta f' \equiv f' - f'_c$ whereby terms up to the fifth order were used in the centre manifold approach [19]. The results are in agreement with approaches done by Elmer [20], who used the singular perturbation theory to describe the developing mode k_c in a system of equations similar to Eqs. (1).

If the parameters of the system are chosen such that there can exist three homogeneous states with only one of them being stable one can vary f' (in the physical system e.g. the voltage U) such that by reaching the critical f'_c the stable homogeneous state is destabilized via the critical mode k_c. This spatially periodic structure grows but cannot be stabilized. Instead of this the system may favour a certain maximum where induced by unavoidable fluctuations the activator develops stronger than in the rest of the system resulting in a well localized high current region, subsequently called current filament. This filament is stabilized because the other maxima become suppressed due to $\varkappa_2 \neq 0$ (the external resistor R_0 in the physical system) and because the inhibitor diffusion constant is large enough to prevent a spreading of the activator into the neighbouring regions due to $\sigma < 1$. The generation of such a stable current filament by increasing the value $\varkappa_1$ (in the physical system the voltage U) is accompanied by a discontinuity in the global current-voltage characteristic indicated by a voltage decrease and a current increase (Fig. 3).

The decrease of the total voltage across the device causes also a decrease of the value of f' corresponding to the low current regions. Increasing $\varkappa_1$ leads to a weak increase of the width of the filaments. If the Turing point is reached again in the homogeneous low current region, i. e. $f' \approx f'_c$, the distributions of v and w are additionally modulated by a periodic structure with wave-vector k_c and an amplitude strongly decaying with growing distance from the filament. Consequently a second filament appears at the maximum nearest to the previous filament in a distance determined by the critical wave-vector. This process recurs with increasing $\varkappa_1$ until the whole system is in the high current state.

In order to explain the periodic modulation near to the critical point regard the behaviour of a disturbance $(\delta v, \delta w)$ of a stable homogeneous state (v_s, w_s) of an infinitely extended system. For the disturbance we make the ansatz:

$$\begin{pmatrix} \delta v \\ \delta w \end{pmatrix} = \begin{pmatrix} v_a \\ w_a \end{pmatrix} \exp(\nu x). \tag{4}$$

The ansatz describes among others a fluctuation which has the amplitude (v_a, w_a) in the arbitrarily chosen origin and decays to zero for $|x| \to \infty$. Especially for $x > 0$ we are interested in solutions with $\mathrm{Re}(\nu) < 0$. Depending on $\mathrm{Im}(\nu) \neq 0$ or $\mathrm{Im}(\nu) = 0$ the asymptotic solution is periodically modulated or not. The insertion of the ansatz (4) into the linearized system of Eqs. (1) leads to the polynomial

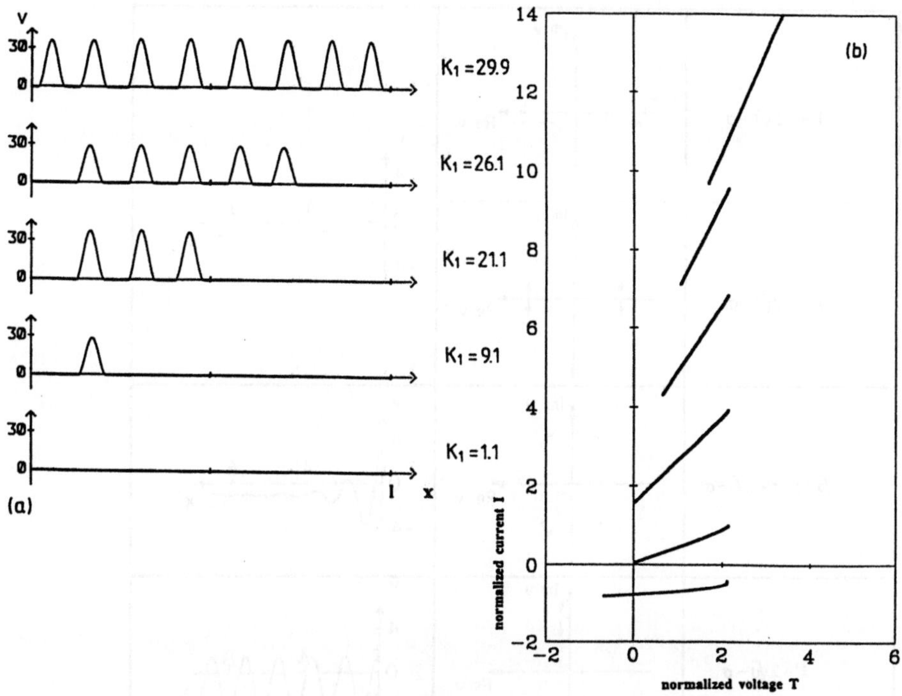

Fig. 3: Results of the numerical calculations obtained from Eqs. (1) by using the parameters $\sigma = 0.1$, $\mu = 0$, $\delta = 1$, $x_2 = 4$, $l = 50$, and

$$f(v) = \begin{cases} -11.5v^3 + 8v & v \leq v_f \\ 0.6\, v + \theta & v \geq v_f \end{cases},$$

with v_f and θ such that $f(v)$ is continuously differentiable:

a) Stationary current density distributions for certain values of the total current $I = (1/l)\, {}_0\int^l v(x)dx$

b) Global current-voltage characteristic $I(T)$ with $T = x_1 - (x_2/l)\, {}_0\int^l v(x)dx$

$$\nu^4 + \left(\frac{f'(v_s)}{\sigma} - 1 \right) \nu^2 + \frac{1 - f'(v_s)}{\sigma} = 0. \tag{5}$$

It delivers four zeros $\nu_i = \nu_i(f',\sigma)$, $i = 1,\dots,4$, two of which have negative real part. For constant σ the four zeros of the polynomial depend merely on the slope f'. Solutions with non-vanishing real parts exist for $f' < f'_c$. If $f' < -2\sigma^{1/2} - \sigma$ all zeros are reals, i. e. a stationary solution

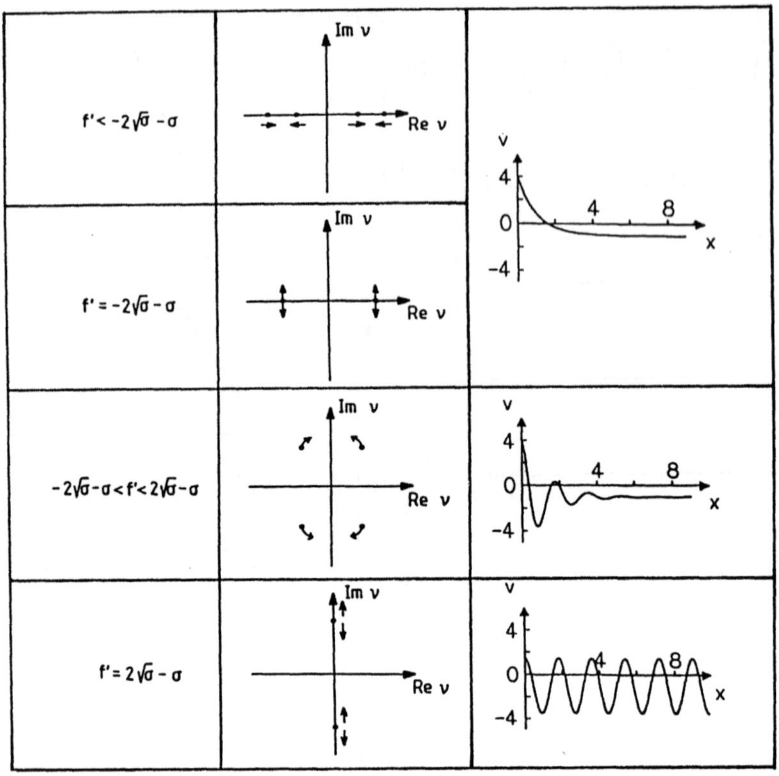

Fig. 4: Schematic diagrams of the eigenvalues ν of the ansatz (4) for different ranges of f'. The arrows indicate the evolution of the eigenvalues with increasing f'. The right column illustrates the corresponding typical distribution of the activator v.

$(v,w) = (v_s,w_s) + (\delta v,\delta w)$ approaches the homogeneous state exponentially (Fig. 4). For f' in the range of $]-2\sigma^{1/2}-\sigma, f'_c[$ the zeros are complex with $Re(\nu_i) \neq 0$ and $Im(\nu_i) \neq 0$ resulting in a spatially periodic modulation of the asymptotic, exponentially decreasing solution. With increasing f' the real part and therefore the damping of this oscillation decreases. For $f' = f'_c$ the real part is zero and the ν_i are purely imaginary. This corresponds to the above mentioned Turing instability; any deflection (v_a,w_a) in the origin leads to a periodic distribution with amplitude (v_a,w_a) and the intrinsic wave-vector $k_c = Im(\nu_i(f'_c))$. An appropriate superposition of the two asymptotic fluctuations delivers an overshoot in the vicinity of a filament.

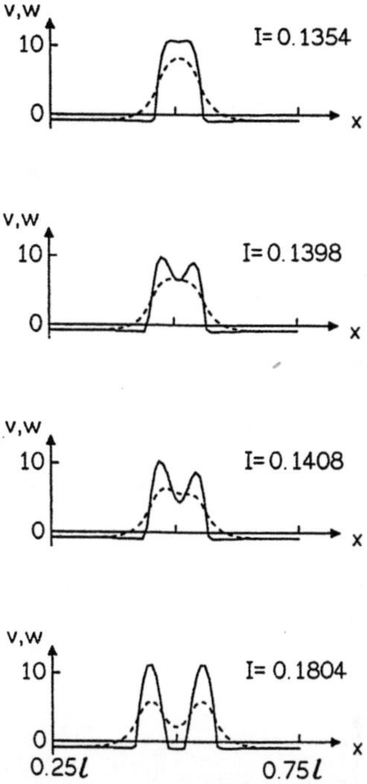

Fig. 5: Numerically calculated stationary activator distributions in the course of the division process for different values of the total current $I = {}_0\int^l v(x)dx$. Up to I = 0.1408 corresponding to a critical width of the distribution the process is reversible. The parameters are σ = 0.1, μ = 0.00002, δ = 1, $\varkappa_2$ = 20, l = 37.8, and f(v) as in Fig. 3.

The filament formation process changes drastically if the drift parameter μ is sufficiently large. Then an additional term proportional to $(\nabla v)(\nabla w)$ has to be taken into account which is particularly efficient at the walls of the current filament. This term causes a rise of the filament walls, and because the activator has a local minimum in the center of the filament whereas the inhibitor has a maximum there, $(\nabla v)(\nabla w)$ becomes negative in the vicinity of the center and therefore reduces the activator. In contrast, at the walls $(\nabla v)(\nabla w)$ is positive and increases the activator. These two processes lead to a splitting of one filament into two independent filaments. In Fig. 5 stationary states of the activator and inhibitor distributions during such a division process by increasing $\varkappa_1$ are shown. Up to a

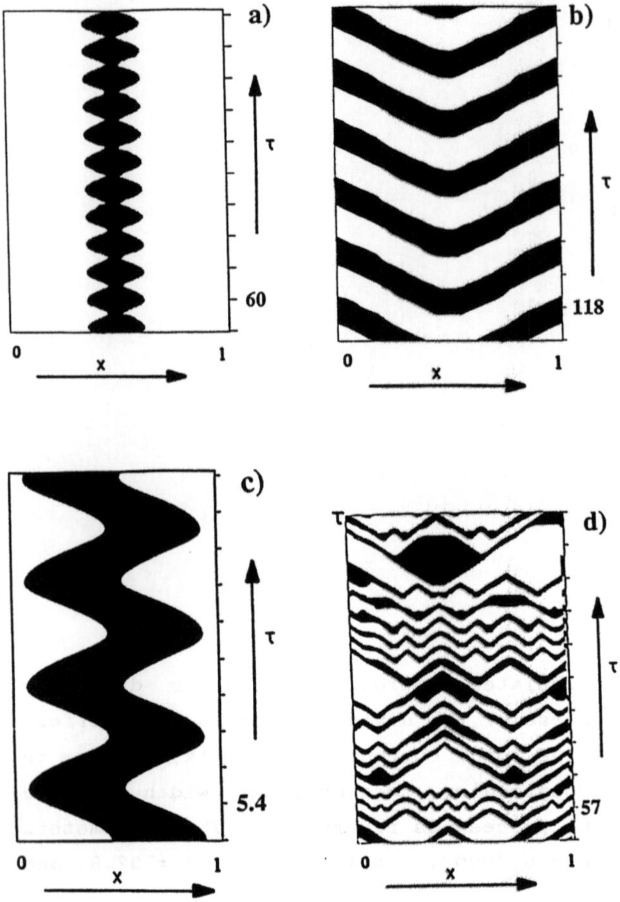

Fig. 6: Numerically calculated time evolution of the spatial distribution of the activator v for different parameter sets. The dark areas mark the parts of the distributions that exceed a certain threshold.

a) A breathing filament for the parameters $\sigma = 0.1$, $\mu = 0$, $\varkappa_1 = -0.5$, $\varkappa_2 = 0$, $\delta = 5$, $l = 22.36$, and $f(v) = -0.01v^3 + 1.37v$

b) Selfcreated travelling pulses for the parameters $\sigma = 10$, $\mu = 0$, $\varkappa_1 = -4$, $\varkappa_2 = 4$, $\delta = 100$, $l = 1000$, and $f(v) = -0.02v^3 + 1.5v$

c) A swaying filament for the parameters $\sigma = 0.1$, $\mu = 0.05$, $\varkappa_1 = -2.15$, $\varkappa_2 = 10$, $\delta = 1, l=4.47$, and $f(v)$ as in Fig. 3b

d) Spatio-temoporal irregular behaviour for parameters as in b) but $\delta = 3$

certain width w_c of the filament the division process is reversible, i. e. reduction of the total current leads to a decreasing width. Exceeding w_c a second filament is generated; if the current is reduced one of the filaments vanishes.

A necessary presupposition for the stability of the above mentioned inhomogeneous structures is that $\sigma < 1$, $\delta < 1$. If the value of δ exceeds 1 the filamentary structure is retained in general, but the boundaries become unstable resulting in a breathing structure, sometimes also called pulsating filament. In Fig. 6a the dark regions mark the breathing filament. For $\delta > 1$, $\varkappa_2 \neq 0$ sufficiently large and $\sigma > 1$, i. e. the diffusion constant of the activator being larger than that of the inhibitor a filament begins to travel to the boundary and vanishes there. Then by a Hopf-bifurcation of the homogeneous low current state a new filament is generated that broadens in the course of time. Finally at a critical width a second Hopf-bifurcation takes place, this time of the homogeneous high current state leading to a division process (Fig. 6b). Both filaments travel to the boundaries and the process recurs. In the case $\delta > 1$, $\sigma < 1$, $\varkappa_2 \neq 0$ and $\mu \geq 0$ another interesting dynamical behaviour emerges. The filament boundaries are oscillating again, but this time in phase, i. e. the filament oscillates around a certain spatial position with definite frequency without changing its width (Fig. 6c). An increase of the total current leads to an increase of the spatial amplitude and of the frequency of the swaying filament. For certain parameter ranges one observes mixtures of breathing, travelling, swaying or splitting filamentary structures resulting in a spatio-temporal irregular behaviour of the activator and inhibitor distributions. An example for such irregular motions is given in Fig. 6d.

3. Spatial and spatio-temporal patterns in a semiconductor structure

Spatial and spatio-temporal patterns similar to those described in chapter 2 can be found in npnp⁻-semiconductor structures where the activator v is essentially the electrical current density and the inhibitor a certain potential in the semiconductor. The samples are prepared from <111>-silicon wafers with a thickness of 500 μm and a weak boron concentration corresponding

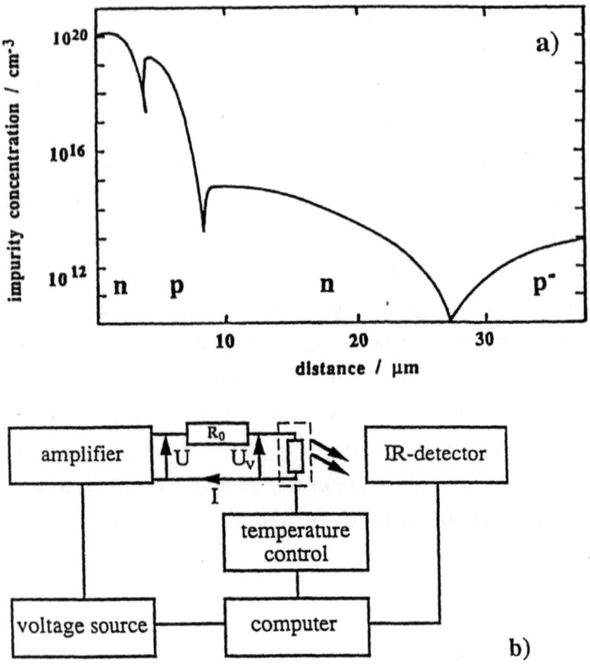

Fig. 7: a) Impurity distributions of the $npnp^-$-structure determined by spreading resistance measurements

b) Experimental arrangement to determine the spatial infrared light distribution of the $npnp^-$-structure

to a specific resistivity of about 1500 Ωcm. In a first step via a predeposition and a subsequent drive-in diffusion a 27 µm wide phosphorus doped layer is prepared. Then by boron diffusion a p-doped region and in a final diffusion process an n-doped layer with widths of 8 µm and 3 µm, respectively, are produced. The first two diffusion profiles are well described by a gaussian distribution whereas the last can be approximated by a complementary error function. The distributions of impurities obtained from spreading-resistance measurements are shown in Fig. 7a.

The samples are ohmically contacted by aluminium layers. For the experimental observation of the current-voltage characteristic and the spatial distribution of the current density small pieces of the wafer with dimensions of about (0.5 x 2.45 x 0.7) mm^3 are prepared. The four edges are polished and the sample is built in a temperature controlled aluminium holder. Via a load resistor R_0 the sample is connected to a voltage source. One of the (2.45 x 0.5) mm^2 surfaces is focused with a lens system to a streak-camera system which serves as an infrared-detector by means of which the emitted infrared recombination radiation can be detected spatially resolved. The IR-detector consists of a photocathode with an area of about (5 x 1) mm^2. Incident light on a certain point of the cathode generates electrons which are accelerated in a high electric field and multiplied in a micro channel plate which stores the spatial information. On a phosphorus screen the electrons create an image in the visible range which can be recorded with a usual vidicon camera. Details about this streak-camera system can be found in [16,21]. The experimental setup is shown in Fig. 7b.

If the high ohmic p^--layer is connected to the positive pole of the voltage source, at small voltage values only a small current is allowed to flow because the p-n junction in the middle is reverse biased. Due to the local S-shaped current-voltage characteristic caused by avalanche breakdown and a positive feedback of carrier injection from the forward biased p-n junctions in the global $j(U_v)$-characteristic (Fig. 8a) at the critical breakdown voltage a discontinuity appears connected with the rise of a current density filament. In regions of large current radiative recombination processes are most likely near p-n junctions and the contacts. Therefore the elevated radiation density in these regions indicates a high current density. In Fig. 8b the measured spatial radiation distribution of the (2.45 x 0.5) mm^2 surface is shown for different values of the total current I. In each picture the device is connected to the positive pole of the voltage source at the top. On the other (2.45 x 0.5) mm^2 surface at the same position a large radiation signal appears, too, and therefore one may conclude that the current flows through the whole sample at this region and not only at the surface.

An increase of the voltage U leads to small changes of the width of the filament and at a critical current an additional current density filament appears of approximately the same size as the first one which is about 500 µm. Again this filament generation process is accompanied by a jump in the $I(U_v)$-characteristic. The filaments are stabilized by the high ohmic p^--layer which acts as inhibitor and prevents the spreading of the filaments.

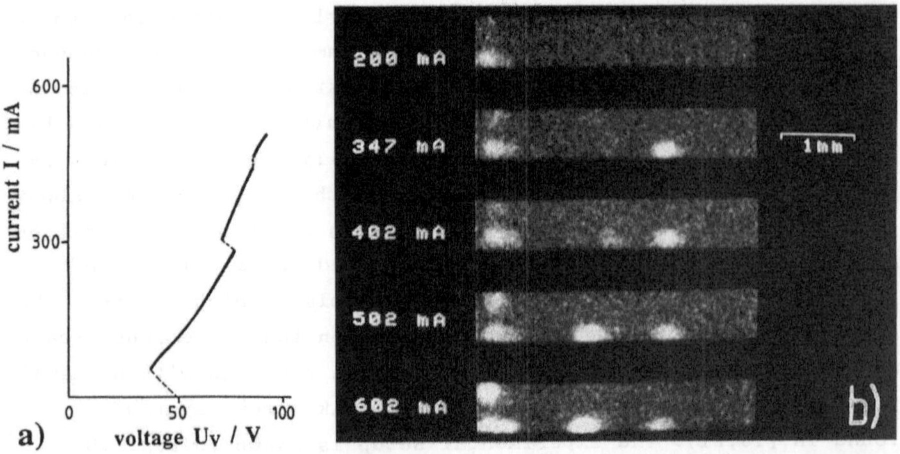

Fig. 8: Measurements obtained from a npnp⁻-semiconductor device with impurity concentrations as shown in Fig. 7a. The size of the sample is (2.45 x 0.7 x 0.5) mm^3 and the measurements were carried out at the temperature T = 123 K.

a) $I(U_V)$-characteristic

b) Spatially resolved infrared radiation distribution for various values of the total current I

In a similar pnpn⁻-structure with impurity distributions as shown in Fig. 9a one also finds the formation of a current filament if the voltage is large enough and the n⁻-layer is connected to the negative pole of the voltage source. But when the voltage is enlarged further instead of the generation of an additional filament the initially stationary filament begins to oscillate around the spatial position of the originally stable filament. Experimentally this behaviour has been observed using the above described streak-camera with an additional electron deflection unit perpendicular to the acceleration field. An electric field increasing linearly in time generates a spatio-temporally resolved picture on the phosporus screen. Fig. 9d shows a measurement of an oscillating filament with a spatial frequency of about 140 Hz. The horizontal and vertical directions correspond to the space and time coordinate, respectively. To get a sufficient time resolution only a 50 µm wide region of the sample surface near the contact has been focused to the photo-cathode.

Due to the spatio-temporal oscillations of the filament an ac-component is added to the dc-voltage drop U_v of the sample (Fig. 9c). The frequency of this oscillation is exactly the same as that of the spatial oscillation observed with the streak-camera. An increase of the total current I causes a weak increase of the spatial amplitude of the oscillating current filament accompanied by an enlargement of the ac-voltage amplitude. At a critical current this behaviour changes abruptly and the current filament oscillates from one boundary of the sample to the other with a somewhat smaller frequency; the ac-voltage amplitude of U_v rises, too, and the frequency of the voltage signal decreases to the same value as that of the spatial oscillation. Fig. 9e shows a filament rocking between the sample boundaries and Fig. 9b the accompanying $I(U_v)$-characteristic of the sample. In the region of oscillations the time averaged values of the voltage U_v and the current I are used for the characteristic. The current I_{c1} marks the instability point at which the stable stationary filament transforms to an oscillating one with small amplitude and I_{c2} marks the current at which the transition to the large amplitude oscillations takes place.

As described above for an adequate nonlinear function f(v) and a suitable choice of the parameters there are solutions of Eqs. (1) which are in qualitative agreement with the experimental results. Therefore it is tentative to identify the n^--layer with the high ohmic layer L of the device in Fig. 1 and the rest of the semiconductor device with the layer N. In Fig. 10 the numerically calculated $I(U_v)$-characteristic for appropriately chosen parameters is depicted. As in the experiment with increasing voltage first a stationary filament appears. Then at a critical current the filament becomes unstable and begins to oscillate around a certain spatial position. The picture in Fig. 6c refers to the normalized average current I = 0.042 of the $I(U_v)$-characteristic in Fig. 10a. In Fig. 10b for a certain moment the activator and inhibitor distributions are depicted. Obviously, the inhibitor distribution is spatially more diffuse and because of the larger time constant persues the activator. The spatial oscillations are accompanied by oscillations of the voltage U_v, too.

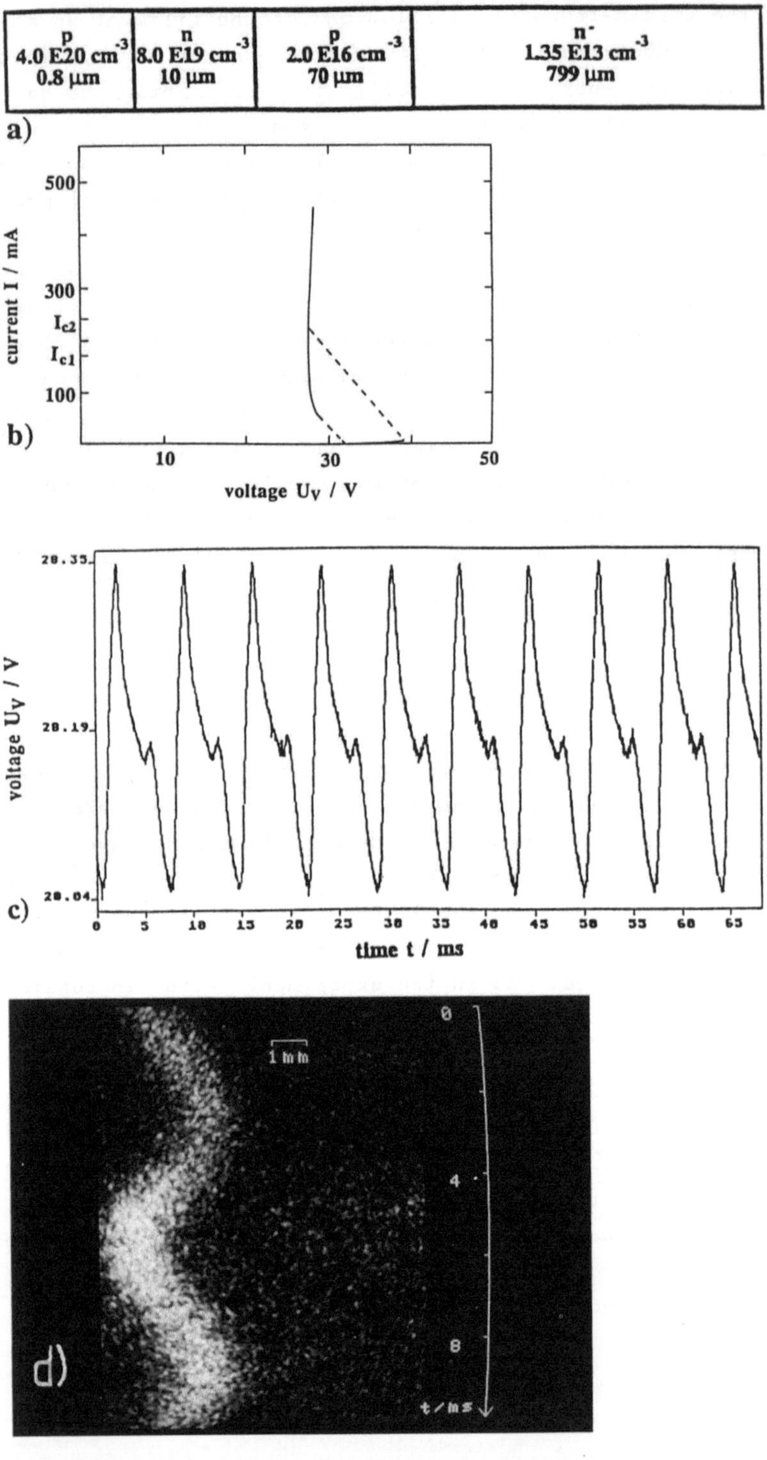
p
4.0 E20 cm-3
0.8 µm
n
8.0 E19 cm-3
10 µm
p
2.0 E16 cm-3
70 µm
n-
1.35 E13 cm-3
799 µm
a)
current I / mA
500
300
Ic2
Ic1
100
b)
10
30
50
voltage Uv / V
voltage Uv / V
28.35
28.19
28.04
c)
0
5
10
15
20
25
30
35
40
45
50
55
60
65
time t / ms
1 mm
0
4
8
t / ms
d)

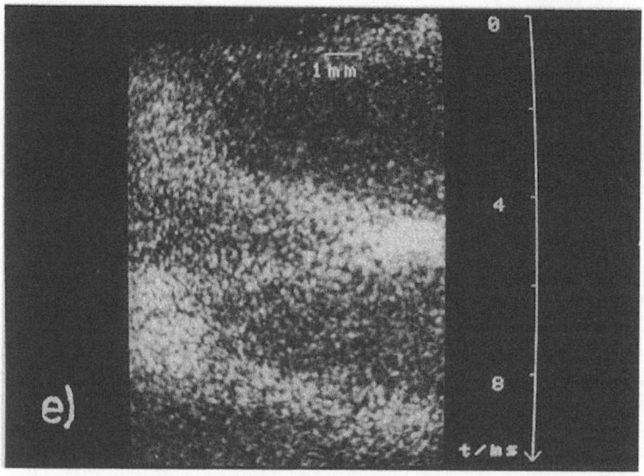

Fig. 9: Measurements obtained fron a pnpn⁻-semiconductor device. The size of the sample is (0.878 x 0.59 x 5.35) mm^3.

a) Material parameters of the investigated semiconductor device specified by the maximum doping concentration and the width of the layers

b) $I(U_V)$-characteristic of the pnpn⁻-structure at room temperature

c) Time series of the voltage drop U_V at the sample for an average total current I = 200 mA corresponding to the spatio-temporal oscillation in d)

d) Time and space resolved infrared measurements of an oscillating current density filament at room temperature. The brightness is a measure for the radiation density, the bright area therefore marks the current filament.

e) Time and space resolved infrared measurements of a filament oscillating from one boundary of the system to the other for an average total current of I = 260 mA at a temperature of 165.6 K

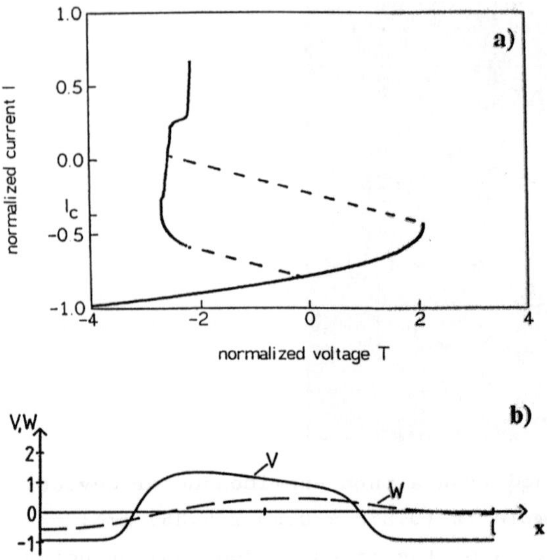

Fig. 10: a) Numerically calculated $I(U_v)$-characteristic using Eqs. (1). The swaying filament shown in Fig. 6c corresponds to a normalized average current total $I = {}_0\int^l v(x)dx = 0.042$. The parameters are as in Fig. 6c. At the currents $I = I_c$ the stable stationary current filament begins to sway.

b) Numerically calculated activator v and inhibitor w distributions at a selected moment which is a horizontal cut of Fig. 6c

4. Spatial and spatio-temporal patterns in a dc-driven gas discharge system

Another experimental example for the two-layer model described in chapter 2 is a special dc-driven gas discharge system. One constituent of the system is formed by a high resistivity silicon electrode which serves as layer with an approximately linear current-voltage characteristic. The material may be p- or n-doped with a specific resistivity of about 1000 Ωcm. The gas between this high ohmic material and a metallic electrode is a highly nonlinear medium with an S-shaped characteristic. The length of the discharge gap is denoted by l, its width by a, the lateral extension of the semiconductor electrode is b and the thickness of the gap is d (Fig. 11a,b). Typical dimensions are l = 40 mm, a = 3 mm, b = 10 mm and d = 0.3 mm. To ensure discharge only

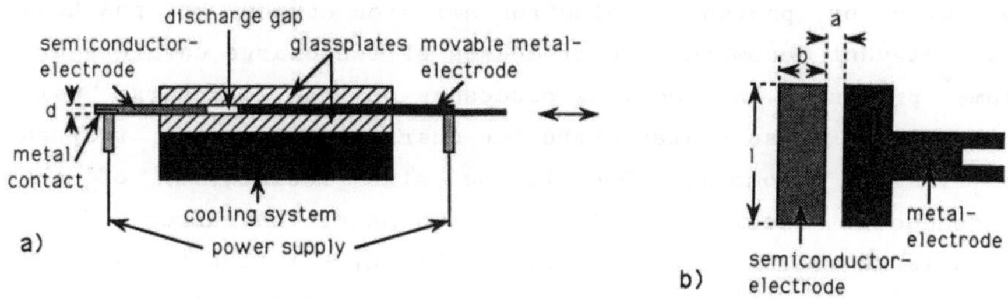

Fig. 11: Experimental arrangement of the gas discharge system
a) Schematic cross-section
b) Top view

between the face plates of the electrodes the whole electrode system is mounted in two glass-plates. The width of the gas discharge gap can be varied by moving the copper electrode. A watercooling system serves to cool the arrangement and to maintain an approximately constant temperature. The experimental investigations take place in a vacuum chamber which contains a gas mixture consisting of an inert gas (helium or argon) and several volume percent of air. The gas pressure is varied between 10 and 250 hPa. The electrodes of the gas discharge system are connected to a high voltage dc-power supply via a load resistor R_0 whereby the semiconductor electrode works as anode.

In terms of the model the semiconductor electrode corresponds to the layer L, the gas represents the nonlinear layer N. The properly normalized current density at the semiconductor-gas interface is identified with the activator v and the properly normalized potential drop across the semiconductor with the inhibitor w.

To characterize the stationary current density patterns the global $I(U_V)$-characteristic of the gas discharge device and the relative radiation density distribution of the gas discharge slit are measured. A cross-section of the gap parallel to the electrode surface is recorded by a vidicon camera. The amplitude of the detected signal is an appropriate measure for the radiation density distribution which in turn is proportional to the current density in the discharge gap. If the applied voltage U is increased a current density filament appears when a certain critical voltage is reached. The generation of secondary electrons is essentially determined by processes at the cathode, especially by photon and electron emission caused by positive ions. Relevant recombination processes of the charge carriers are volume processes as well as processes on the discharge wall formed by the glass-plates where the charge carriers are trapped and finally recombine. Due to the high resistivity of the semiconductor electrode in the surrounding of the current filament a large voltage drop lateral to the current flow takes place and consequently the filaments have a finite width. An enlargement of the voltage U results in a small broadening of the filament and in an increase of the current density and finally in the generation of a second filament. In Fig. 12a stationary radiation density distributions for different values of the voltage U are depicted.

Obviously, all filaments have approximately the same size and are carrying the same current which indicates their solitary character. The generation of new filaments is entailed by discontinuities in the global $I(U_V)$-characteristic (Fig. 12b). While the first filament develops from a homogeneous nonconductive state in a localized conductive nonequilibrium state the spatial position of which depends on unavoidable system inhomogeneities as, e. g., the roughness of the electrode surfaces or statistical electron generation processes, the additional filaments arise from division processes of already existing filaments. A more detailed investigation of these splitting processes using the above mentioned streak-camera system reveals that there are different stages when the applied voltage is increased. First the current filament only broadens. Then at a critical width the stationary filament becomes unstable and a second filament branches off and removes from the original one but as a rule the total current is too small to allow the existence of two independent stationary filaments. Therefore after a short time - in the considered case about a few milliseconds - the second filament disappears. This process of splitting and extinction recurs periodically thereby developing a spatio-temporal radiation pattern. With growing total current the oscillation becomes more complex, e. g. a period-doubling occurs. If the total current exceeds a critical value, approximately double the minimum current necessary to maintain one filament, the second filament is able to exist independently so that the splitting process is accomplished. Up to a critical current value the

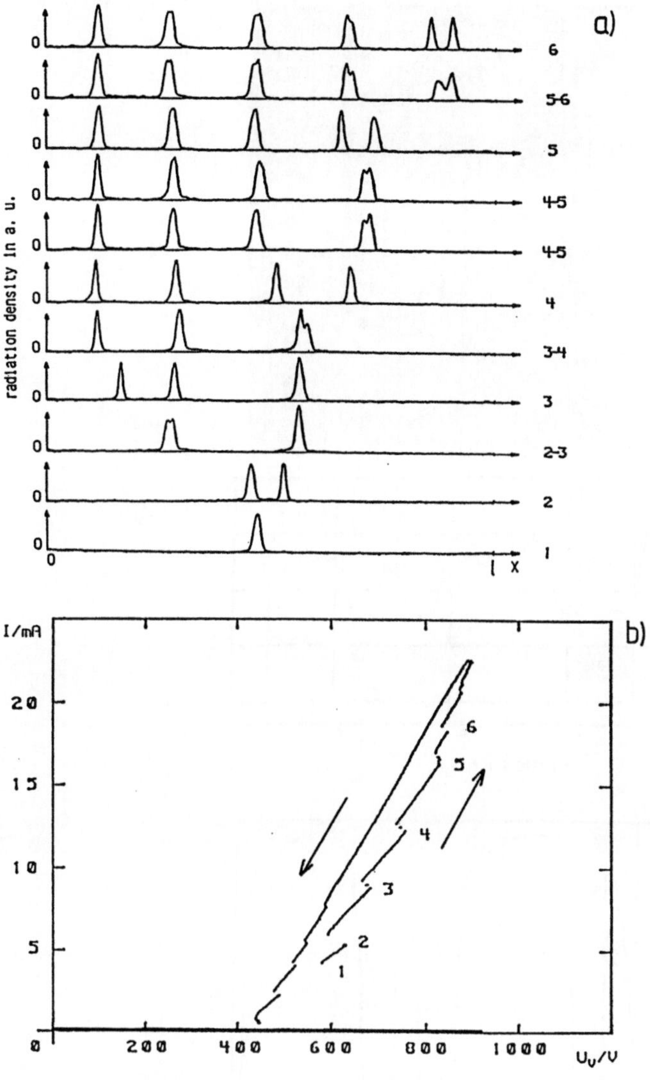

Fig. 12: Experimental results on a gas discharge system with the parameters R_o = 164 kΩ, ϱ = 1.7 kΩcm, p = 200 hPa, l = 32 mm, b = 10 mm, a = 0.3 mm, and d = 3.4 mm. The gas mixture is argon with 10 vol. percent of air.

a) Scenario of filament generation measured with increasing applied voltage U. The numbers refer to those in b) indicating the point (I,U_V) at which the radiation density distribution is recorded.

b) Measured $I(U_V)$-characteristic corresponding to the scenario of a)

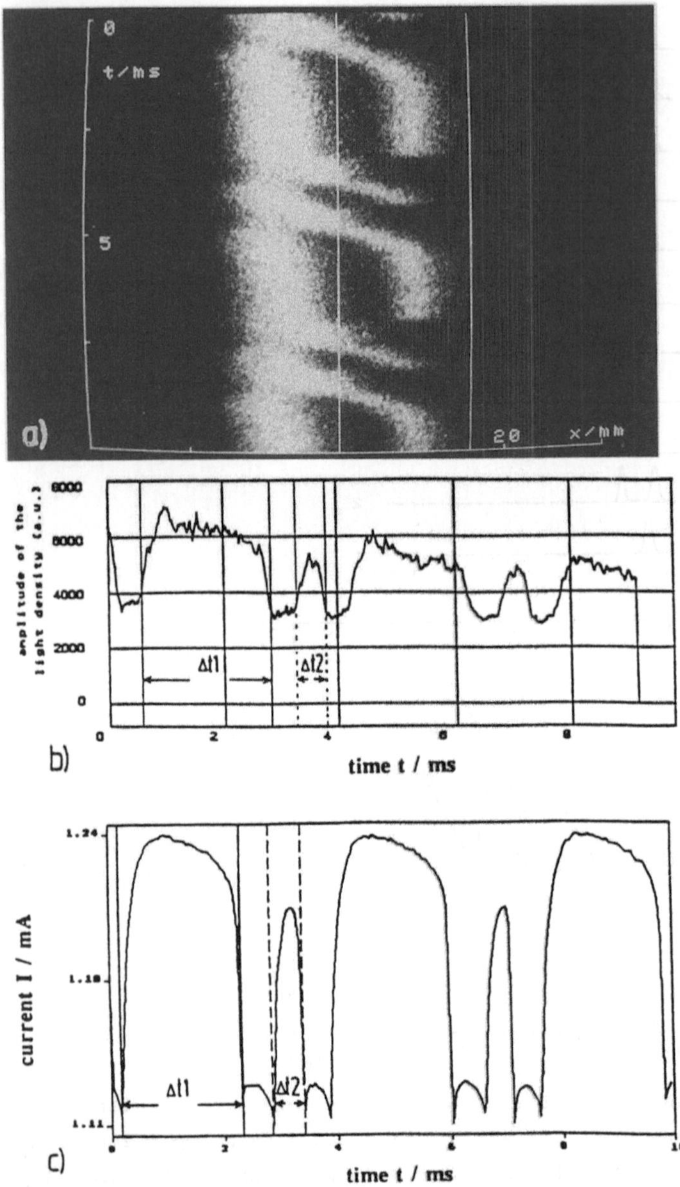

Fig. 13: Experimental results on a gas discharge system with the parameters R_0 = 200 kΩ, ϱ = 1.7 kΩcm, p = 30 hPa, l = 30 mm, b = 8 mm, and a = 3.5 mm. The gas mixture is argon with 10 vol. percent of air.

a) Streak-camera record of a splitting filament

process is reversible, i. e. if the total current is reduced the oscillation vanishes and the original filament is stable. Above the critical value a decrease of the current results in an extinction of one of the filaments without any oscillations.

Associated with the spatio-temporal oscillations of the radiation density distribution (Fig. 13a) during the splitting process one observes oscillations of the total current I in the external circuit which correlate directly with the spatio-temporal oscillations. To illustrate this in Fig. 13b,c the temporal behaviour of the total current I and the integral of the radiation density in the area between the two vertical solid lines in Fig. 13a are drawn. The splitting processes occur in a wide range of parameters. Somewhat different forms of oscillations are found for gas pressures between 30 and 300 hPa and distances of the electrodes in the range of 2 and 5 mm. However, in any case the observed oscillations are attributed to the splitting and extinction of filamentary structures which seem to be typical for all filament generation processes in the system.

For appropriately chosen parameters besides the described solitary structures the gas discharge system self-organizes also stable spatially periodic light patterns transversal to the current flow. Starting from a high current homogeneously lightening gas discharge and decreasing the voltage U below a certain value the homogeneous discharge becomes unstable and finally bifurcates into a spatially periodic radiation density distribution. In most cases the periodic patterns rise at the boundary of the system and with decreasing U spread out to the center of the system. A typical fully developed periodic distribution is shown in Fig. 14a. A further decrease of the bifurcation parameter U leads to an increase of the amplitude of the periodic distribution. Near the bifurcation point $U_c \approx 1230$ V the amplitude grows proportional to the square root of U as shown in Fig. 14b. This

Fig. 13 b) Total light intensity from the area marked in a) as function of time where $\Delta t_1 = 2.17$ ms and $\Delta t_2 = 0.54$ ms

c) Corresponding time series of the total current with $\Delta t_1 = 2.19$ ms and $\Delta t_2 = 0.55$ ms

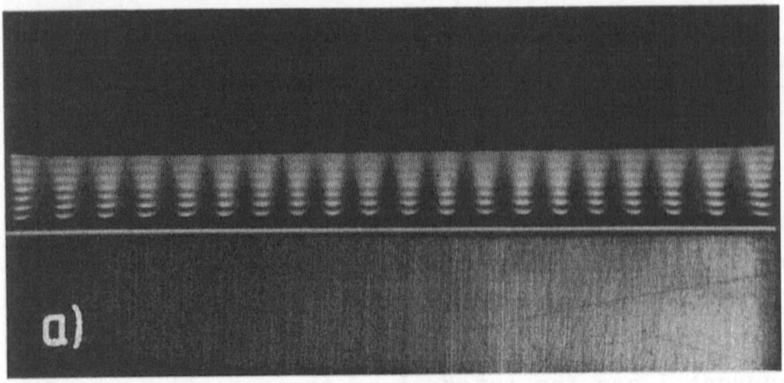

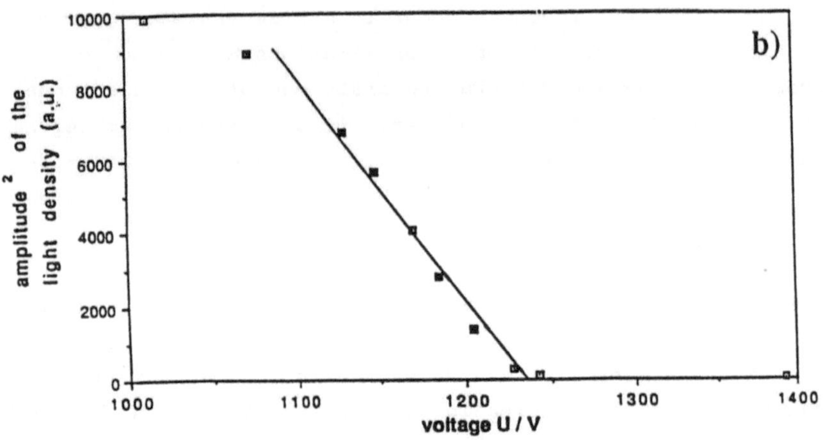

Fig. 14: Experimental results on a gas discharge system with the parameters R_0 = 20 kΩ, ϱ = 0.9 kΩcm, p = 90 hPa, l = 45 mm, b = 12 mm, a = 4.6 mm, and d = 0.3mm. The gas mixture is 80 hPa He and 10 hPa air.

a) Photograph of a typical periodic spatial radiation density distribution. The semiconductor is placed at the top and the metallic electrode at the bottom of the picture. The small bright horizontal region corresponds to the cathode fall.

b) Dependence of the amplitude square of the developing periodic distribution on the voltage U result is in good agreement with the findings of the centre manifold theory applied to Eqs. (1) at the Turing point which leads to the same dependence of the amplitude of the developing mode on the bifurcation parameter. Similar periodic patterns occur if the load resistor R_0, the width a of the gas gap or the gas pressure p is varied appropriately.

5. Conclusion

The presented model is the result of a phenomenological consideration which reduces a given physical system to a layer system with characteristic electric properties. Necessary presuppositions are on one side a layer with a nonlinear S-shaped current-voltage characteristic and a layer with a single valued monotonically increasing one on the other side. Using one of the simplest approximations for an S-shaped characteristic, namely an algebraic function containing only terms up to the third power and a linear characteristic for the other layer the numerical and analytical analysis of the model gives insight into the rich behaviour of self-organized localized nonequilibrium structures of such systems. Besides solitary filamentary structures the system is able to organize spatially periodic structures and a wide variety of dynamical patterns. Because of the rough approximation made deducing the equations an agreement between experiment and calculation can only be qualitative. The two expounded experimental systems fulfill two important conditions. Both contain a high ohmic layer realized by a semiconductor layer and a region with nonlinear electric properties realized by several p-n junctions (chapter 3) or a gas (chapter 4) with an S-shaped characteristic. Many patterns predicted by the model can be found in these systems. Current filaments the generation of which is accompanied by discontinuities in the global $I(U_V)$-characteristic appear in both systems. Whereas the generation process in the semiconductor device is not investigated in detail up to now in the gas discharge system it was found to consist in division processes. Periodic spatial patterns observed in the gas discharge system are also well described by the model. Concerning the dynamical behaviour of these systems a qualitative agreement between the experimentally investigated swaying filaments in the semiconductor structure and the numerically calculated ones becomes apparent. Therefore one may conclude that the described model with its rough approximation contains the essential features to allow for an explanation of various complex self-organized nonequilibrium structures in different physical systems.

Acknowledgement

The authors would like to thank M. van dem Brink who carried out some measurements. Also, the financial support of the Stiftung Volkswagenwerk is gratefully acknowledged.

References

[1] A. V. Nedospasov, Sov. Phys. Uspekhi 11 (1968) 174

[2] Ch. Radehaus, T. Dirksmeyer, H. Willebrand, and H.-G. Purwins, Phys. Lett. A 125 (1987) 92

[3] H.-G. Purwins, G. Klempt, and J. Berkemeier, Festkörperprobleme 27 (1987) 27

[4] H. Willebrand, Ch. Radehaus, F.-J. Niedernostheide, R. Dohmen, and H.-G. Purwins, Phys. Lett. A 149 (1990) 131

[5] K. G. Müller, Phys. Rev. A 37 (1988) 4836

[6] D. G. Boyers and W. A. Tiller, Appl. Phys. Lett. 41 (1982) 28

[7] P. G. Gennes, "The Physics of Liquid Cristals" (Clarendon, Oxford 1974)

[8] A. Newzella and H.-G. Purwins, to be published

[9] B. S. Kerner and V. F. Sinkevich, JETP Lett. 36 (1982) 436

[10] K. M. Mayer, J. Parisi, and R. P. Huebener, Z. Phys. B - Condensed Matter 71 (1988) 171

[11] D. Jäger, H. Baumann, and R. Symanczyk, Phys. Lett. A 117 (1986) 141

[12] B. S. Kerner and V. V. Osipov, Sov. Phys. Usp. 32 (1989) 101

[13] Ch. Radehaus, K. Kardell, H. Baumann, D. Jäger, and H.-G. Purwins, Z. Phys. B 65 (1987) 515

[14] K. M. Mayer, R. Gross, J. Parisi, J. Peinke, and R. P. Huebener, Sol. State Commun. 63 (1987) 55

[15] A. Brandl and W. Prettl, Festkörperprobleme 30 (1990) 371

[16] H. Willebrand, F.-J. Niedernostheide, E. Ammelt, R. Dohmen, and H.-G. Purwins, "Spatio-Temporal Oscillations during Filament Splitting in Gas Discharge Systems", accepted for publication in Phys. Lett. (1991)

[17] Ch. Radehaus, R. Dohmen, H. Willebrand, and F.-J. Niedernostheide, Phys. Rev. A 42 (1990) 7426

[18] J. Carr, "Applications of the Centre Manifold Theory" (Springer, Berlin 1981)

[19] Ch. Radehaus, H. Willebrand, R. Dohmen, F.-J. Niedernostheide, G. Bengel, and H.-G. Purwins, to be published in J. Appl. Phys.

[20] F.-J. Elmer, Physica D 30 (1988) 321

[21] Y. Tsuchiya, IEEE J. QE 20 (1984) 1516

Index of Contributors

Springer Proceedings in Physics

Managing Editor: H. K. V. Lotsch

1 *Fluctuations and Sensitivity in Nonequilibrium Systems*
Editors: W. Horsthemke and D. K. Kondepudi

2 *EXAFS and Near Edge Structure III*
Editors: K. O. Hodgson, B. Hedman, and J. E. Penner-Hahn

3 *Nonlinear Phenomena in Physics* Editor: F. Claro

4 *Time-Resolved Vibrational Spectroscopy*
Editors: A. Laubereau and M. Stockburger

5 *Physics of Finely Divided Matter*
Editors: N. Boccara and M. Daoud

6 *Aerogels* Editor: J. Fricke

7 *Nonlinear Optics: Materials and Devices*
Editors: C. Flytzanis and J. L. Oudar

8 *Optical Bistability III*
Editors: H. M. Gibbs, P. Mandel, N. Peyghambarian, and S. D. Smith

9 *Ion Formation from Organic Solids (IFOS III)*
Editor: A. Benninghoven

10 *Atomic Transport and Defects in Metals by Neutron Scattering*
Editors: C. Janot, W. Petry, D. Richter, and T. Springer

11 *Biophysical Effects of Steady Magnetic Fields*
Editors: G. Maret, J. Kiepenheuer, and N. Boccara

12 *Quantum Optics IV*
Editors: J. D. Harvey and D. F. Walls

13 *The Physics and Fabrication of Microstructures and Microdevices*
Editors: M. J. Kelly and C. Weisbuch

14 *Magnetic Properties of Low-Dimensional Systems*
Editors: L. M. Falicov and J. L. Morán-López

15 *Gas Flow and Chemical Lasers*
Editor: S. Rosenwaks

16 *Photons and Continuum States of Atoms and Molecules*
Editors: N. K. Rahman, C. Guidotti, and M. Allegrini

17 *Quantum Aspects of Molecular Motions in Solids*
Editors: A. Heidemann, A. Magerl, M. Prager, D. Richter, and T. Springer

18 *Electro-optic and Photorefractive Materials*
Editor: P. Günter

19 *Lasers and Synergetics*
Editors: R. Graham and A. Wunderlin

20 *Primary Processes in Photobiology*
Editor: T. Kobayashi

21 *Physics of Amphiphilic Layers*
Editors: J. Meunier, D. Langevin, and N. Boccara

22 *Semiconductor Interfaces: Formation and Properties*
Editors: G. Le Lay, J. Derrien, and N. Boccara

23 *Magnetic Excitations and Fluctuations II*
Editors: U. Balucani, S. W. Lovesey, M. G. Rasetti, and V. Tognetti

24 *Recent Topics in Theoretical Physics*
Editor: H. Takayama

25 *Excitons in Confined Systems*
Editors: R. Del Sole, A. D'Andrea, and A. Lapiccirella

26 *The Elementary Structure of Matter*
Editors: J.-M. Richard, E. Aslanides, and N. Boccara

27 *Competing Interactions and Microstructures: Statics and Dynamics*
Editors: R. LeSar, A. Bishop, and R. Heffner

28 *Anderson Localization*
Editors: T. Ando and H. Fukuyama

29 *Polymer Motion in Dense Systems*
Editors: D. Richter and T. Springer

30 *Short-Wavelength Lasers and Their Applications*
Editor: C. Yamanaka

31 *Quantum String Theory*
Editors: N. Kawamoto and T. Kugo

32 *Universalities in Condensed Matter*
Editors: R. Jullien, L. Peliti, R. Rammal, and N. Boccara

33 *Computer Simulation Studies in Condensed Matter Physics: Recent Developments*
Editors: D. P. Landau, K. K. Mon, and H.-B. Schüttler

34 *Amorphous and Crystalline Silicon Carbide and Related Materials*
Editors: G. L. Harris and C. Y.-W. Yang

35 *Polycrystalline Semiconductors: Grain Boundaries and Interfaces*
Editors: H. J. Möller, H. P. Strunk, and J. H. Werner

36 *Nonlinear Optics of Organics and Semiconductors*
Editors: T. Kobayashi

37 *Dynamics of Disordered Materials*
Editors: D. Richter, A. J. Dianoux, W. Petry, and J. Teixeira

38 *Electroluminescence*
Editors: S. Shionoya and H. Kobayashi

39 *Disorder and Nonlinearity*
Editors: A. R. Bishop, D. K. Campbell, and S. Pnevmatikos

40 *Static and Dynamic Properties of Liquids*
Editors: M. Davidović and A. K. Soper

41 *Quantum Optics V*
Editors: J. D. Harvey and D. F. Walls

42 *Molecular Basis of Polymer Networks*
Editors: A. Baumgärtner and C. E. Picot

43 *Amorphous and Crystalline Silicon Carbide II: Recent Developments*
Editors: M. M. Rahman, C. Y.-W. Yang, and G. L. Harris

MIX
Papier aus verantwortungsvollen Quellen
Paper from responsible sources
FSC® C105338

If you have any concerns about our products,
you can contact us on
ProductSafety@springernature.com

In case Publisher is established outside the EU,
the EU authorized representative is:
Springer Nature Customer Service Center GmbH
Europaplatz 3, 69115 Heidelberg, Germany

Printed by Libri Plureos GmbH
in Hamburg, Germany